Resource and Security Management in Electronic Communications and Networks

Resource and Security Management in Electronic Communications and Networks

Edited by **Bob Tucker**

WILLFORD PRESS

New York

Published by Willford Press,
118-35 Queens Blvd., Suite 400,
Forest Hills, NY 11375, USA
www.willfordpress.com

Resource and Security Management in Electronic Communications and Networks
Edited by Bob Tucker

International Standard Book Number: 978-1-68285-076-3 (Hardback)

Printed in the United States of America.

Contents

Preface

Electronic communications and networks play a major part in management of industries and businesses in the current scenario. However, the use of technology has also increased the need for management and security of different databases, networks and information systems. This book provides a detailed analysis of different security protocols and measures, allocation of various information resources and databases, their assessment and evaluation, etc. It discusses some of the significant concepts, frameworks and techniques of resource and security management in electronic communications and networks. Students, professionals and academicians would find comprehensive insights on the subject in this book.

Significant researches are present in this book. Intensive efforts have been employed by authors to make this book an outstanding discourse. This book contains the enlightening chapters which have been written on the basis of significant researches done by the experts.

Finally, I would also like to thank all the members involved in this book for being a team and meeting all the deadlines for the submission of their respective works. I would also like to thank my friends and family for being supportive in my efforts.

Editor

Fair Secure Computation with Reputation Assumptions in the Mobile Social Networks

Yilei Wang,[1,2] Chuan Zhao,[1] Qiuliang Xu,[1] Zhihua Zheng,[3] Zhenhua Chen,[4] and Zhe Liu[5]

[1]School of Computer Science and Technology, Shandong University, Jinan 250101, China
[2]School of Information and Electrical Engineering, Ludong University, Yantai 264025, China
[3]School of Information Science and Engineering, Shandong Normal University, Jinan 250014, China
[4]School of Computer Science, Shaanxi Normal University, Xi'an 710062, China
[5]Laboratory of Algorithmics, Cryptology and Security (LACS), 1359 Luxembourg, Luxembourg

Correspondence should be addressed to Qiuliang Xu; xql@sdu.edu.cn

Academic Editor: David Taniar

With the rapid development of mobile devices and wireless technologies, mobile social networks become increasingly available. People can implement many applications on the basis of mobile social networks. Secure computation, like exchanging information and file sharing, is one of such applications. Fairness in secure computation, which means that either all parties implement the application or none of them does, is deemed as an impossible task in traditional secure computation without mobile social networks. Here we regard the applications in mobile social networks as specific functions and stress on the achievement of fairness on these functions within mobile social networks in the presence of two rational parties. Rational parties value their utilities when they participate in secure computation protocol in mobile social networks. Therefore, we introduce reputation derived from mobile social networks into the utility definition such that rational parties have incentives to implement the applications for a higher utility. To the best of our knowledge, the protocol is the first fair secure computation in mobile social networks. Furthermore, it finishes within constant rounds and allows both parties to know the terminal round.

1. Introduction

Mobile computing and telecommunications are areas of rapid growth. A mobile social network connects individuals or organizations using off-the-shelf, sensor-enabled mobile phones with sharing of information through social networking applications such as Facebook, MySpace, and scientific collaboration networks [1]. A mobile social network plays an important role as the spread of information and influence in the form of "word of mouth" [2]. The advantages of wireless communications are that they can provide many new services that will revolutionize the way that society handles information [3]. The most significant property for mobile users in the mobile social network lies in the fact that they have reputation when they interact in the network [4–6], which can be utilized to boost cooperation in secure two-party computation. Secure two-party computation [7] means that two distributed parties wish to correctly compute some functionality using their private inputs while disclosing nothing except for the output. The computation should suffice three basic requirements: (i) privacy: nothing is learned from the protocol other than the output, (ii) correctness: the output is distributed according to the prescribed functionality, and (iii) independence: parties cannot make their inputs depending on other parties' inputs. Another requirement is fairness which means that either all parties learn the results or none of them does. Plenty of researchers delve into implementing fairness among parties. Unfortunately, Cleve [8] shows that fairness cannot be achieved in two-party settings. So, the accepted folklore is that nothing nontrivial can be computed with fairness. The usual treatment of secure two-party computation [9] wakens the ideal world to the one where fairness is not guaranteed at all.

In the setting of two-party games under incomplete information, two selfish parties wish to maximize their utilities with their private information. Each party has a set of strategies and certain private information-like types. Both parties take their strategies simultaneously or alternately in each round (maybe just in one shot) and the last round leads to an outcome which assigns each party a utility. Cryptography and game theory are both concerned with understanding interactions among mutually distrusted parties with conflicting interests. Cryptographic protocols are designed to protect the private inputs of each party against arbitrary behaviors, while game theory protocols are designed to reach various Nash equilibria against rational deviations.

1.1. Related Works. Research shows great increases in communications through mobile phone call, text messages, and the spatial reach of social networks [10–12]. People frequently have ties at a distance and they socialize with these ties through mobile phones and so forth. Larsen et al. [13] consider how mobile phones are used to coordinate face-to-face meetings between distanced friends and family members. Wang et al. [14] deal with the problem of influence maximization in a mobile social network where users in the network communicate through mobile phones. A mobile social network can be extracted from call logs and is modeled as a weighted directed graph. A mobile phone user corresponds to a node. The weight of one node is its reputation and is established when it interacts with other nodes in the network. Miluzzo et al. [15] discuss the design, implementation, and evaluation of the cenceme application on the basis of mobile social networks. González et al. [16] represent a model of mobile agents to construct social networks on the basis of a system of moving particles by keeping track of the collisions during their permanence in the system. Beach et al. [17] discuss the security and privacy issues in mobile social network when users in the network share their IDs or handles.

On the other hand, users in mobile social networks are assumed as rational parties who care about their utilities as those in game theory. Wang et al. [18] propose social rational secure multiparty computation protocol when rational parties belong to a social network. Rational parties, introduced by Halpern and Teague [19], behave neither like honest parties who always follow the protocol nor like malicious parties who arbitrarily violate the protocol. Rational parties only adopt the strategies which maximize their utilities. Halpern and Teague [19] prove the impossible result with rational parties and then give a random solution for rational multiparty computation. However, given at least three malicious parties, their protocol cannot achieve fairness at all.

1.2. Motivations and Contributions. Rational parties in secure computations are expected to cooperate with each other. However, they have no incentives to cooperate according to traditional utility definition. Therefore new utility definition must be considered assigning incentives to rational parties. With the motivation that reputation derived from mobile social networks can boost cooperation among users, we consider rational secure computation in mobile social works such that rational parties can utilize the reputation in the

networks. In particular, users in the mobile social networks are willing to cooperate with those who have good reputation of cooperation. Furthermore, the good reputation can be transmitted among friends in the networks. For example, if Alice cooperated with Bob once, then Bob's friends are willing to cooperate with Alice or Bob will cooperate with Alice when they meet again. Therefore, reputation is a useful tool to encourage mutual cooperation.

In this paper, we only consider two rational parties to securely compute a function. The parties come from a mobile social network, where they both have reputation value and use Tit-for-Tat (TFT) strategies to boost cooperation. Note that reputation affects the way parties achieve their utilities. The rational computation protocol in the presence of such rational parties is divided into several iterations. At the end of each iteration both parties gain some utilities and update their reputations. This process is similar to repeated games with stage games. Maleka et al. [20] first introduce repeated games into secret sharing scheme and get positive/negative results in infinitely/finitely repeated games. They discuss repeated games under complete information scenarios and conclude that parties cannot reconstruct secret when they know the terminal iteration. In this paper, we introduce the TFT strategy, reputation assumption, and incomplete information in order to facilitate mutual cooperation between both parties. Thus, it is possible for parties to achieve fairness in constant rounds.

Our settings are approximately similar to those of Groce and Katz [21] with the exception of the TFT strategy [22], the reputation assumption, and incomplete information scenarios. The main contributions of this paper are the introduction of the TFT strategy and reputation assumptions.

(i) The main target of rational two-party computation in the mobile social networks is how to facilitate cooperation among parties in order to complete the protocol (like the prisoners' dilemma game). In game theory scenario (especially in repeated games), TFT is an efficient strategy to promote cooperation. In fact, this seemingly simple and quite natural strategy defeats other strategies in Axelrod's prisoners' dilemma tournament [23]. The main intuition of the TFT strategy is that parties implement cooperation at the first round to make an attempt to elicit mutual cooperation from their opponents and copy the opponent's last action in the next round. In other words, a TFT party (who adopts the TFT strategy) cooperates with parties who cooperate and finks with parties who fink. Nowak and Sigmund [24] design experiments based on Axelrod's tournament to simulate the role of reciprocity in societies. In rational secure two-party computation, parties participate in the computation using the TFT strategy.

(ii) In previous works, parties in rational multiparty computation have no private types. Namely, the fact that parties are rational is common knowledge (common knowledge about an event between two parties means that one party knows the event and he knows the other party knows the event too, and vice versa [25]) and parties run the protocol under complete information scenario. Consequently, parties execute the protocol according to the Nash equilibrium. However, feasibility condition is that parties may have their own private type. For example, some people are kind, some others are vicious, and still others may be revengeful. Everybody knows exactly his own type and only has a priori probability on the private type of other parties. We call this incomplete information scenario.

Under this scenario, parties adopt their strategies consulting the preceding actions when executing the protocol. The preceding actions form a reputation for a certain type. For example, in the mobile social networks people who often help others have a good reputation, while people who often deceive others have a bad reputation.

In rational computation under incomplete information scenario, parties need to build a good reputation if they want to obtain the computation results. On the other hand, parties should show their private type to others through their actions. Otherwise, other parties may always adopt their dominating strategies which may lead to lower utilities.

(iii) Traditional utility assumptions in rational multiparty computation include two sides: (i) *correctness*, parties wish to compute the functionality correctly and (ii) *exclusivity*, parties wish that other parties do not obtain the correct result. Following the results of [19], parties have no incentives to participate in the protocol, not to mention how to realize fairness among them. Therefore, new assumptions should be introduced such that parties are willing to participate in the protocol. Other than the above utility assumptions, we introduce a new reputation assumption when parties come from a mobile social network. Namely, parties value and form their reputation in the network. We note that parties with a good reputation can inspire other parties to cooperate with them and boost their ultimate utilities. Reputation exists in many business-related, financial, political, and diplomatic settings and a good reputation is of great concern. Sometimes, companies, institutions, and individuals involved cannot afford the embarrassment, loss of reputation.

(iv) In this paper, there are two private types of parties: rational parties who always adopt their dominating strategies and TFTer parties who follow the TFT strategies. Each party knows his own private type and has a prior probability γ on the type of the other party. We stress that the prior probability (corresponding to their reputation) is not static, and it is updated after each round of the protocol.

Loosely speaking, we assume that there are two parties (each has his private type), say P_0 and P_1, wishing to jointly compute a function f with their private inputs x_0 and x_1, where the distributions of them are common knowledge.

Following [26–28], our protocol consists of two stages, where the first stage is regarded as a "preprocessing" stage and the second stage includes several iterations.

1.3. Paper Outline. Section 2 presents some preliminaries in our protocol, such as the TFT strategy, utility assumptions, and the reputation assumption. Section 3 presents the description of our protocol in the ideal-real world paradigm. Then Section 4 proves how to construct a fair protocol with constant rounds. In the last section, we conclude this paper and anticipate some open problems.

2. Preliminaries

2.1. Utility Assumptions. We first introduce the concept of the *stage game*, a building block of repeated games and our protocol. Let $\Gamma(P, A, U)$ denote a stage game, where $P = \{P_b\}_{b\in\{0,1\}}$. In the following section, we denote by $-b$ the complementary of b. Furthermore, let $A = A_0 \times A_1$, where A_b includes the strategy *fink* (F) and *cooperate* (C). Let $U = \{u_b\}$ be the utility set of parties. Let $\mu_b(o)$ be the utility of P_b with the outcome o, and let $\delta_b(o)$ be an indicator denoting the notion whether P_b learns the output of the function, and let $\text{num}(o) = \sum_b \delta_b(o)$ denote the aggregated number of parties who learn the output of the function. According to [19], we make the utility function assumptions as follows.

(a) Correctness. If $\delta_b(o) > \delta_b(o')$, then $\mu_b(o) > \mu_b(o')$; that is, parties prefer to learn the output of the function.

(b) Exclusivity. If $\delta_b(o) = \delta_b(o')$ and $\text{num}(o) < \text{num}(o')$, then $\mu_b(o) > \mu_b(o')$; that is, P_b hopes the other party does not learn the output of the function.

For simplicity, we define the following outcomes:

(i) $u_b = a$ if P_b learns the output of the function, while P_{-b} does not;

(ii) $u_b = 1$ if both P_b and P_{-b} learn the output of the function;

(iii) $u_b = 0$ if neither P_b nor P_{-b} learns the output of the function;

(iv) $u_b = c$ if P_{-b} learns the output of the function, while P_b does not.

Here $a > 1$, $c < 0$, and $a + c < 2$ hold (if $a + c < 2$, the strategy where both parties take cooperation is Pareto-dominated by the strategy where both parties alternately take fink and cooperate); otherwise parties have no incentives to participate in the protocol (this is very much like the scenario of prisoner's dilemma game [23]).

In repeated games, parties interact in several periods and take actions simultaneously or nonsimultaneously in each stage game $(\Gamma_1, \Gamma_2, \ldots, \Gamma_T)$, where T is a finite number. The total utility of P_b in the repeated games is

$$U_b = \sum_{t=0}^{T} u_b. \tag{1}$$

TABLE 1: Reputation $R_i^j(t+1)$ updating rules.

$R_i^j(t)$	Cooperation by j	Fink by j
>0	$R_i^j(t) + \alpha(1 - R_i^j(t))$	$(R_i^j(t) + \beta)/(1 - \min\{|R_i^j(t)|, |\beta|\})$
<0	$(R_i^j(t) + \alpha)/(1 - \min\{|R_i^j(t)|, |\alpha|\})$	$R_i^j(t) + \beta(1 + R_i^j(t))$
$=0$	α	β

2.2. Reputation Assumption. Rational parties are allowed to have utilities, and then we might as well regard the rational parties as parts of a social network and endow them with an additional property like reputation. Reputation plays an important role when distrusted parties interact under incomplete information scenarios, where parties only have a prior probability on the types of other parties. The famous prisoner's dilemma game under incomplete information accounts for how reputation encourages reciprocal cooperation in multistage games. In this paper we use reputation effects for our purpose. Put differently, a rational party values his reputation, because a high reputation can attract other parties to cooperate with him and boost his total utilities. That is, reputation makes a difference to the utilities. Precisely for this reason, we introduce another assumption on utility that rational parties under incomplete information think highly of their reputations. The definition of reputation in this paper is in accordance with [29]. (Although there is an improvement in [30] on the definition of reputation, we still use the original definition in this paper. Since the reputation equals trust when there are only two parties, we do not distinguish these two notions in this paper.)

Definition 1. Let $R_i^j(t)$ denote the reputation of party P_j assigned by P_i in period t such that $R_i^j(t) \in (-1, 1)$ and $R_i^j(0) = 0$, where period 0 denotes the initial period of the protocol.

The reputation is not static. If there are no specific instructions, in the following sections, we denote by P_j the other parties except for P_i. Party P_i adjusts his reputation of the $(t+1)$th period according to P_j's action in the tth period (Definition 2). To prevent parties from maliciously finking, we set $|\alpha| < |\beta|$, where α is the positive evidence to reward the parties who cooperate and β is the negative evidence to punish the parties who fink. In other words, the reputation grows slowly when parties cooperate while it drops quickly once parties fink.

Definition 2. After the tth stage game, reputation $R_i^j(t+1)$ is updated according to the rules in Table 1.

Under incomplete information scenarios, each party has a private type. Here, we assume that parties have two types: rational parties maximizing their utilities and TFTer parties adopting the TFT strategy. It is obvious that the utility is higher when they both obtain the correct value than when they do not. Parties would be apt to cooperate with other parties with high reputation. The more frequently parties cooperate, the higher utilities they obtain. Thus parties have incentives to cooperate with others in order to maintain

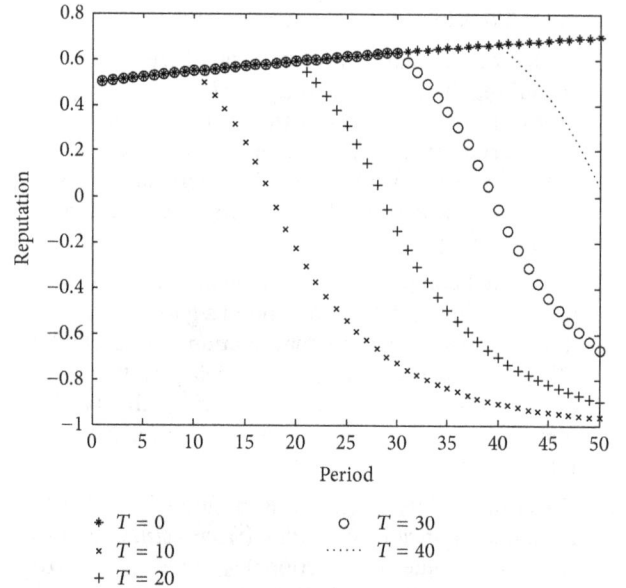

FIGURE 1: Reputation for each party when $\alpha = 0.01$ and $\beta = -0.1$.

a higher reputation. Meanwhile, high reputation will in turn make it easier for the other party to cooperate. This forms a virtuous cycle. We simulate the reputation value of Definition 2 in Figure 1, where the horizontal axis denotes the total periods (50 times here) and T denotes the outset of the deviating period.

We observe that the reputation will decrease once parties deviate, so parties have no incentives to deviate in each stage game if they want to preserve a higher reputation. Thus far, we give the third assumption if the protocol is considered to be a long-term process.

(c) Reputation. Each party has incentives to preserve a higher reputation for the sake of inducing reciprocally cooperation and the mutual cooperation consequently promotes parties' whole utilities in the long run.

In fact, the reputation assumption is a virtual part in the definition of utilities. Its main role is to warn other parties not to fink. Otherwise, the protocol will consequently enter into mutual fink. If so, all parties will not get the correct results and their utilities will decrease in the long run. Namely, although reputation does not intervene with the direct utilities in the current iteration, it actually affects the future utilities. In the future work, we will add reputation assumption as a real part into the definition of utilities.

3. Ideal-Real World Paradigm

3.1. Execution in the Ideal World. In the ideal world where a third trusted party (TTP) exists, it is trivial to achieve fairness. For completeness we represent the two-party (P_0, P_1) protocol in a natural way.

(1) Each party knows his private type and the other party only has a prior probability on the private type of his opponent.

(2) P_0 (*resp.,* P_1) randomly chooses his input value x_0 (*resp.,* x_1) according to a joint probability distribution D over input pairs.

(3) Each party P_b sends its value x_b' to the TTP. In the fail-stop setting, x_b' is restricted to a special symbol $\perp$ and x_b.

(4) If $x_b' = \perp$ then the TTP sends $\perp$ to both parties and the protocol ends. Otherwise, the TTP sends $f(x_0, x_1)$ to both parties.

(5) Each party outputs some values and the protocol ends.

At the end of the protocol, both parties either get utility 1 (when both parties follow the protocol) or get utility 0 (when at least one party sends $\perp$ to the TTP). Since utility function is common knowledge, both parties will follow the protocol in the fail-stop setting. We assume that $f(x_0, x_1)$ has full support, if for every x_0 and x_1 the distribution $f(x_0, x_1)$ puts nonzero probability on one unique element in the range of f. In other words, there does not exist any element $x_0' \neq x_0$ (*resp.,* $x_1' \neq x_1$) such that $f(x_0, x_1) = f(x_0', x_1)$ (*resp.,* $f(x_0, x_1') = f(x_0, x_1)$). If one party sends a fictitious value to the TTP, both parties get utility zero, which is absolutely smaller than 1, so both parties have incentives to send their true values to the TTP.

3.2. Execution in the Real World. It is more complex to construct a protocol completing the computation without a TTP. A hybrid protocol including two stages is first proposed as a transition. The first stage is an ideal functionality *ShareGen* and the second stage consists of several rounds (Section 1.3). In each round, they communicate with each other and exchange some messages. Each rational party satisfies the utility assumptions (a) and (b) and reputation assumption (c). In the ideal world, the TTP is the arbitrator which restrains both parties from deviating. In the hybrid world, the TFT strategy, the reputation assumption, and the incomplete information stimulate both parties to comply with the protocol.

In the second stage, one party, say P_0, is not sure whether P_1 is a TFTer party. We assume that P_0 has a priori probability γ on the type of P_1. In other words, P_0 considers that P_1 is a rational party with probability $1 - \gamma$ and a TFTer party with a small probability γ. According to the TFT strategy, parties are required to begin with cooperation and keep it if the other party cooperated at the preceding stages. The fact that one party has a good/bad reputation means that the party has a reputation to cooperate/fink. From the utility definition,

utility 1 when both cooperate is higher than utility 0 when both fink. So each party hopes to cooperate in each round. Mutual cooperation is easily realized when both parties are TFTers. However, parties will get into a dilemma when both parties are rational. Because the strategy profile, where both rational parties fink reach Nash equilibrium. Incomplete information can solve this dilemma to some extent. On the conditions that each party is not sure about his opponent true type, even rational parties have incentives to establish a good reputation hoping to obtain higher utilities in the future.

As the results of [21], the protocol is assumed to be in the hybrid world, where *ShareGen* is directed by a trusted dealer. We will remove this limitation here. Fortunately, Canetti [31] proves that a secure-with-abort protocol for *ShareGen* in the real world exists if there are enhanced trapdoor permutations. Therefore, a protocol in real world can be established using the composable theorem [32].

The protocols in this paper have finite rounds and parties know the last round when the protocols terminate. We will prove that mutual cooperation is a sequential equilibrium. To demonstrate a sequential equilibrium especially in the last round is cumbersome. Nevertheless, such a sequential equilibrium does exist [33]. We stress that the length of shares is the total iterations in the second stage, so the protocol will end in constant rounds. For clarity, we redescribe the lemma in [18].

Lemma 3. *Given* $m^* = 1 + (2a - 4c + 2\gamma)/\gamma$, *where* m^* *denotes the remained rounds in the protocol, there exists a sequential equilibrium such that both parties cooperate before* $n - m^*$ *rounds in the protocol, where* n *denotes the total rounds of the protocol.*

4. Fairness with Constant Rounds

In complete information scenario, there is no two-party protocol to compute functionality f on account of backward induction [25]. When the last round n is reached, parties no longer fear the future punishment and prefer to fink. As we know that fink is dominating strategy with respect to cooperation, consequently, round $n - 1$ is now the last round, and players will take strategy fink as before. This process continues in this way backwards in times and shows that parties are better off finking in rounds $n-2, n-3, \ldots, 1$ as well. If we release this condition, the predicament will be broken. We assume that each party has a private type like rational party or TFTer party. According to Lemma 3, both parties cooperate before the last "few" rounds. Inspired by this result, we construct a protocol with fairness between two parties. The informal description is given in Section 1.3, and now we give particular representations of the protocol.

4.1. The Fail-Stop Setting. Just as Groce and Katz [21], our protocol $\Pi^{ShareGen}$ consists of two stages. In the fail-stop setting, the first stage is a functionality *ShareGen* (see Box 1). The second stage includes the protocol Π (see Box 2) where both parties exchange their shares under incomplete information.

Functionality *ShareGen*

(1) **Inputs**: *ShareGen* takes as input a value x_0 (resp. x_1) form P_0 (resp. P_1).
 If either input is of no vail, then *ShareGen* returns $\perp$ to both parties.
(2) **Computation**: It includes the following steps:
(a) Choose a value n such that $n = t + m^*$ (t is a constant and is the threshold of Shamir's secret sharing scheme), so $t = n - m^*$ (see Lemma 3).
(b) Generate two shares $s_0 \neq 0$, $s_1 \neq 0$ of $f(x_0, x_1)$ such that $s_0 \oplus s_1 = f(x_0, x_1)$.
(c) Randomly select two $t - 1$ degree polynomials g_0 and g_1, where $g_0(0) = s_0$ and $g_1(0) = s_1$.
(3) **Outputs**: *ShareGen* sends $g_0(i)$ to P_0 and $g_1(i)$ to P_1, where $i \in \{1, 2, \ldots, n\}$.

Box 1: The description of functionality *ShareGen* in the fail-stop setting.

Protocol II

(1) **Step one**: Both parties run functionality *ShareGen* to receive $g_0(i)$ and $g_1(i)$,
 where $i \in \{1, 2, \ldots, n\}$.
(2) **Step two**: For $i = 1$ to n, each party decides whether to exchange his shares with the other party using the TFT strategy. We highlight two premises.
(a) Each party satisfies assumptions (a)–(c).
(b) Meanwhile, parties do not know exactly whether his opponent is a TFTer party.
Note: The utility assumptions and the incomplete information compel cooperation before round m^*.
(3) **Outputs**: Parties decide the outputs according to messages they have received.

Box 2: The description of protocol II in the fail-stop setting.

4.2. Positive Results

Theorem 4 (main theorem). *Given the utility assumptions (a) and (b) and reputation assumption (c), there exists a completely fair protocol Π with $n > 1 + (2a - 4c + 2\gamma)/\gamma$ constant rounds to compute f under incomplete information in fail-stop setting, where a party is a TFTer party with probability γ. If enhanced trapdoor permutations exist, the completely fair protocol Π also is established in the real world.*

Proof. We will first analyze the protocol $\Pi^{ShareGen}$ in a hybrid world where there is a trusted dealer computing *ShareGen*. Then following [32], if the protocol $\Pi^{ShareGen}$ is computational in the hybrid world, it is also established in the real world when enhanced trapdoor permutations exist. The correctness and privacy of the protocol are guaranteed by the ideal functionality *ShareGen*. We omit the formal definitions and straightforward proofs here. We prove fairness in the fail-stop setting.

(i) When *step one* of Box 2 finishes, it is obvious that party P_b can obtain s_b using Lagrange's interpolation after he receives all $g_b(i)$. The rest to do is to exchange shares with his opponent and recover s_{-b} using Lagrange's interpolation. Then at last he gets $f(x_0, x_1)$.

(ii) When *step two* of Box 2 finishes, we know that even if the parties know the value m^*, they still cooperate at the first t rounds (Lemma 3). Therefore both parties have no incentives to deviate before the previous t rounds, where $t = n - 1 - (2a - 4c + 2\gamma)/\gamma$ is the threshold of Shamir's secret sharing scheme. Under

this circumstance, both parties will receive at least t shares from their opponents. In other words, party P_b may retrieve s_{-b} using Lagrange's interpolation and finally learn $f(x_0, x_1)$.

To sum up, fairness is achieved in both settings. The round complexity is $O(1)$ which is more efficient than $O(1/p)$ in [21]. We stress that our conclusion of Nash equilibrium is stronger than that of [21], where only computational Nash equilibrium is established. Here, a sequential equilibrium in the fail-stop setting is met according to Lemma 3. $\square$

4.3. The Applications of Our Protocol.

The most important property of our protocol is the achievement of fairness in rational secure two-party computations. Although fairness is achieved in previous works, this is the first time that it is achieved through reputation assumptions, where parties in the protocol adopt TFT strategy. The property of fairness is essential in most secure multiparty computations, such as electronic voting and electronic auction. Take electronic voting; for instance, voters vote for candidates and wish to receive a fair and correct result. That is, the result cannot be biased by adversaries and should truly reflect their opinions. Traditional secure multiparty computations cannot achieve the property of fairness. Therefore, they cannot prevent adversaries from biasing the result. Fortunately, rational secure multiparty computations can realize fairness. On one hand, our rational protocols guarantee that each party may receive the same voting result. On the other hand, the adversary cannot bias the result.

The application of protocol $\Pi^{ShareGen}$ in electronic voting is present as follows. We describe the electronic voting protocol in the fail-stop setting using the protocol $\Pi^{ShareGen}$. Suppose that voters who participate in the voting may meet in the future to participate in other voting. When they meet again, they will evaluate each other through previous interactions. After several meetings, each voter win a reputation about his type. The type indicates that voters are rational or that they may adopt TFT strategy. As mentioned above, there is a probability γ to describe the prior probability about the type. So far, voters in electronic voting have the same features as those in $\Pi^{ShareGen}$. Next we will describe the process of electronic voting in which the voters mentioned above have participated.

(i) Voters run *ShareGen* using their specific inputs and receive their outputs, respectively (Box 1).

(ii) Voters run protocol Π according to their types and update reputation after each step (Box 2).

(iii) Voters output what they received in the protocol.

We prove that, given proper parameters, fairness can be achieved in protocol $\Pi^{ShareGen}$. Since voters have the same features as parties in $\Pi^{ShareGen}$, Theorem 4 can be applied rightly into electronic voting, where fairness is also achieved.

5. Conclusions

The importance of security guarantee in mobile social networks and telecommunication services is rapidly increasing since the applications in mobile social networks are more and more popular. The property of fairness is becoming an eye-catching aspect in secure computation especially between two rational parties. Game theory opens up another avenue to intensively study fairness of secure multiparty computation. Asharov et al. [34] give negative results based on improper utility assumptions. They conclude that no parties have incentives to cooperate with others. Groce and Katz [21] amend the deficiencies with new utility assumptions and two modifications which bring some new troubles. Consequently, the protocol in [21] has large round complexity and the trust dealer is required to participate in the protocol Π even in the real world.

Inspired by the fact that parties in mobile social networks value their reputation, which can boost cooperation between two rational parties, we modify the utility definition and allow parties to consider the effect of reputation derived from mobile social networks when they interact in the protocol. The results show that cooperation appears before the last "few" rounds even when they know the terminal round in finitely repeated games under incomplete information. Then we construct a protocol just like Groce and Katz [21]. Finally, with the help of the TFT strategy and the reputation from mobile social networks, the protocol Π in this paper can achieve fairness and sequential equilibrium.

Disclosure

An abstract of this paper has been presented in the INCOS2013 conference, pages 309–314, 2013 [35].

Conflict of Interests

The authors declare that there is no conflict of interests regarding the publication of this paper.

Acknowledgments

This work was supported by the Natural Science Foundation of China under Grant nos. 61173139 and 61202475, Natural Science Foundation of Shandong Province under Grant no. BS2014DX016, Ph.D. Programs Foundation of Ludong University under Grant no. LY2015033.

References

[1] M. Kimura and K. Saito, "Tractable models for information diffusion in social networks," in *Knowledge Discovery in Databases: PKDD 2006*, vol. 4213, pp. 259–271, Springer, Berlin, Germany, 2006.

[2] H. Ma, H. Yang, M. R. Lyu, and I. King, "Mining social networks using heat diffusion processes for marketing candidates selection," in *Proceedings of the 17th ACM Conference on Information and Knowledge Management (CIKM '08)*, pp. 233–242, ACM, October 2008.

[3] A. B. Waluyo, W. Rahayu, D. Taniar, and B. Scrinivasan, "A novel structure and access mechanism for mobile data broadcast in digital ecosystems," *IEEE Transactions on Industrial Electronics*, vol. 58, no. 6, pp. 2173–2182, 2011.

[4] J. Goh and D. Taniar, "Mining frequency pattern from mobile users," in *Knowledge-Based Intelligent Information and Engineering Systems*, pp. 795–801, Springer, 2004.

[5] D. Taniar and J. Goh, "On mining movement pattern from mobile users," *International Journal of Distributed Sensor Networks*, vol. 3, no. 1, pp. 69–86, 2007.

[6] J. Y. Goh and D. Taniar, "Mobile data mining by location dependencies," in *Intelligent Data Engineering and Automated Learning—IDEAL 2004*, vol. 3177 of *Lecture Notes in Computer Science*, pp. 225–231, Springer, Berlin, Germany, 2004.

[7] A. Yao, "Protocols for secure computation," in *Proceedings of the 23rd Annual Symposium on Foundations of Computer Science (FOCS '82)*, pp. 160–164, IEEE Computer Society, Chicago, Ill, USA, November 1982.

[8] R. Cleve, "Limits on the security of coin flips when half the processors are faulty," in *STOC 1986*, J. Hartmanis, Ed., pp. 364–369, ACM, Berkeley, Calif, USA, 1986.

[9] O. Goldreich, *Foundations of Cryptography*, vol. 2, Cambridge University Press, 2004.

[10] J. Urry, "Social networks, travel and talk," *British Journal of Sociology*, vol. 54, no. 2, pp. 155–175, 2003.

[11] K. W. Axhausen, "Social networks and travel: some hypotheses," in *Social Dimensions of Sustainable Transport: Transatlantic Perspectives*, pp. 90–108, 2005.

[12] B. Wellman, B. Hogan, K. Berg et al., "Connected lives: the project1," in *Networked Neighbourhoods*, pp. 161–216, Springer, 2006.

[13] J. Larsen, J. Urry, and K. Axhausen, "Coordinating face-to-face meetings in mobile network societies," *Information Communication & Society*, vol. 11, no. 5, pp. 640–658, 2008.

[14] Y. Wang, G. Cong, G. Song, and K. Xie, "Community-based greedy algorithm for mining top-k influential nodes in mobile

social networks," in *Proceedings of the 16th ACM SIGKDD International Conference on Knowledge Discovery and Data Mining*, pp. 1039–1048, ACM, 2010.

[15] E. Miluzzo, N. D. Lane, K. Fodor et al., "Sensing meets mobile social networks: the design, implementation and evaluation of the CenceMe application," in *Proceedings of the 6th ACM Conference on Embedded Networked Sensor Systems (SenSys '08)*, pp. 337–350, ACM, New York, NY, USA, November 2008.

[16] M. C. González, P. G. Lind, and H. J. Herrmann, "System of mobile agents to model social networks," *Physical Review Letters*, vol. 96, no. 8, Article ID 088702, 2006.

[17] A. Beach, M. Gartrell, S. Akkala et al., "WhozThat? Evolving an ecosystem for context-aware mobile social networks," *IEEE Network*, vol. 22, no. 4, pp. 50–55, 2008.

[18] Y. Wang, Z. Liu, H. Wang, and Q. Xu, "Social rational secure multi-party computation," *Concurrency Computation Practice and Experience*, vol. 26, no. 5, pp. 1067–1083, 2014.

[19] J. Halpern and V. Teague, "Rational secret sharing and multiparty computation: extended abstract," in *Proceedings of the Symposium of Theory of Computing (STOC '04)*, pp. 623–632, ACM, Chicago, Ill, USA.

[20] S. Maleka, A. Shareef, and C. P. Rangan, "Rational secret sharing with repeated games," in *Information Security Practice and Experience: Proceedings of the 4th International Conference, ISPEC 2008 Sydney, Australia, April 21–23, 2008*, L. Chen, Y. Mu, and W. Susilo, Eds., vol. 4991 of *Lecture Notes in Computer Science*, pp. 334–346, Springer, Berlin, Germany, 2008.

[21] A. Groce and J. Katz, "Fair computation with rational players," in *Advances in Cryptology—EUROCRYPT 2012*, D. Pointcheval and T. Johansson, Eds., vol. 7237, pp. 81–98, Springer, Cambridge, UK, 2012.

[22] Y. Wang, Q. Xu, and Z. Liu, "Fair computation with tit-for-tat strategy," in *Proceedings of the 5th IEEE International Conference on Intelligent Networking and Collaborative Systems (INCoS '13)*, pp. 309–314, Xi'an, China, September 2013.

[23] R. Axelrod, *The Evolution of Cooperation*, Penguin Press, London, UK, 1990.

[24] M. A. Nowak and K. Sigmund, "Tit for tat in heterogeneous populations," *Nature*, vol. 355, no. 6357, pp. 250–253, 1992.

[25] D. Fudenberg and J. Tirole, *Game Theory, 1991*, MIT Press, Cambridge, Mass, USA, 1991.

[26] S. Gordon, C. Hazay, J. Katz, and Y. Lindell, "Complete fairness in secure two-party computation," in *Proceedings of the Symposium on Theory of Computing Conference (STOC '08)*, C. Dwork, Ed., pp. 413–422, ACM, Victoria, Canada, May 2008.

[27] J. Katz, "On achieving the best of both worlds in secure multiparty computation," in *Proceedings of the 39th Annual ACM Symposium on Theory of Computing (STOC '07)*, pp. 11–20, ACM, San Diego, Calif, USA, 2007.

[28] T. Moran, M. Naor, and G. Segev, "An optimally fair coin toss," in *Theory of Cryptography*, O. Reingold, Ed., vol. 5444 of *Lecture Notes in Computer Science*, pp. 1–18, San Francisco, Calif, USA, 2009.

[29] B. Yu and M. P. Singh, "A social mechanism of reputation management in electronic communities," in *Cooperative Information Agents IV—The Future of Information Agents in Cyberspace*, pp. 154–165, Springer, Boston, MA, USA, 2000.

[30] M. Nojoumian and T. C. Lethbridge, "A new approach for the trust calculation in social networks," in *E-Business and Telecommunication Networks*, pp. 64–77, Springer, Berlin, Germany, 2008.

[31] R. Canetti, "Security and composition of multiparty cryptographic protocols," *Journal of Cryptology*, vol. 13, no. 1, pp. 143–202, 2000.

[32] R. Canetti, "Universally composable security: a new paradigm for cryptographic protocols," in *Proceedings of the 42nd IEEE Symposium on Foundations of Computer Science (FOCS '01)*, pp. 136–145, IEEE Computer Society, Las Vegas, Nev, USA, October 2001.

[33] D. M. Kreps and R. Wilson, "Reputation and imperfect information," *Journal of Economic Theory*, vol. 27, no. 2, pp. 253–279, 1982.

[34] G. Asharov, R. Canetti, and C. Hazay, "Towards a game theoretic view of secure computation," in *Proceedings of the 30th Annual International Conference on the Theory and Applications of Cryptographic Techniques (EUROCRYPT '11)*, K. G. Paterson, Ed., pp. 426–445, Springer, Tallinn, Estonia, 2011.

[35] Y. Wang, Q. Xu, and Z. Liu, "Fair computation with tit-for-tat strategy," in *Proceedings of the 5th IEEE International Conference on Intelligent Networking and Collaborative Systems (INCoS '13)*, pp. 309–314, IEEE, Los Alamitos, Calif, USA, September 2013.

Quantal Response Equilibrium-Based Strategies for Intrusion Detection in WSNs

Shigen Shen,[1,2] Keli Hu,[1] Longjun Huang,[1,3] Hongjie Li,[2] Risheng Han,[2] and Qiying Cao[4]

[1]Department of Computer Science and Engineering, Shaoxing University, Shaoxing 312000, China
[2]College of Mathematics, Physics and Information Engineering, Jiaxing University, Jiaxing 314001, China
[3]College of Computer Science and Technology, Zhejiang University of Technology, Hangzhou 310014, China
[4]College of Computer Science and Technology, Donghua University, Shanghai 201620, China

Correspondence should be addressed to Shigen Shen; shigens@126.com

Academic Editor: Laurence T. Yang

This paper is to solve the problem stating that applying Intrusion Detection System (IDS) to guarantee security of Wireless Sensor Networks (WSNs) is computationally costly for sensor nodes due to their limited resources. For this aim, we obtain optimal strategies to save IDS agents' power, through Quantal Response Equilibrium (QRE) that is more realistic than Nash Equilibrium. A stage Intrusion Detection Game (IDG) is formulated to describe interactions between the Attacker and IDS agents. The preference structures of different strategy profiles are analyzed. Upon these structures, the payoff matrix is obtained. As the Attacker and IDS agents interact continually, the stage IDG is extended to a repeated IDG and its payoffs are correspondingly defined. The optimal strategies based on QRE are then obtained. These optimal strategies considering bounded rationality make IDS agents not always be in *Defend*. Sensor nodes' power consumed in performing intrusion analyses can thus be saved. Experiment results show that the probabilities of the actions adopted by the Attacker can be predicted and thus the IDS can respond correspondingly to protect WSNs.

1. Introduction

Recently, Wireless Sensor Networks (WSNs) have attracted considerable concerns owing to their broad applications. Typical examples exist in environment monitoring, health monitoring, earthquake monitoring, objects tracking, and so on [1]. One of the major issues that we must face is how to guarantee security of WSNs before they are widely applied. Similar to traditional networks, there are to realize secure WSNs and prevention- and detection-based mechanisms [2–8]. The prevention-based mechanism, which aims to prevent any attack before it occurs, includes cryptography, key management, and authentication. On the contrary, the detection-based mechanism is to identify specifically those compromised nodes after they have broken down the measures taken by the prevention step. This mechanism is generally applied using Intrusion Detection System (IDS) as the second line of defense, while the prevention-based mechanism is referred to as the first line of defense. With an IDS, key data such as

intruder identification, intrusion time, and intrusion activity are provided to mitigate and remedy attack influences.

Currently, lots of IDSs [9, 10] have been proposed for various WSNs structures to provide an important security mechanism against both insider and outsider attacks. However, applying an IDS to WSNs is challenging since sensor nodes have resources limited in terms of energy, memory, computation, and communication capacities. Generally, different methods including anomaly-, misuse-, and specification-based detection are computationally expensive, which are particularly costly for small sensor nodes. This situation motivates us to seek optimal strategies of intrusion detection to possibly save sensor nodes' resources.

As a formal and mathematical tool that studies competition among involved individuals, game theory has provided us with an efficient method to explore optimal strategies in the field of intrusion detection of WSNs [11–14]. Nevertheless, game-theoretic approaches have a common assumption in which players are completely rational and the solutions to games are based on Nash Equilibrium (NE). In real-world applications, however, all Attackers (a player in game theory) may not be always rational and they do not even care about being detected. Therefore, NE-based solutions are not suitable for such circumstances and we need a more appropriate method to solve Intrusion Detection Games.

Nowadays, Quantal Response Equilibrium (QRE) has turned into a popular alternative to the traditional NE in behavior game theory. The QRE model maintains the assumption that individuals have beliefs that are supported in equilibrium by the strategies that players choose, but with the assumption that players make systematic mistakes or deviations in their choices [15]. There are two reasons resulting in the deviation behavior. One is called bounded rationality. The other is that players' payoffs are influenced by social preference in which subjects appear altruistic or fair or seek to reciprocate fairness or seek to limit inequality in payoffs [15].

In this paper, QRE is adopted to seek optimal strategies of saving IDS agents' power in WSNs. Considering the characteristics of sensor nodes, we construct a stage Intrusion Detection Game to describe interactions between the Attacker (a player) and IDS agents (the other player). The preference structures for the Attacker and IDS agents are defined, which lead to form payoffs of players. As the stage game evolves (the Attacker and IDS agents interact continually), we extend it to a repeated game and define the corresponding payoffs. We further obtain QRE-based strategies to show how the Attacker and IDS agents will select their actions.

To the best of our knowledge, this paper is the first work to focus on exploring QRE-based strategies for intrusion detection in WSNs. The main contributions of this paper are summarized as follows:

(1) we formulate a stage Intrusion Detection Game according to Binmore's method to study strategies of malicious sensor nodes and IDS agents, which is able to reflect interactions between the Attacker and IDS agents as well as their preferences;

(2) we extend the stage Intrusion Detection Game to a repeated Intrusion Detection Game by redefining the corresponding payoffs, which is able to reflect the reality that malicious sensor nodes and IDS agents interact continually;

(3) instead of NE-based strategies, we obtain QRE-based strategies of the Attacker and IDS agents, which satisfies such a situation to the point that the Attacker and IDS agents always make their decisions with bounded rationality;

(4) we realize an implementation of applying the repeated Intrusion Detection Game to WSNs based on the algorithm of calculating QRE-based strategies that can predict the Attacker's future behavior.

The rest of this paper is organized as follows. In Section 2, we overview related work to distinguish the difference between our work and other related works. In Section 3, we construct our stage Intrusion Detection Game for WSNs and

extend it to a repeated game. Further, we give a method of calculating QRE-based strategies. In Section 4, we implement an intrusion detection mechanism based on QRE-based strategies and give the corresponding algorithm. In Section 5, we perform experiments to show how the repeated Intrusion Detection Game is actually played. Finally, conclusions are provided in Section 6.

2. Related Work

IDSs in WSNs have attracted considerable attention. In the good survey, Butun et al. [5] presented detailed information about IDSs and the applicability of IDSs to WSNs, which are followed by the analysis and comparison of each scheme along with their advantages and disadvantages. Al-Hamadi and Chen [16] considered an optimization problem for the case where a voting-based distributed intrusion detection algorithm is employed to detect and isolate malicious nodes in WSNs. They then can dynamically determine the best redundancy level to apply to multipath routing for achieving the case of intrusion tolerance. In another paper [17], they analyzed dynamic redundancy management of integrated intrusion detection and tolerance, which is to maximize the lifetime of homogeneous clustered WSNs. To cope with potential Denial of Service attacks in WSNs, Cho et al. [18] proposed a partially distributed intrusion detection system with low memory and power requirements. In [19], Farooqi et al. proposed a novel intrusion detection mechanism including online prevention and offline detection for securing WSNs from routing attacks. To obtain efficient performance under limited computation resources of sensor nodes, Kim et al. [20] developed a Wu-Manber algorithm-based network intrusion detection system. By integrating system monitoring modules and intrusion detection modules in WSNs, Sun et al. [21] proposed an extended Kalman filter-based mechanism to detect false injected data. They further combine cumulative summation and generalized likelihood ratio to increase detection sensitivity. Shamshirband et al. [22] developed a cooperative-based fuzzy artificial immune system, in which the Cooperative-Decision-Making Module incorporates the danger detector module with the fuzzy Q-learning vaccination module to produce optimum defense strategies for detecting intrusion in WSNs. In addition, Riecker et al. [23] proposed a lightweight, energy-efficient IDS, where mobile agents are used to detect intrusions based on the energy consumption of the sensor nodes.

Since selecting the profitable detection strategy is able to lower resources consumption, game theory has been widely applied to obtain these optimal strategies. For example, the optimal strategies of launching IDS agents installed in sensor nodes are obtained by the signaling game in [24]. To determine the best defense strategies, Huang et al. [25] proposed a Markovian IDS incorporating game theory with anomaly and misuse detection, where Markov decision processes are employed with an attack-pattern-mining algorithm to predict future attack patterns. Moosavi and Bui [26] considered non-zero-sum discounted stochastic games to formally formulate and analyze the intrusion detection problem in WSNs. They assumed that the game data are not to be fully known to the players and achieved a robust optimization approach to address this data uncertainty. On the contrary, a zero sum stochastic game is applied in [27] to predict malicious behavior of Attackers. In [28], Shen et al. formulate a malware-defense differential game, in which the system can dynamically choose its strategies to minimize the overall cost whereas the malware intelligently varies its strategies over time to maximize this cost, to obtain optimal dynamic strategies for the system. In addition, cooperative games are also applied to formulate intrusion detection problem in WSNs. Shamshirband et al. [29] combined the game-theoretic approach and the fuzzy Q-learning algorithm to implement cooperative defense counter-attack scenarios for the sink node and the base station. The game is composed of three players consisting of sink nodes, a base station, and an Attacker and performs when a victim node in WSNs receives a flooding packet as a DDoS attack beyond a specific alarm event threshold. To obtain secure and reliable defenses of virtual sensor services in cloud-assisted WSNs, Liu et al. [30] proposed a stochastic evolutionary coalition game which is able to decide how evolutionary coalitions should be dynamically formed for reliable virtual-sensor-service composites to deliver data and how to adaptively defend in the face of uncertain attack strategies.

Among various game types, a repeated game consists of some number of repetitions of a stage game. Such a game is generally divided into two categories: finitely and infinitely repeated game, depending on whether interactions among players are finite or infinite. Players in a repeated game must consider the effects produced by their current chosen strategies on the opponents' strategies in subsequent rounds [31]. The same stage game, when played repeatedly, may result in different equilibriums. Therefore, each player must take optimal reactions against the opponent, which will affect one's payoffs in future.

Some applications of repeated game have been devoted to various aspects in wireless networks. Agah and Das [32] formulated a repeated game between IDS and sensor nodes to prevent Denial of Service (DoS) attacks in WSNs. Upon their proposed game, a protocol was proposed to category different sensor nodes based on their behavior. In [33], Pandana et al. proposed a self-learning repeated game framework to overcome selfishness and noncooperation of autonomous nodes in wireless ad hoc networks. The framework ensures the cooperation among nodes for the current packet forwarding and finds the better cooperation probabilities by self-learning algorithms. Chen et al. [34] constructed a repeated game model based on reputation for wireless networks to fully utilize the scarce spectrum resource. The model is able to help multiple primary and secondary users coexist and share the spectrum. Using a repeated game to enforce cooperation among nodes in wireless networks, Kong and Kwok [35] proposed an efficient packet-scheduling algorithm that leads to an equilibrium. Upon the algorithm, the wireless channel resources are fully utilized. The other typical cooperation applications of repeated game are composed of cooperative multicast [36], network selection [37], and power trading [38]. In addition, Sagduyu et al. [39] formulated a repeated game

under network uncertainty to deal with jamming attacks in wireless networks. A multiattacker repeated colluding game is proposed in [40] to find subgame equilibriums that indicate the optimal strategies of Attackers. Upon these equilibriums, a security policy is established to detect malicious nodes that collude with each other to launch the selective forwarding attacks. Moreover, cognitive radio users, using a repeated game in [41], can adapt their power by observing the interference from the feedback signals of primary users and transmission rates obtained in the previous stage. Zhu and Martínez [42] developed a repeated game to solve the coverage optimization problem of mobile sensors. To defend against multistage attacks, Luo et al. [43] modeled a two-player non-zero-sum noncooperative dynamic multistage game with incomplete information to find the best actions for defenders. Sun et al. [44], considering inherent uncertainty of nodes in ad hoc networks, proposed a power control mechanism with a dynamic repeated game-theoretic framework. Smith et al. [45] proposed a dynamic noncooperative repeated game for transmitting power control across multi-source-destination distributed wireless networks. Recently, the "zero-determinant strategies" [46–50] of a repeated game have attracted much attention in scientific world. In particular, Farraj et al. [48] employed a repeated game-theoretic formulation to describe the interactions of the parties in cyber-enabled power systems. Transient stabilization over time using zero-determinant strategies is obtained to indicate the potential of the constrained controller.

Based on the repeated game, QRE developing the concept of NE considers bounded rationality and thus is profitable to describe the dilemma of security source allocation. To fit the bounded rationality of human adversaries in security game, Yang et al. [51] modeled human behavior of adversaries and provided new mathematical models based on prospect theory and stochastic discrete choice model. A modification of QRE is proposed to develop algorithms that are efficient to compute the best response of the security forces when playing against the different adversaries. In [52], QRE is used to capture players' bounded rationality and to model internal Attackers' behavior. The results are able to predict how an internal Attacker will act in future. Then, a detailed game-based detection algorithm taking advantage of these results is described in detail.

3. Constructing Intrusion Detection Game for WSNs

3.1. Network Model. According to classification based on the installation location of IDS agents, there are purely distributed, purely centralized, and distributed-centralized structures [53]. For the purely distributed situation, each sensor node has been equipped with an IDS agent that locally examines malicious actions from neighboring sensor nodes. On the contrary, for the purely centralized situation, the base station (BS) has been equipped with the IDS agent, where a special protocol is necessary to gather information from sensor nodes to examine the behavior of sensor nodes. In addition, for the distributed-centralized situation, monitor sensor nodes are introduced and have been equipped with

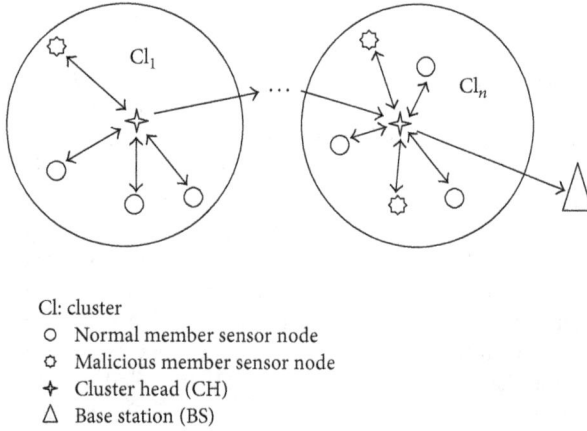

Cl: cluster
○ Normal member sensor node
✿ Malicious member sensor node
✛ Cluster head (CH)
△ Base station (BS)

FIGURE 1: Network model [24].

IDS agents. Not only do these monitor nodes perform activities like normal nodes, but also they check for intrusion detection.

Our network model adopts the same one in [24], as depicted in Figure 1. This model belongs to the distributed-centralized case. However, under clustering hierarchy, each sensor node has been equipped with an IDS agent, which is not the same as the situation where IDS agents are only installed in monitor sensor nodes. When an energy-abundant sensor node is elected as a cluster head (CH), the deployed IDS agent will launch simultaneously while the IDS agents in member sensor nodes are in sleep. Therefore, a CH executes the task of intrusion detection by the IDS agent in addition to aggregating and sending data.

3.2. Stage Intrusion Detection Game

Definition 1. The stage Intrusion Detection Game (IDG) is a 3-tuple $\mathbb{G} = (\mathcal{N}, \mathcal{A}, \mathcal{U})$, where

(1) $\mathcal{N} = \{\text{Attacker } i, \text{IDS agent } j\}$ is a set of players;

(2) $\mathcal{A} = \mathcal{A}_i \times \mathcal{A}_j$, where $\mathcal{A}_i = \{Cooperate (C), Preattack (P), Attack (A)\}$ and $\mathcal{A}_j = \{Sleep (S), Grant (G), Defend (D)\}$ are the sets of actions adopted by players i and j, respectively;

(3) $\mathcal{U} = \mathcal{U}_i \times \mathcal{U}_j$, where $\mathcal{U}_i = \{u_i(s_i) : \mathcal{A}_i \mapsto \mathbb{R}\}$ and $\mathcal{U}_j = \{u_j(s_j) : \mathcal{A}_j \mapsto \mathbb{R}\}$ are the sets of payoffs of players i adopting strategy s_i and j adopting strategy s_j, respectively.

In Definition 1, we consider that the game is played by the Attacker (i) versus the IDS agents (j). Player i is in fact referred to malicious sensor nodes that have such purposes as to listen to sensor information, devastate a sensor node's communication abilities, or entirely disable a sensor node. On the other hand, player j is referred to IDS agents that are initially installed in CHs. The goal of our Intrusion Detection Game is, from the view of game theory, to supply optimal

strategies for IDS agents in response to the Attacker selecting its strategies dynamically.

As an Attacker, player i has three possible actions. It may take the action *Cooperate* (C), meaning that it acts normally during communications among other sensor nodes. This action disguises it to avoid being captured by its opponent. However, the intentions of player i are hostile, and therefore its aim is to systematically arrange methods so that it can attack other sensor nodes for its own profits. Generally, it might disclose private information of other sensor nodes for obtaining other information required for it to finish an attack. These actions are known as reconnaissance attacks and can be summarized as the action *Preattack* (P). Moreover, an Attacker finally achieve the phase in which the action *Attack* (A) is made to obtain its expected profit. This action is without doubt the most threatening action among all. It raises and strengthens the seriousness of the problem and leads to many unexpected results such as a network unavailable for its legitimate sensor nodes, inaccurate sensing information, and leaking private data. In summary, the set of actions of player i is $\{C, P, A\}$.

To confront Attackers, player j also has three actions. Due to limited resources in sensor nodes, the strategy that IDS agents are always in *Defend* is not optimal. Otherwise, cluster heads installed IDS agents will consume their power quickly since processing intrusion detection is generally costly. Player j may therefore take the action *Sleep* (S) for saving energy. After launching IDS agents, it may grant sensor nodes to continue when no malicious behavior has been discovered. Note that two cases result in the fact that player j takes this action *Grant* (G). One case is that the Monitored Events are truly normal. The other is that IDS agents cannot detect the malicious events since any IDS has the false negative rate. In addition, player j will take the action *Defend* (D) to stop the work of malicious sensor nodes once violations are detected. In a summary, the set of actions of player j is $\{S, G, D\}$.

Based on the above analyses, there are nine possible combinations between the Attacker's actions and the IDS agent's actions. For example, strategy profile (C, S) means that player i acts normally and player j is in sleep for saving energy. (P, G) means that player i acts in a preattack step and player j grants its opponent to continue for not detecting reconnaissance attacks. (A, D) means that player i performs attacking behavior and player j prevents its opponent from its malicious work to protect sensor nodes.

Finally, let us quantify preferences and payoffs of players in the stage IDG. Let the symbols $\succ$ and $\sim$ be the preference and indifference, respectively. For example, if $x \succ y$, then it is said that x is preferred to y.

For the player Attacker, it is most profitable to attack successfully the WSNs without being defended. Since the IDS agents taking the action *Sleep* or *Grant* cannot defend the Attacker, the preference of strategy profile (A, G) is indifferent to that of (A, S). Its next choice is to take the action *Cooperate* without being defended. The following preference action is *Preattack* without any deterrence. The action *Attack* that is defended follows the Attacker's favorite, which is more preferable than the action *Cooperate* that is defended. Finally, the worst choice is the action *Preattack* responded by the

TABLE 1: Payoff matrix.

	S	G	D
C	(6, 8)	(6, 7)	(1, 4)
P	(4, 3)	(4, 2)	(0, 5)
A	(8, 1)	(8, 0)	(2, 6)

action *Defend*. The above analyses result in the following preference structure:

$$(A, S) \sim (A, G) \succ (C, S) \sim (C, G) \succ (P, S) \sim (P, G) \tag{1}$$
$$\succ (A, D) \succ (C, D) \succ (P, D).$$

With respect to the player IDS agents, the most preferable profile is the action *Cooperate* followed by the action *Sleep*. The following is the action *Cooperate* followed by the action *Grant* since taking the action *Grant* spends more power for detection than taking *Sleep*. When it takes the action *Defend*, it prefers orderly the actions *Attack*, *Preattack*, and *Cooperate*. It next prefers the action *Preattack* followed orderly by the actions *Sleep* and *Grant*. The least preferable profile is the action *Attack* followed by the action *Grant*. Therefore, the preference structure attained is

$$(C, S) \succ (C, G) \succ (A, D) \succ (P, D) \succ (C, D) \succ (P, S) \tag{2}$$
$$\succ (P, G) \succ (A, S) \succ (A, G).$$

According to Binmore's method [54], rational numbers are assigned to reflect players' preferences ranked in (1) and (2). Then, after being multiplied with their least common factor, the values of payoff functions u_i and u_j, free of fractions, can be formed in Table 1.

3.3. Repeated Intrusion Detection Game. In the realistic WSNs, interactions between players Attacker and IDS agents are continually performed. Therefore, the stage IDG will be played more than once and it is reasonable to model these interactions as a repeated game. Generally, a repeated game is a particular style of an extensive form game in which each stage is a repetition of the same strategic-form game. The times of playing a repeated game may be finite or infinite. If the game never ends (Attacker and IDS agent interact forever) or players (Attacker and IDS agent) do not know when the game ends, it is called an infinitely repeated game, which will be employed in this paper. In a repeated game, a strategy is an entire plan of action described in the stage game. When each stage ends, all players are able to observe the consequence of the stage game and make a choice to select the future actions depending on the history of actions. The overall payoff in a repeated game is denoted by a normalized discounted aggregate of the payoff at each stage game. Our repeated Intrusion Detection Game (RIDG) can be defined as follows.

Definition 2. The infinite δ-discounted RIDG is composed of repeated game $\mathbb{G}$, which is denoted by $\mathbb{G}(\infty, \delta)$, where

(1) the set of players is $\mathcal{N}$ defined in Definition 1;

(2) for every player $x \in \{i, j\}$, its overall strategy at the tth stage IDG is $s_x^t = [s_x(h_0), s_x(h_1), \ldots, s_x(h_t)]$, where

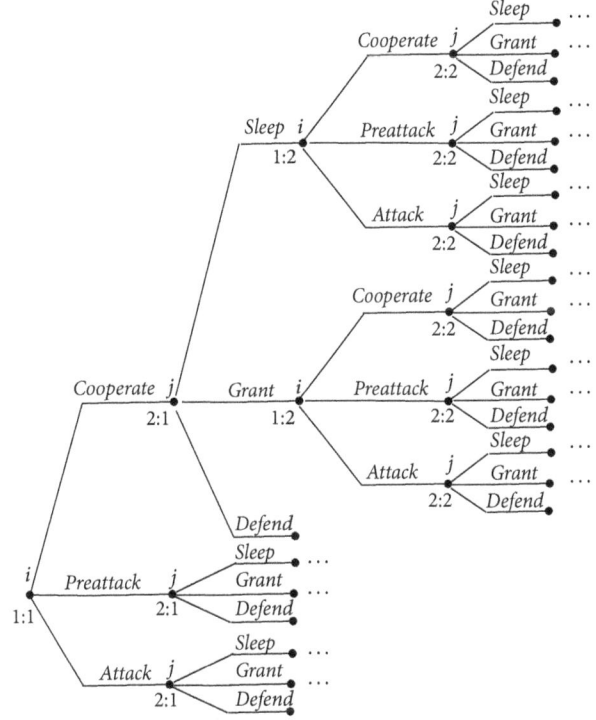

FIGURE 2: Repeated Intrusion Detection Game.

$h_y, y \in \{0, 1, \ldots, t\}$, denotes the yth history stage and $s_x(h_y)$ denotes the strategy adopted by player x at the yth history stage;

(3) for every player $x \in \{i, j\}$, its overall payoff is the δ-discounted average of instant payoffs from each round of the repeated Intrusion Detection Game.

Figure 2 shows a representation of the RIDG in extensive form. In fact, an Attacker is perfectly aware of IDS agents' past actions because IDS agents exert actions on the Attacker. In other words, player j's (IDS agents') actions are perfectly known by player i (Attacker). On the other hand, player j is imperfectly aware of player i's past choices because player j judges its opponent's actions with uncertainty. Consequently, the RIDG belongs to a repeated dynamic game with imperfect information.

From Figure 2, player i first takes an action at the beginning node. It may select action *Cooperate*, *Preattack*, or *Attack*. Next, player j responds to its opponent with action *Sleep*, *Grant*, or *Defend*. As soon as it selects action *Defend*, the game ends. Except for this case, the game will be played repeatedly.

Now, let us define players' payoffs for the repeated IDG. Players Attacker and IDS agents strive to maximize their expected payoffs over multiple rounds of the stage IDG. The expected payoff is generally described as a sum of per-period payoffs, multiplied by a discount factor δ, $\delta \in [0, 1)$. If the discount factor is not too high, the players then are interested enough in future outcomes of the game. Both players therefore put more weight on the current payoff than

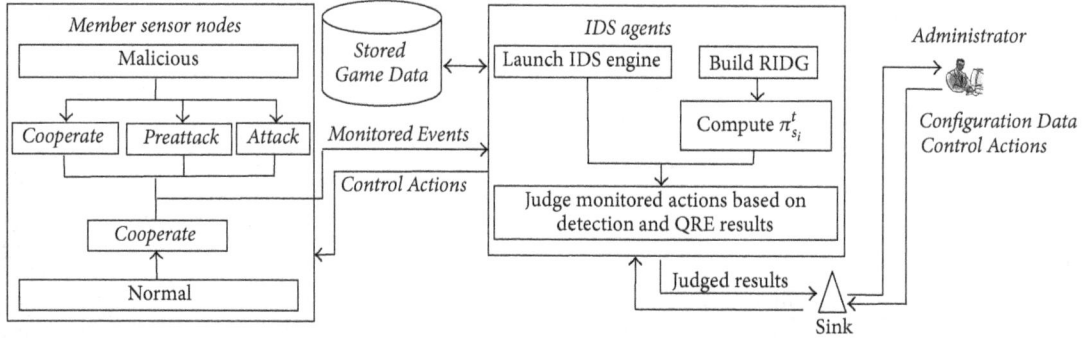

FIGURE 3: Implementation of applying our RIDG to WSNs.

on the future payoffs. The total payoff for player x, $x \in \{i, j\}$, is given as

$$u_x^t (s_x) = \sum_{y=0}^{t} \delta^y u_x^y (s_x), \quad (3)$$

where $u_x^y(s_x)$ denotes the payoff obtained by player x, $x \in \{i, j\}$, adopting strategy s_x at slot time, y, $y = 0, 1, 2, \ldots, t$. Further, in the repeated game with infinite rounds, the total payoff in (3) is often averaged. Therefore, the average discounted payoff for player x, $x \in \{i, j\}$, can be expressed as

$$\overline{u}_x^t (s_x) = (1 - \delta) \sum_{y=1}^{t} \delta^y u_x^y (s_x). \quad (4)$$

Next let us analyze the total number of strategy profiles in our RIDG. Generally, as an infinitely repeated game, the total number of strategy profiles at the tth stage is computed by multiplying the number of history strategy profiles at all stages $0, 1, \ldots, t - 1$, with the number of actions to be played at the tth stage. However, in our RIDG, the action *Defend* adopted by player j means that the game ends. In this sense, the number of combined actions excluding the terminal action *Defend* is $3 \times 2 = 6$. Therefore, the total number of strategy profiles at the tth stage, n_t, can be computed as

$$n_t = 6 \times n_{t-1}, \quad t = 1, 2, 3, \ldots, \quad (5)$$

where $n_0 = 9$.

From (5), we can see that the total number of strategy profiles of our RIDG will increase quickly as the number of all repeated stage IDGs grows. As a result, complexity to predict the future behavior of player i by computing the NE of a subgame becomes higher and higher, which motivates us to find an optimal alternative, QRE.

3.4. QRE-Based Strategies. QRE for extensive form games is first defined by McKelvey and Palfrey [55], which provides an equilibrium notion with bounded rationality. QRE is not an equilibrium refinement, and it can obtain significantly different results from NE. It is only defined for games with separate strategies, regardless of the fact that there are repeated-strategy analogues. In particular, it is developed as

a probabilistic extension of NE and can be used to give reasons why players might systematically deviate from the NE path. This is because players in QRE are assumed to make errors in selecting which strategy to play. The probability of any particular strategy being picked is positively related to the highest expected payoff from that strategy. Therefore, strategy choices in QRE are probabilistic rather than deterministic.

The characteristic that QRE provides equilibrium with bounded rationality is realized by introducing a rationality parameter to the payoff. The rationality parameter denoted by λ is changed during the process of QRE converging to the NE. When $\lambda = 0$, players are completely irrational. This case means that even though a player cannot obtain greater payoff, players Attacker and IDS agents will select another strategy other than the one indicated by NE. On the contrary, when $\lambda \to \infty$, players will follow NE since they become completely rational in this case. So far, the QRE can be calculated by

$$\pi_{s_x}^t = \frac{\exp\left(\lambda \cdot \overline{u}_x^t (s_x)\right)}{\sum_{y \in \mathcal{A}_x} \exp\left(\lambda \cdot \overline{u}_x^t (y)\right)}, \quad (6)$$

where $\pi_{s_x}^t$ is in fact the probability of player x, $x \in \{i, j\}$, selecting strategy s_x. From (6), QRE-based strategies of players Attacker and IDS agents can be obtained, respectively. In essence, QRE-based strategies are based on the introduction of payoff perturbations associated with actions adopted by players Attacker and IDS agents. The probability of a strategy profile is positively related to the average discounted payoffs held by players. The set of QRE can be regarded as a correspondence mapping the rationality parameter λ into a set of mixed strategy (the probability that each action of IDS agents will be selected by IDS agents) in $\mathcal{A}$.

4. QRE-Based Intrusion Detection for WSNs

As given in Figure 3, we realize an implementation of applying our RIDG to WSNs. The data flow begins with member sensor nodes that are being monitored by the IDS agents installed in the corresponding CH. These member sensor nodes may be normal or malicious, so they take possible actions including *Cooperate*, *Preattack*, and *Attack*. As soon as the IDS agent is woken by events of member sensor nodes,

```
(1)   t ← 1;
(2)   Initialize game parameters required in Definition 2;
(3)   Do UNTIL the end of interactions between players Attacker and the corresponding IDS agent
(4)       Woken by Monitored Events;
(5)       Judge whether Monitored Events are normal or malicious with the known intrusion detection techniques;
(6)       IF the output of detection is malicious THEN
(7)           IF the RIDG is not existed THEN
(8)               Construct the first stage RIDG with game parameters including 𝒩, 𝒜, and 𝒰;
(9)           ELSE
(10)              Obtain the current stage RIDG from the Stored Game Data;
(11)          ENDIF
(12)          Compute π_{s_x}^t according to (6);
(13)          Compute ū_x^t(s_x) according to (4) and store it into the Stored Game Data for the next stage RIDG;
(14)          Combine IDS results and π_{s_x}^t, and send them to Administrator;
(15)      ENDIF
(16)      t ← t + 1;
(17) ENDDO
```

ALGORITHM 1: QRE-based intrusion detection algorithm for IDS agents.

it filters the Monitored Events and employs an IDS engine to judge whether an event is normal or not.

Generally, IDS agents have been previously configured to make them more accurate and reliable, through Configuration Data sent by *Administrator*. Upon completion of events detection, the relevant results will be temporarily stored for the final decision. On the other hand, the IDS agent starts to initialize game parameters required in Definition 2. It accepts the results of events detection and formulates the RIDG. When the RIDG is constructed at the first stage, preferences and payoffs of two players, which have been stored in the *Stored Game Data*, are manually set by *Administrator*. The IDS agent then, employing (6), calculates the QRE probabilities with the events detection results and the stage RIDG. The QRE probabilities attained will be combined with the IDS results, and this combination will be sent to Administrator who may take Control Actions on member sensor nodes through the IDS agent. After one round of RIDG is played, the game parameters will be updated to the *Stored Game Data*. In particular, the payoffs of two players are adjusted according to (4), which will be used in the next stage RIDG. The above process will then be repeated until the IDS agent selects action *Defend*. In fact, reaching this point means the end of interactions between players Attacker and IDS agents. Next, we describe the algorithm for the process of QRE-based intrusion detection (Algorithm 1).

5. Experiments

With Gambit [56], QRE-based strategies are calculated to show us how the RIDG is actually played, as illustrated in Table 2 and Figures 4 and 5. These illustrations show that we are able to predict the Attacker's actions, so that the corresponding IDS agent can adopt the appropriate action in advance. Calculations begin with equal probabilities for each action. In this manner, every action has a probability of 0.3333

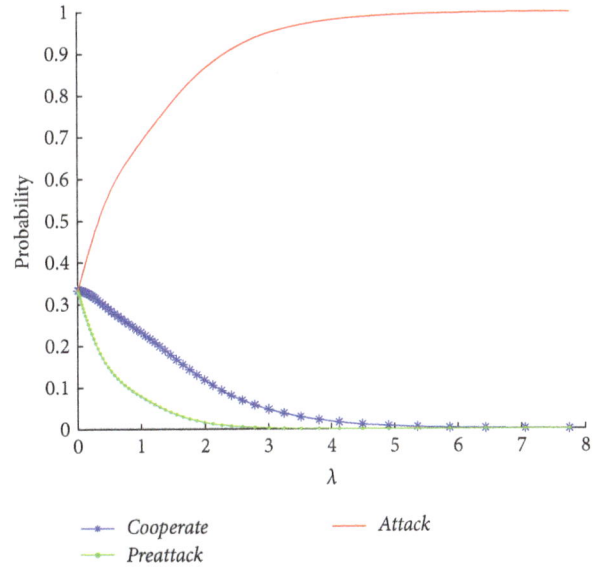

FIGURE 4: QRE-based strategies for the player Attacker.

or so, since there are three actions for each player. In addition, the rationality parameter λ starts with $\lambda = 0$ at step 1.

The trend of actions adopted by the Attacker is shown in Figure 4, where y-axis represents the probability the Attacker will select a certain strategy for a given λ. It is remarkable that the probability of the Attacker adopting the action *Cooperate* or *Preattack* is gradually decreasing while the probability of the action *Attack* is increasing. From Table 2, when $\lambda \approx 3.238326$, the probability of the action *Preattack* becomes zero approximately. This case means the action *Preattack* has been eliminated from this step. Adapting the Attacker's strategies continually, it is $\lambda \approx 161.049147$, when the selection by the Attacker of the action *Attack* becomes certain. This means

TABLE 2: QRE calculations for the players Attacker and IDS agents in the RIDG.

Step	λ	Attacker			IDS agents		
		Cooperate	Preattack	Attack	Sleep	Grant	Defend
1	0	0.333333	0.333333	0.333333	0.333333	0.333333	0.333333
2	0.010248	0.333301	0.327668	0.339031	0.333302	0.329903	0.336795
3	0.021502	0.333192	0.321514	0.345294	0.333192	0.326104	0.340703
4	0.033854	0.332985	0.314843	0.352172	0.332978	0.321894	0.345128
5	0.047402	0.332655	0.307631	0.359715	0.332625	0.317226	0.35015
6	0.062251	0.332172	0.299857	0.367971	0.332088	0.312046	0.355866
7	0.078512	0.331504	0.291508	0.376987	0.331314	0.306297	0.36239
8	0.0963	0.330612	0.28258	0.386808	0.330231	0.299913	0.369856
$\vdots$	$\vdots$	$\vdots$	$\vdots$	$\vdots$	$\vdots$	$\vdots$	$\vdots$
47	3.238326	0.0376928	0.00147867	0.960829	$2.83E-07$	$1.11E-08$	1
48	3.502404	0.0292182	0.000880193	0.969902	$6.29E-08$	$1.89E-09$	1
49	3.797081	0.021933	0.000492093	0.977575	$1.21E-08$	$2.71E-10$	1
$\vdots$	$\vdots$	$\vdots$	$\vdots$	$\vdots$	$\vdots$	$\vdots$	$\vdots$
89	161.049147	$1.14E-70$	$1.30E-140$	1	0	0	1
90	177.142263	$1.17E-77$	$1.37E-154$	1	0	0	1
91	194.844691	$2.40E-85$	$5.76E-170$	1	0	0	1
$\vdots$	$\vdots$	$\vdots$	$\vdots$	$\vdots$	$\vdots$	$\vdots$	$\vdots$
104	672.367387	$9.88E-293$	0	1	0	0	1
105	739.592327	$6.27463E-322$	0	1	0	0	1
106	813.539761	0	0	1	0	0	1

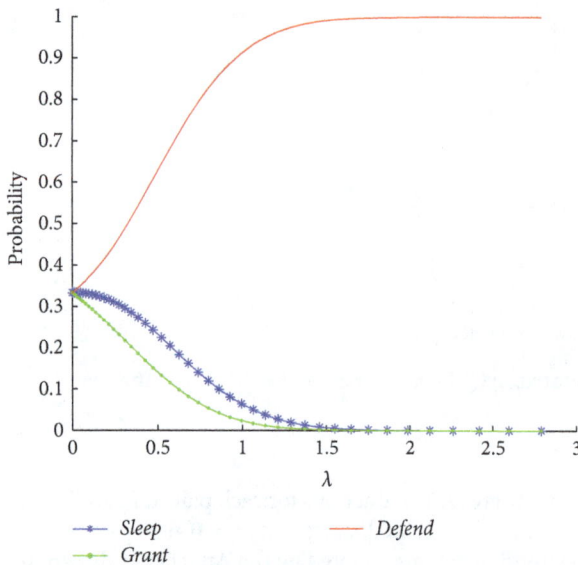

FIGURE 5: QRE-based strategies for the player IDS agents.

that if λ is greater than 161.049147, then the Attacker always selects the action *Attack* that is in fact the NE of the stage IDG.

Figure 5 shows the trend of actions adopted by IDS agents, where the probability of the action *Sleep* or *Grant* is decreasing and the probability of the action *Defend* is increasing. However, compared to the changeable trend of the selection by the Attacker, the action adopted by the IDS agents converges quickly to the action *Defend* that is the NE of the stage IDG. This point, from Table 2, is obtained when $\lambda \approx 3.238326$ for the IDS agents while λ is 161.049147 or so for the Attacker.

6. Conclusion

To save sensor nodes' power, we have put forward a method based on QRE to make IDS agents not always be in *Defend*. A stage IDG that is able to reflect interactions between the Attacker and IDS agents has been formulated, where we have thoroughly considered players' preferences and have assigned payoffs of players according to Binmore's method. To reflect the reality that the Attacker and IDS agents interact continually, we have extended the stage IDG to a repeated IDG and have defined the corresponding payoffs. Further, we have given the method of calculating QRE-based strategies that predict the Attacker's future behavior. As a result, optimal reactions can be suggested to the IDS agents to protect WSNs.

In the future, to extend the current game model RIDG when taking into account multiple Attackers that may collude is an interesting work.

Conflict of Interests

The authors declare that there is no conflict of interests regarding the publication of this paper.

Acknowledgments

This work was supported by National Natural Science Foundation of China under Grant no. 61272034, by Zhejiang Provincial Natural Science Foundation of China under Grants LY13F030012 and LY13F020035, and by Science Foundation of Shaoxing University under Grants 20145021 and 2014LG1009.

References

[1] J. Yick, B. Mukherjee, and D. Ghosal, "Wireless sensor network survey," *Computer Networks*, vol. 52, no. 12, pp. 2292–2330, 2008.

[2] A. Ramos and R. H. Filho, "Sensor data security level estimation scheme for wireless sensor networks," *Sensors*, vol. 15, no. 1, pp. 2104–2136, 2015.

[3] A. Derhab, A. Bouras, M. R. Senouci, and M. Imran, "Fortifying intrusion detection systems in dynamic Ad Hoc and wireless sensor networks," *International Journal of Distributed Sensor Networks*, vol. 2014, Article ID 608162, 15 pages, 2014.

[4] R. Mitchell and I.-R. Chen, "A survey of intrusion detection in wireless network applications," *Computer Communications*, vol. 42, pp. 1–23, 2014.

[5] I. Butun, S. D. Morgera, and R. Sankar, "A survey of intrusion detection systems in wireless sensor networks," *IEEE Communications Surveys and Tutorials*, vol. 16, no. 1, pp. 266–282, 2014.

[6] C.-F. Hsieh, R.-C. Chen, and Y.-F. Huang, "Applying an ontology to a patrol intrusion detection system for wireless sensor networks," *International Journal of Distributed Sensor Networks*, vol. 2014, Article ID 634748, 14 pages, 2014.

[7] S.-H. Seo, J. Won, S. Sultana, and E. Bertino, "Effective key management in dynamic wireless sensor networks," *IEEE Transactions on Information Forensics and Security*, vol. 10, no. 2, pp. 371–383, 2015.

[8] R. Soosahabi, M. Naraghi-Pour, D. Perkins, and M. A. Bayoumi, "Optimal probabilistic encryption for secure detection in wireless sensor networks," *IEEE Transactions on Information Forensics and Security*, vol. 9, no. 3, pp. 375–385, 2014.

[9] A. Abduvaliyev, A.-S. K. Pathan, J. Zhou, R. Roman, and W.-C. Wong, "On the vital areas of intrusion detection systems in wireless sensor networks," *IEEE Communications Surveys and Tutorials*, vol. 15, no. 3, pp. 1223–1237, 2013.

[10] N. A. Alrajeh, S. Khan, and B. Shams, "Intrusion detection systems in wireless sensor networks: a review," *International Journal of Distributed Sensor Networks*, vol. 2013, Article ID 167575, 7 pages, 2013.

[11] M. H. Manshaei, Q. Zhu, T. Alpcan, T. Basar, and J.-P. Hubaux, "Game theory meets network security and privacy," *ACM Computing Surveys*, vol. 45, no. 3, article 25, 39 pages, 2013.

[12] S. Shen, G. Yue, Q. Cao, and F. Yu, "A survey of game theory in wireless sensor networks security," *Journal of Networks*, vol. 6, no. 3, pp. 521–532, 2011.

[13] X. Liang and Y. Xiao, "Game theory for network security," *IEEE Communications Surveys and Tutorials*, vol. 15, no. 1, pp. 472–486, 2013.

[14] H.-Y. Shi, W.-L. Wang, N.-M. Kwok, and S.-Y. Chen, "Game theory for wireless sensor networks: a survey," *Sensors*, vol. 12, no. 7, pp. 9055–9097, 2012.

[15] M. D. McCubbins, M. Turner, and N. Weller, "Testing the foundations of quantal response equilibrium," in *Social Computing, Behavioral-Cultural Modeling and Prediction*, vol. 7812 of *Lecture Notes in Computer Science*, pp. 144–153, Springer, Berlin, Germany, 2013.

[16] H. Al-Hamadi and I.-R. Chen, "Redundancy management of multipath routing for intrusion tolerance in heterogeneous wireless sensor networks," *IEEE Transactions on Network and Service Management*, vol. 10, no. 2, pp. 189–203, 2013.

[17] H. Al-Hamadi and I. R. Chen, "Integrated intrusion detection and tolerance in homogeneous clustered sensor networks," *ACM Transactions on Sensor Networks*, vol. 11, no. 3, article 47, 24 pages, 2015.

[18] E. J. Cho, C. S. Hong, S. Lee, and S. Jeon, "A partially distributed intrusion detection system for wireless sensor networks," *Sensors*, vol. 13, no. 12, pp. 15863–15879, 2013.

[19] A. H. Farooqi, F. A. Khan, J. Wang, and S. Lee, "A novel intrusion detection framework for wireless sensor networks," *Personal and Ubiquitous Computing*, vol. 17, no. 5, pp. 907–919, 2013.

[20] I. Kim, D. Oh, M. K. Yoon, K. Yi, and W. W. Ro, "A distributed signature detection method for detecting intrusions in sensor systems," *Sensors*, vol. 13, no. 4, pp. 3998–4016, 2013.

[21] B. Sun, X. Shan, K. Wu, and Y. Xiao, "Anomaly detection based secure in-network aggregation for wireless sensor networks," *IEEE Systems Journal*, vol. 7, no. 1, pp. 13–25, 2013.

[22] S. Shamshirband, N. B. Anuar, M. L. M. Kiah et al., "CoFAIS: cooperative fuzzy artificial immune system for detecting intrusion in wireless sensor networks," *Journal of Network and Computer Applications*, vol. 42, pp. 102–117, 2014.

[23] M. Riecker, S. Biedermann, R. El Bansarkhani, and M. Hollick, "Lightweight energy consumption-based intrusion detection system for wireless sensor networks," *International Journal of Information Security*, vol. 14, no. 2, pp. 155–167, 2015.

[24] S. Shen, Y. Li, H. Xu, and Q. Cao, "Signaling game based strategy of intrusion detection in wireless sensor networks," *Computers & Mathematics with Applications*, vol. 62, no. 6, pp. 2404–2416, 2011.

[25] J.-Y. Huang, I.-E. Liao, Y.-F. Chung, and K.-T. Chen, "Shielding wireless sensor network using Markovian intrusion detection system with attack pattern mining," *Information Sciences. An International Journal*, vol. 231, pp. 32–44, 2013.

[26] H. Moosavi and F. M. Bui, "A game-theoretic framework for robust optimal intrusion detection in wireless sensor networks," *IEEE Transactions on Information Forensics and Security*, vol. 9, no. 9, pp. 1367–1379, 2014.

[27] S. Shen, R. Han, L. Guo, W. Li, and Q. Cao, "Survivability evaluation towards attacked WSNs based on stochastic game and continuous-time Markov chain," *Applied Soft Computing*, vol. 12, no. 5, pp. 1467–1476, 2012.

[28] S. Shen, H. Li, R. Han, A. V. Vasilakos, Y. Wang, and Q. Cao, "Differential game-based strategies for preventing malware propagation in wireless sensor networks," *IEEE Transactions on Information Forensics and Security*, vol. 9, no. 11, pp. 1962–1973, 2014.

[29] S. Shamshirband, A. Patel, N. B. Anuar, M. L. M. Kiah, and A. Abraham, "Cooperative game theoretic approach using fuzzy Q-learning for detecting and preventing intrusions in wireless sensor networks," *Engineering Applications of Artificial Intelligence*, vol. 32, pp. 228–241, 2014.

[30] J. Liu, S. Shen, G. Yue, R. Han, and H. Li, "A stochastic evolutionary coalition game model of secure and dependable virtual service in Sensor-Cloud," *Applied Soft Computing*, vol. 30, pp. 123–135, 2015.

[31] A. B. MacKenzie and L. A. DaSilva, *Game Theory for Wireless Engineers*, Morgan & Claypool Publishers, San Rafael, Calif, USA, 2006.

[32] A. Agah and S. K. Das, "Preventing DoS attacks in wireless sensor networks: a repeated game theory approach," *International Journal of Network Security*, vol. 5, no. 2, pp. 145–153, 2007.

[33] C. Pandana, Z. Han, and K. J. R. Liu, "Cooperation enforcement and learning for optimizing packet forwarding in autonomous wireless networks," *IEEE Transactions on Wireless Communications*, vol. 7, no. 8, pp. 3150–3163, 2008.

[34] J. Chen, S. Lian, C. Fu, and R. Du, "A hybrid game model based on reputation for spectrum allocation in wireless networks," *Computer Communications*, vol. 33, no. 14, pp. 1623–1631, 2010.

[35] Z. Kong and Y.-K. Kwok, "Efficient wireless packet scheduling in a non-cooperative environment: game theoretic analysis and algorithms," *Journal of Parallel and Distributed Computing*, vol. 70, no. 8, pp. 790–799, 2010.

[36] B. Niu, H. V. Zhao, and H. Jiang, "A cooperation stimulation strategy in wireless multicast networks," *IEEE Transactions on Signal Processing*, vol. 59, no. 5, pp. 2355–2369, 2011.

[37] R. Trestian, O. Ormond, and G.-M. Muntean, "Reputation-based network selection mechanism using game theory," *Physical Communication*, vol. 4, no. 3, pp. 156–171, 2011.

[38] S. Kandeepan, S. K. Jayaweera, and R. Fedrizzi, "Power-trading in wireless communications: a cooperative networking business model," *IEEE Transactions on Wireless Communications*, vol. 11, no. 5, pp. 1872–1880, 2012.

[39] Y. E. Sagduyu, R. A. Berry, and A. Ephremides, "Jamming games in wireless networks with incomplete information," *IEEE Communications Magazine*, vol. 49, no. 8, pp. 112–118, 2011.

[40] D. Hao, X. Liao, A. Adhikari, K. Sakurai, and M. Yokoo, "A repeated game approach for analyzing the collusion on selective forwarding in multihop wireless networks," *Computer Communications*, vol. 35, no. 17, pp. 2125–2137, 2012.

[41] P. Zhou, Y. Chang, and J. A. Copeland, "Reinforcement learning for repeated power control game in cognitive radio networks," *IEEE Journal on Selected Areas in Communications*, vol. 30, no. 1, pp. 54–69, 2012.

[42] M. Zhu and S. Martínez, "Distributed coverage games for energy-aware mobile sensor networks," *SIAM Journal on Control and Optimization*, vol. 51, no. 1, pp. 1–27, 2013.

[43] Y. Luo, F. Szidarovszky, Y. Al-Nashif, and S. Hariri, "A fictitious play-based response strategy for multistage intrusion defense systems," *Security and Communication Networks*, vol. 7, no. 3, pp. 473–491, 2014.

[44] Y. Sun, Y. Guo, Y. Ge, S. Lu, J. Zhou, and E. Dutkiewicz, "Improving the transmission efficiency by considering non-cooperation in ad hoc networks," *Computer Journal*, vol. 56, no. 8, pp. 1034–1042, 2013.

[45] D. B. Smith, M. Portmann, W. L. Tan, and W. Tushar, "Multi-source-destination distributed wireless networks: pareto-efficient dynamic power control game with rapid convergence," *IEEE Transactions on Vehicular Technology*, vol. 63, no. 6, pp. 2744–2754, 2014.

[46] A. A. Daoud, G. Kesidis, and J. Liebeherr, "Zero-determinant strategies: a game-theoretic approach for sharing licensed spectrum bands," *IEEE Journal on Selected Areas in Communications*, vol. 32, no. 11, pp. 2297–2308, 2014.

[47] D. Hao, Z.-H. Rong, and T. Zhou, "Zero-determinant strategy: an underway revolution in game theory," *Chinese Physics B*, vol. 23, no. 7, Article ID 078905, 2014.

[48] A. Farraj, E. Hammad, A. Al Daoud, and D. Kundur, "A game-theoretic analysis of cyber switching attacks and mitigation in smart grid systems," *IEEE Transactions on Smart Grid*, 2015.

[49] H. Zhang, N. Dusit, L. Song, T. Jiang, and Z. Han, "Zero-determinant strategy in cheating management of wireless cooperation," in *Proceedings of the IEEE Global Communications Conference (GLOBECOM '14)*, pp. 4382–4386, Austin, Tex, USA, December 2014.

[50] D. Hao, Z. Rong, and T. Zhou, "Extortion under uncertainty: zero-determinant strategies in noisy games," *Physical Review E—Statistical, Nonlinear, and Soft Matter Physics*, vol. 91, no. 5, Article ID 052803, 8 pages, 2015.

[51] R. Yang, C. Kiekintveld, F. Ordóñez, M. Tambe, and R. John, "Improving resource allocation strategies against human adversaries in security games: an extended study," *Artificial Intelligence*, vol. 195, pp. 440–469, 2013.

[52] I. Kantzavelou and S. Katsikas, "A game-based intrusion detection mechanism to confront internal attackers," *Computers & Security*, vol. 29, no. 8, pp. 859–874, 2010.

[53] A. H. Farooqi and F. A. Khan, "A survey of intrusion detection systems for wireless sensor networks," *International Journal of Ad Hoc and Ubiquitous Computing*, vol. 9, no. 2, pp. 69–83, 2012.

[54] K. Binmore, *Playing for Real : A Text on Game Theory*, Oxford University Press, New York, NY, USA, 2007.

[55] R. D. McKelvey and T. R. Palfrey, "Quantal response equilibria for extensive form games," *Experimental Economics*, vol. 1, no. 1, pp. 9–41, 1998.

[56] R. D. McKelvey, A. M. McLennan, and T. L. Turocy, "Gambit: Software Tools for Game Theory," Version 14.1.0, 2014, http://www.gambit-project.org/.

3

Authenticated Diffie-Hellman Key Agreement Scheme that Protects Client Anonymity and Achieves Half-Forward Secrecy

Hung-Yu Chien

Department of Information Management, National Chi-Nan University, 470 University Road, Puli, Nantou, Taiwan

Correspondence should be addressed to Hung-Yu Chien; hychien@ncnu.edu.tw

Academic Editor: Francesco Gringoli

Authenticated Diffie-Hellman key agreement (D-H key) is the de facto building block for establishing secure session keys in many security systems. Regarding the computations of authenticated D-H key agreement, the operation of modular exponentiation is the most expensive computation, which incurs a heavy loading on those clients where either their computational capacities or their batteries are limited and precious. As client's privacy is a big concern in several e-commerce applications, it is desirable to extend authenticated D-H key agreement to protect client's identity privacy. This paper proposes a new problem: the modified elliptic curves computational Diffie-Hellman problem (MECDHP) and proves that the MECDHP is as hard as the conventional elliptic curves computational Diffie-Hellman problem (ECDHP). Based on the MECDHP, we propose an authenticated D-H key agreement scheme which greatly improves client computational efficiency and protects client's anonymity from outsiders. This new scheme is attractive to those applications where the clients need identity protection and lightweight computation.

1. Introduction

Authentication is an essential security; however, many existent authentication schemes either did not provide key agreement during authentication [1–3], did not protect client privacy [4], or required many expensive modular exponentiations [4–16]. Authenticated key agreement aims at simultaneously providing authentication of communicating parties and establishing secure session keys. Nowadays it is very easy for any network operators or even outsiders to collect users' behavior and violate their privacy when they surf the Internet if key agreement schemes do not protect users' identities. It is, therefore, imperative to design key agreement schemes which protect client's identity. With such schemes, users' privacy can be protected when they enjoy Internet activities, e-commerce or m-commerce transactions.

Conventionally, the computational Diffie-Hellman problem over Galis field (called CDHP) and the same problem over elliptic curves (called ECDHP) are the most popular building blocks for many authenticated key agreement schemes [4–8, 10–19], owing to their hardness. However, the modular exponentiation computations over Galis field or the point multiplications over elliptic curves impose a heavy computational stress on those clients where either

their computing capacities or their batteries are limited. Such clients are called thin clients in the rest of this paper. Even a native D-H key scheme (not including authentication of communicating parties) that uses the CDHP requires of each party two modular exponentiations, and the corresponding version over elliptic curves would require of each party two point multiplications. Generally an authenticated version of D-H key scheme or an extended version of D-H key scheme which protects client's anonymity would require more modular exponentiations or more point multiplications [9–12, 14–16, 20], respectively. *It is, therefore, important to reduce the number of exponentiation computations/point multiplications for those thin clients.* In 2014, Chien [4] formulated the modified computational D-H problem (MCDHP) and proposed an authenticated D-H key agreement scheme, using the MCDHP. The scheme [4] effectively reduced the number of modular exponentiations but it did not protect client's anonymity.

Conventionally, key agreement schemes transmit participants' identities in plaintext. As privacy has been a big concern in many applications, it is desirable to protect participants' identities during the key agreement process. Such kinds of key agreement schemes are called authenticated key agreement schemes with client anonymity (or simply an anonymous authenticated key agreement). Chien [12] classified four types of two-party key agreement schemes, according to the protection of participants' anonymity. Type 1: the privacy of identities of two communicating parties is not protected; this type corresponds to those conventional two-party key agreement schemes. Type 2: the client's identity is protected from the outsiders, but the identity of the server is not protected. Type 3: the client's identity is protected from the outsiders, but the anonymity of the server is only protected from unregistered entities. The scenarios for such a type are like that; in a mission-oriented ad hoc network, the clients and the servers want to protect their identities from outsiders, while all the preregistered clients know the IP address or MAC address of the servers. Type 4: both the identity of the client and the identity of the server are protected from outsiders. Type 2 is the most popular one for authenticated key agreement schemes that protect participant's anonymity, because it corresponds to the cases where clients want to protect their anonymity from outsiders. In this paper, we focus on authenticated two-party key agreement for Type 2 cases and aim to improve client's computational efficiency.

Authentication with anonymity protection is a popular topic, and there are some popular techniques for achieving client's anonymity during authentication process. Chien [2] classifies existent techniques into four categories: probabilistic encryption-based scheme like the OAKley protocol [13], pseudonym-based schemes like [20], hash-chain-based scheme like Ohkubo et al.'s scheme [1], and error-correction-codes-based scheme like [2, 3]. It is challenging to extend the existent CDHP-based key agreement schemes like [5–7, 13] (or ECDHP-based schemes like [8]) to their anonymous versions and reduce the modular exponentiation load (or the point multiplication load). In this paper, we first formulate a modified ECDHP (MECDHP) and prove its security. Then we propose a new authenticated two-party key agreement scheme with client anonymity, based on the MECDHP problem and Ohkubo et al.'s hashing chain technique [1]. *The new scheme protects client anonymity, effectively reduces the computational load for client, and preserves the strong security of the ECDHP.*

The rest of this paper is organized as follows. Section 2 introduces the MECDHP and proves its security. Section 3 proposes our new scheme. Section 4 examines its security and evaluates its performance. Section 5 states our conclusions and discussions.

2. The MECDHP Problem and the Security Requirements

In this section, we first propose the MECDHP and prove its security. Then we introduce the model and discuss the security requirements of an authenticated key agreement scheme with client anonymity.

Elliptic Curves over GF(p). A nonsupersingular elliptic curve $E(Fp)$ is the set of points $P = (x, y)$ and the point O (called

the *point at infinity*), where $x, y \in Z_p$ satisfy the equation $y^2 \equiv x^3 + ax + b \pmod{p}$ ($a, b \in Z_p$ are constants, such that $4a^3 + 27b^2 \neq 0 \bmod p$). Two points $P = (x_1, y_1)$ and $Q = (x_2, y_2)$ on the elliptic curve E can be added together using the following rule: if $x_2 = x_1$ and $y_2 = -y_1$, then $P + Q = O$; otherwise, $P + Q = (x_3, y_3)$, where $x_3 = \lambda^2 - x_1 - x_2 \bmod p$, $y_3 = \lambda(x_1 - x_3) - y_1 \bmod p$, and $\lambda = (y_2 - y_1)/(x_2 - x_1)$ if $P \neq Q$ or $\lambda = (3x_1^2 + a)/(2y_1)$ if $P = Q$.

Definition 1. The computational elliptic curve Diffie-Hellman problem (ECDHP) is as follows: given an elliptic curve over a finite field F_p, a point $P \in E(F_p)$ of order q and points $A = aP$ and $B = bP \in \langle P \rangle$ find the point $C = abP$.

Now we formulate a new problem called the modified computational elliptic curve Diffie-Hellman problem (MECDHP) as follows.

Definition 2. The *modified computational elliptic curve Diffie-Hellman problem* (the MECDHP) is as follows: given an elliptic curve over a finite field F_p, a point $P \in E(F_p)$ of order q, $a + x$, and points $A = xP$ and $B = bP \in \langle P \rangle$ find the point $C = abP$.

We prove the hardness of the MECDHP to be as hard as the ECDHP as follows.

Theorem 3. *The MECDHP is as hard as the ECDHP.*

Proof. We prove this by reduction.

(1) The MECDHP Is Reduced to the ECDHP. Given an instance of the MECDHP problem $(x + a, P, xP, \text{ and } bP)$, we can compute $(x + a)P - xP = aP$ and get the instance $(P, aP, \text{ and } bP)$ for the ECDHP. Assume there is one oracle that can answer the ECDHP. Now we input the instance $(P, aP, \text{ and } bP)$ to the oracle, and we get the answer abP.

(2) The ECDHP Is Reduced to the MECDHP. Assume there is one oracle that can answer the MECDHP: given $(x + a, P, xP, \text{ and } bP)$, it outputs abP.

Now given an instance of the CDHP $(P, aP, \text{ and } bP)$, we then choose a random value t and input the instance $(t, P, aP, \text{ and } bP)$ to the MECDHP oracle. The oracle will answer $b(tP - aP) = b(t - a)P = tbP - abP$. Using the response, we can derive $-(tbP - abP - t(bP)) = -(-abP) = abP$. That is, we get the answer for the ECDHP problem $(P, aP, \text{ and } bP)$.

Based on the above arguments, we prove the theorem.   □

Now we introduce the model and discuss the security requirements of authenticated D-H key agreement scheme with client anonymity as follows. Our model consists of three kinds of entities: clients, servers, and outsiders. Clients would like to establish secure session keys with servers via key agreement schemes and their identities should not be learned by any outsiders. An outsider can actively manipulate the communications via replay, modification, or interception.

(1) Consider Mutual authentication of client and server and resistance to various attacks like replay attack, impersonation attack, man-in-the-middle attack, and known key attack.

(2) Consider Partial forward secrecy and perfect forward secrecy. Here partial forward secrecy requires that even if we assume one party's long-term private key is disclosed someday, the previous communications (the session keys before the disclosure) are still secure. If partial forward secrecy is preserved only when one specific party's private key is compromised but it does not hold for the other party, then it is called half forward secrecy. While perfect forward requires that even if we assume both of the two parties' long-term private keys are disclosed someday, the previous communications (the session keys before the disclosure) are still secure.

(3) Anonymity of the client: the identities of clients should be well protected from outsiders.

3. New Authenticated D-H Key Agreement Scheme with Client Anonymity

We introduce the following notations. We will omit the mod q operation in the rest of this paper to simplify the presentation when the context is clear.

The Notations.

$E(F_p)$, P, and q are as follows: $E(F_p)$ is an elliptic curve over F_p, $P \in E(F_p)$ is a generator point for a group over $E(F_p)$.

$h()$, $g()$ are two cryptographic hash functions. Two different hash functions are chosen here to hinder outsiders from tracking users, using correlated data. This idea is inspired by [1]. These functions could be implemented using a pseudorandom function (PRF) with distinct paddings.

C, S are as follows: C and S, respectively, denote client and server.

ID_C, ID_S are as follows: ID_C and ID_S, respectively, denote the identity of the client and that of the server, where ID_S is static while ID_C is dynamically updated.

c, s are as follows: $c, s \in Z_q^*$, respectively, denote the private key of C and that of S.

P_C, P_S are as follows: $P_C = cP$ and $P_S = sP$, respectively, denote their corresponding public keys.

x, y are as follows: $x, y \in Z_q^*$, respectively, denote ephemeral private keys.

X, Y are as follows: $X = xP$, and $Y = yP$ denotes their corresponding public keys.

$\oplus$, $\|$ are as follows: $\oplus$ denotes the exclusive OR operation. $\|$ denotes concatenation. Here we abusively use the notation $\oplus$ between two elliptic curve points to represent $(x1, y1) \oplus (x2, y2) = (x1 \oplus x2, y1 \oplus y2)$.

The scheme is depicted in Figure 1 and is described as follows.

The scheme consists of two phases: initial phase and key agreement phase.

The Initialization Phase. Each registered client C shares its identity ID_C and public key P_C with the server S, where S keeps two values $\text{ID}_{C,\text{old}}$ and $\text{ID}_{C,\text{new}}$ to track the dynamically changing identity of the client. Both $\text{ID}_{C,\text{old}}$ and $\text{ID}_{C,\text{new}}$ are initialized as ID_C. Each client keeps the identity and the public key P_S of the server.

The Key Agreement Phase.

(1) $S \rightarrow C: \text{ID}_S, Y$.

The server randomly chooses an integer $y \in_R Z_q^*$, computes $Y = yP$, and sends (ID_S, Y) to the client.

(2) $C \rightarrow S: g(\text{ID}_C), c + x, \text{Auth}_C$.

The client chooses a random integer $x \in_R Z_q^*$ and computes $K_{C,S} = xY$, $BL = xP_S$, and $\text{Auth}_C = h((P_C + K_{C,S}) \oplus BL)$. He sends $(g(\text{ID}_C), c + x, \text{Auth}_C)$ to the server.

(3) $S \rightarrow C: \text{Auth}_S$.

Upon receiving the client's data, the server first searches its data base to look up the entry that satisfies either $g(\text{ID}_C) \stackrel{?}{=} g(\text{ID}_{C,\text{old}})$ or $g(\text{ID}_C) \stackrel{?}{=} g(\text{ID}_{C,\text{new}})$. The matched entry ($\text{ID}_{C,\text{old}}$ or $\text{ID}_{C,\text{new}}$) is referred as $\text{ID}_{C,\text{matched}}$. The server uses the matched entry to get the corresponding public key P_C and computes $X = (c + x)P - P_C = xP$, $BL = sX = sxP$, and $K_{C,S} = yX = xyP$. It then checks whether the equation $h((P_C + K_{C,S}) \oplus BL) \stackrel{?}{=} \text{Auth}_C$ holds. If the verification succeeds, then it updates $\text{ID}_{C,\text{old}} = \text{ID}_{C,\text{matched}}$ and $\text{ID}_{C,\text{new}} = h(\text{ID}_{C,\text{matched}})$ and computes $\text{Auth}_S = h(P_C \oplus K_{C,S} \oplus BL)$.

(4) C: the client uses its local values to verify the validity of Auth_S. If the verification succeeds, then it updates its identity ID_C as $h(\text{ID}_C)$.

The final session key of the session is computed using any secure hash function: one example is like $h(P_C, P_S, x + c, Y, K_{C,S})$.

4. Security Analysis and Performance Evaluation

4.1. Security Analysis. In this section, we examine the security of the proposed authenticated D-H key agreement scheme with client anonymity. The formulated MECDHP used in the scheme has been proved to be as hard as the conventional ECDHP. Our new scheme improves the computational efficiency of client while preserving the strong security properties of conventional secure D-H key agreement schemes *under the assumption that the communicating parties' long-term secret keys are secure.* But, the security properties might

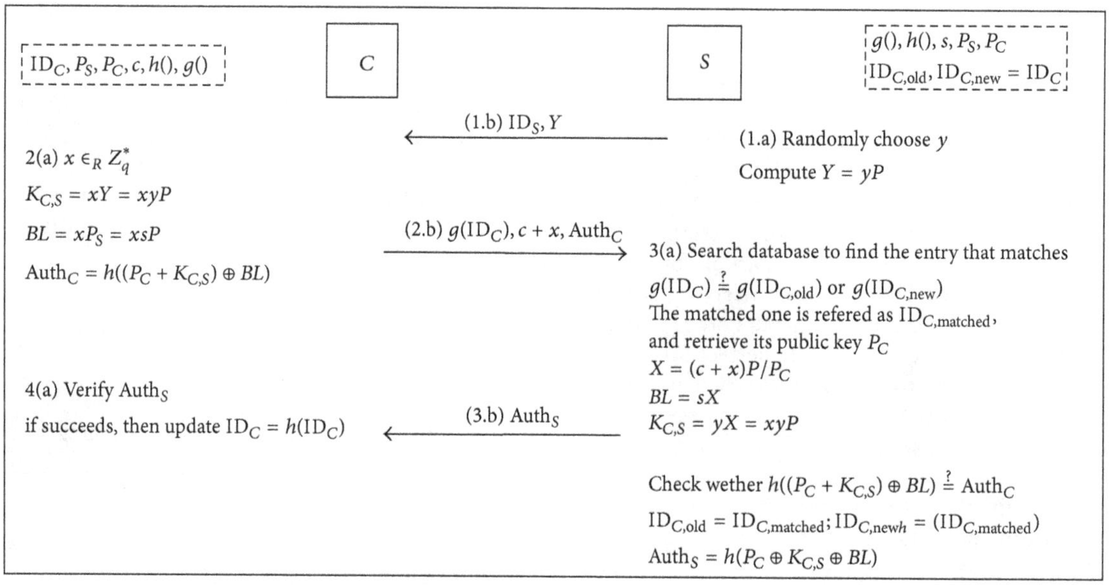

FIGURE 1: Authenticated D-H key agreement with client anonymity.

change when client's long-term secret keys are compromised. We examine these properties as follows.

Authentication of the Client. The authentication of the client depends on its ability of providing a valid $\text{Auth}_C = h((P_C + K_{C,S}) \oplus BL)$ in response to server's challenge Y. To have a valid Auth_C, the client should be able to provide *well-formed "x+c,"* on which the server can correctly derive the corresponding $X = (c + x)P - P_C = xP$, $BL = sX = sxP$, and $K_{C,S} = yX = xyP$ to verify the Auth_C. A well-formed "$x+c$" implies that the client knows the private key c. This ensures the authentication of the client.

Authentication of the Server. The authentication of server depends on server's ability of providing a valid $\text{Auth}_S = h(P_C \oplus K_{C,S} \oplus BL)$, which depends on the server's ability of computing $BL = sX$, using $X = (c + x)P - P_C = xP$. To compute $BL = sX$, it requires the knowledge of the server's private key s. This ensures the authentication of the server.

Resistance to Replay Attack and Man-in-the-Middle Attack. The client and the server, respectively, issue new challenges $x + c$ and Y in each session. The server is expected to generate the corresponding value $BL = sX = s((x + c)P - P_C)$, where the value is derived from the well-formed challenge $x+c$. The client is expected to have the ability of computing the value $K_{C,S} = xY$, where Y is the challenge. All of the computations involve new challenges. As long as the challenges are random and fresh, this ensures the resistance to replay attack and man-in-the-middle attack.

Half Forward Secrecy. Partial forward secrecy concerns the privacy of session keys corresponding to those sessions before one party's private key was compromised.

(1) If we assume that the server's long-term secret s is compromised, then the forward privacy of the session

key $h(P_C, P_S, x + c, Y, K_{C,S})$ is still preserved, because the computation of the session key involves $K_{C,S}$. Here the computation of $K_{C,S}$ needs the knowledge of either x or y. This ensures the partial forward secrecy against server's compromise.

(2) If we assume the client's long-term secret c is compromised, then the attacker can derive the ephemeral secret x from $x + c$ and derive the session key; that is, our scheme does not preserve partial forward secrecy against client's compromise. So we call that our scheme owns *half forward secrecy* property but not perfect forward secrecy.

Client Anonymity. The client-identity-related data transmitted in our protocol include $g(\text{ID}_C)$, $\text{Auth}_C = h((P_C + K_{C,S}) \oplus BL)$, and $\text{Auth}_S = h(P_C \oplus K_{C,S} \oplus BL)$. The identity ID_C is updated as $h(\text{ID}_C)$ per successful authentication, and only its hashed value $g(\text{ID}_C)$ is transmitted. Therefore, an attacker could not learn clients' identities and could not track any client using eavesdropped data. The other identity-related data is client's public key P_C. The calculations of Auth_C and Auth_S first obscure P_C with random value BL and then apply hashing: this ensures that attackers cannot link the transmissions to any specific client.

4.2. Performance Evaluation. Now we examine the performance of our scheme in terms of communication cost and computational cost. Regarding communication cost, we concern the number of message steps and the length of message length. Our scheme requires three message steps. It demands the message length $3L_h + 2L_{point} + 1L_q$, where L_h denotes the length of hash function ($|h()|$ and $|g()|$), L_{point} denotes the length of string representation of one elliptic curve point, and L_q denotes the bit length of q.

TABLE 1: Summary of performance comparison among two-party key agreement schemes with client anonymity.

Scheme	Ours.	Yoon-You [11]	Chien [12]	Yang et al. [10]	Wu-Hsu [14]	Wang et al. [15]
Number of steps	3	3	3	3	3	3
Length of message	$3L_h + 2L_{\text{point}} + 1L_q$	$3L_N + 1L_{\text{Time}} + 1L_{\text{ID}}$	$3L_N + 1L_{\text{Time}} + 1L_{\text{ID}}$	$3L_N + 1L_{\text{Time}} + 1L_{\text{ID}}$	$3L_N + 1L_{\text{Time}}$	$5L_{\text{nonce}} + 4L_{\text{point}} + 1L_{\text{ID}}$
Computational cost of client	$2T_{\text{PS}} + T_{\text{PA}} + 3T_{\text{XOR}} + 3T_h$	$5T_{\text{exp}} + 2T_{\text{multi}} + 1T_{\text{enc}}$	$5T_{\text{exp}} + 1T_{\text{multi}} + 1T_{\text{enc}} + 2T_h$	$5T_{\text{exp}} + 3T_{\text{multi}} + 1T_{\text{enc}} + 1T_h$	$4T_{\text{exp}} + 3T_{\text{multi}} + 1T_h$	$4T_{\text{PS}} + 2T_{\text{enc}} + 3T_h$
Computational cost for server	$4T_{\text{PS}} + 2T_{\text{PA}} + 3T_{\text{XOR}} + (n+2)T_h$	$5T_{\text{exp}} + 2T_{\text{multi}} + 1T_{\text{enc}}$	$5T_{\text{exp}} + 2T_{\text{multi}} + 1T_{\text{enc}} + 2T_h$	$5T_{\text{exp}} + 2T_{\text{multi}} + 1T_{\text{enc}} + 1T_h$	$4T_{\text{exp}} + 2T_{\text{multi}} + 1T_h$	$4T_{\text{PS}} + 2T_{\text{enc}} + 3T_h$
Security properties	Mutual auth.	Mutual auth.	Mutual auth.	Client/server impersonation [11]	Client/server impersonation [11]	Mutual auth.
Client anonymity	Yes	Fail (Note 5)	Yes	Yes	Yes	Yes
Forward secrecy	Half forward secrecy	Yes	Yes	Yes	Yes	Note 3
NP problem	MCDHP	RSA, CDHP	RSA, CDHP	RSA, CDHP	RSA, CDHP	Note 3

Note 1: L_N: the bit length of RSA modulus N; L_h: the length of hash function ($|h()|$ and $|g()|$); L_{point}: the length of string representation of one elliptic curve point, and L_q denotes the bit length of q. L_{Time}: the bit length of timestamp; L_{ID}: the length of identity; L_{nonce}: the length of nonce.

Note 2: T_{PS}: time complexity of one elliptic curve point multiplication; T_{PA}: time complexity for one elliptic curve point addition; T_{XOR}: time complexity for one XOR operation; T_h: time complexity for one hash operation; T_{exp}: time complexity of one modular exponentiation; T_{multi}: time complexity of one modular multiplication; T_{enc}: time complexity of one symmetric encryption.

Note 3: The session key security was based on the privacy of nonce and the fixed D-H key (the fixed key $sP_C = cP_C = csP$). If one private key of either the server or the client is compromised, then the long-term, fixed D-H key is compromised and one of the nonce is compromised; therefore, its security is not strong as the ECDHP.

Note 4: Our scheme applies hashing to protect the dynamic identity while other schemes use encryption of identity to protect anonymity; therefore, we can simplify the comparison by assuming $L_h = L_{\text{ID}}$.

Note 5: Chien [12] reported the failure of client anonymity of Yoon-You's scheme [11].

Next we examine the computational cost of our scheme. The client in our scheme demands $2T_{\text{PS}} + 1T_{\text{PA}} + 3T_{\text{XOR}} + 3T_h$, where T_{PS} denotes the time complexity of one elliptic curve point multiplication, T_{PA} denotes that of one elliptic curve point addition, T_{XOR} denotes that of one XOR operation, and T_h denotes that of one hash operation. The server averagely needs $4T_{\text{PS}} + 2T_{\text{PA}} + 3T_{\text{XOR}} + (n+2)T_h$ to find the matched client (each trial for one potential client takes two hash operations), where n denotes the number of entries of potential clients in the database.

Table 1 summarizes the performance comparison of the proposed scheme with its counterparts. The notations used in the table are introduced in the bottom of the table. The security of the schemes [10–12, 14] is based on RSA and the CDHP problem, and our scheme is based on the MECDHP problem. *Please note that even though the authors [15] claimed that their scheme was based on the ECDHP problem but actually its session key security was based on the privacy of nonce and the fixed D-H key (the fixed key $sP_C = cP_C = csP$), if one private key of either the server or the client is compromised, then the fixed D-H key and one nonce are compromised; therefore, its security is not strong as the ECDHP.*

Regarding the communicational performance, we focus on the number of message steps and the length of the messages. According to National Security Agency of the USA [18], to have an equivalent security of an 80-bit symmetric encryption, RSA algorithm and Diffie-Hellman algorithm need 1024-bit key while ECC only needs 160-bit key. From the table, we can see that all the schemes require three message steps. Because the schemes [9–12, 14] used RSA and the

CDHP, their message length is larger than our scheme and the scheme [15]. And, the message length of our scheme is shorter than its ECC-based counterpart [15].

Now we compare the computational performance. Recently there have been several publications like [21–24] discussing the implementation performances of ECC standard and other public-key standards (like RSA and El-Gamal public key), where [21] compared various software implementations of point multiplication, [22] aimed at improving hardware implementation of point multiplication, [23] discussed the hardware implementation performance of RSA digital signature and ECDSA, and [24] compared implementation performance of El-Gamal elliptic curve encryption method using two different libraries. These works focused on the implementation performances of elliptic curve point multiplication and related standards (like ECC standards, RSA digital signature, and El-Gamal elliptic curve encryption). For a fair comparison of custom-designed algorithms that consist of underlying field operations (like field multiplication and exponentiation in Z_p, field multiplication in Z_q, and elliptic curve point multiplication), the figures in NSA [18] and the algebra equations of elliptic curve operations in terms of underlying field operations from [19] have been widely adopted [15]. We adopt the same principle in the following evaluation, since it gives a fair comparison of custom-designed algorithms.

The security of ECC with 160-bit key is equivalent to that of RSA with 1024-bit key or D-H algorithm with 1024-bit key. Under the above figures, $T_{\text{multi},p}$ (the time complexity of a field multiplication in Z_p, where p is 1024-bit) is 41 times

$T_{\text{multi},q}$ (the time complexity of field multiplication in Z_q, where q is 160-bit), $T_{\text{PS}} \sim= 29T_{\text{multi},p}$, $T_{\text{exp}} \sim= 240T_{\text{multi},p} \sim= 8T_{\text{PS}}$, $T_{\text{PA}} \sim= 0.12T_{\text{multi},p}$, and $T_{\text{PS}} \sim= 241\,T_{\text{PA}}$, where $\sim=$ means "roughly equal." To simplify the comparison and get an insight of the computational performance, we can focus on the number of ECC point multiplications, point addition, modular exponentiation, and modular multiplication only because the other operations are not computationally significant. In this simplification, the client in our scheme needs $2T_{\text{PS}} + T_{\text{PA}} \sim= 58.12T_{\text{multi},p}$, the client in [11] needs $5T_{\text{exp}} + 2T_{\text{multi},p} \sim= 1202T_{\text{multi},p}$, the client in [12] takes $5T_{\text{exp}} + 1T_{\text{multi},p} \sim= 1201T_{\text{multi},p}$, the client in [10] requires $5T_{\text{exp}} + 3T_{\text{multi},p} \sim= 1203T_{\text{multi},p}$, the client in [14] demands $4T_{\text{exp}} + 3T_{\text{multi},p} \sim= 9633T_{\text{multi},p}$, and the client in Wang et al.'s scheme [15] requires $4T_{\text{PS}} \sim= 116T_{\text{multi},p}$. Based on these figures, we can get an insight that the client in our scheme only takes roughly 1/20 computational complexity of those RSA-based schemes like [10–12, 14] and takes only 1/2 time complexity of Wang et al.'s ECC-based scheme [15].

5. Conclusions and Further Investigations

This paper has formulated a new problem called the MECDHP and has proved its security being equivalent to the conventional ECDHP. Based on the MECDHP, we have proposed a new authenticated D-H key agreement scheme which protects client anonymity and greatly improves client's computational efficiency. Based on the figures of 160-bit ECC key and 1024-bit RSA key (or 1024-bit D-H algorithm), the client of our scheme requires only 1/20 time complexity of its RSA-based counterparts [10–12, 14] and takes only 1/2 time complexity of Wang et al.'s ECC-based scheme [15]. Security analysis shows that the scheme achieves half forward secrecy: it achieves forward secrecy if server's private key is compromised but the session keys of a compromised client would be disclosed. Here we note that it only affects the session keys of compromised clients. These excellent performances make our scheme quite attractive to those thin clients that require anonymity protection. One interesting open issue is whether we can improve the computational efficiency of existent authenticated D-H key agreement schemes while still preserving the perfect forward secrecy.

Conflict of Interests

The author declares that there is no conflict of interests regarding the publication of this paper.

Acknowledgment

This project is partially supported by the National Science Council, Taiwan, under Grant no. MOST 103-2221-E-260-022.

References

[1] M. Ohkubo, K. Suzki, and S. Kinoshita, "Cryptographic approach to 'privacy-friendly' tags," in *Proceedings of the RFID Privacy Workshop*, Massachusetts Institute of Technology, November 2003.

[2] H.-Y. Chien and C.-S. Laih, "ECC-based lightweight authentication protocol with untraceability for low-cost RFID," *Journal of Parallel and Distributed Computing*, vol. 69, no. 10, pp. 848–853, 2009.

[3] H.-Y. Chien, "Combining Rabin cryptosystem and error correction codes to facilitate anonymous authentication with untraceability for low-end devices," *Computer Networks*, vol. 57, no. 14, pp. 2705–2717, 2013.

[4] H. Y. Chien, "Provably secure authenticated Diffie-Hellman key exchange for resource-limited smart card," *Journal of Shanghai Jiaotong University (Science)*, vol. 19, no. 4, pp. 436–439, 2014.

[5] A. Brusilovsky, I. Faynberg, Z. Zeltsan, and S. Patel, "RFC683-Password-Authenticated Key (PAK) Diffie-Hellman Exchange," 2010, http://tools.ietf.org/html/rfc5683.

[6] V. Boyko, P. MacKenzie, and S. Patel, "Provably secure password-authenticated key exchange using Diffie-Hellman," in *Advances in Cryptology—EUROCRYPT 2000*, vol. 1807 of *Lecture Notes in Computer Science*, pp. 156–171, Springer, Berlin, Germany, 2000.

[7] ISO/IEC 9798-3 Authentication SASL Mechanism, http://www.faqs.org/rfcs/rfc3163.html.

[8] S. Blake-Wilson, N. Bolyard, V. Gupta, C. Hawk, and B. Moeller, "Elliptic curve cryptography (ECC) cipher suites for transport layer security (TLS)," RFC 4492, Internet Engineering Task Force (IETF), 2006.

[9] W.-B. Lee and C.-C. Chang, "User identification and key distribution maintaining anonymity for distributed computer networks," *Computer Systems Science and Engineering*, vol. 15, no. 4, pp. 113–116, 2000.

[10] Y. Yang, S. Wang, F. Bao, J. Wang, and R. H. Deng, "New efficient User identification and key distribution scheme providing enhanced security," *Computers and Security*, vol. 23, no. 8, pp. 697–704, 2004.

[11] E.-J. Yoon and K.-Y. Yoo, "Cryptanalysis of two user identification schemes with key distribution preserving anonymity," in *Information and Communications Security: 7th International Conference, ICICS 2005, Beijing, China, December 10–13, 2005. Proceedings*, vol. 3783 of *Lecture Notes in Computer Science*, pp. 315–322, Springer, Berlin, Germany, 2005.

[12] H.-Y. Chien, "Practical anonymous user authentication scheme with security proof," *Computers & Security*, vol. 27, no. 5-6, pp. 216–223, 2008.

[13] P. Szalachowski and Z. Kotulski, "Enhancing the Oakley key agreement protocol with secure time information," in *Proceedings of the International Symposium on Performance Evaluation of Computer and Telecommunication Systems (SPECTS '12)*, pp. 1–8, Genoa, Italy, July 2012.

[14] T.-S. Wu and C.-L. Hsu, "Efficient user identification scheme with key distribution preserving anonymity for distributed computer networks," *Computers and Security*, vol. 23, no. 2, pp. 120–125, 2004.

[15] R.-C. Wang, W.-S. Juang, and C.-L. Lei, "Provably secure and efficient identification and key agreement protocol with user anonymity," *Journal of Computer and System Sciences*, vol. 77, no. 4, pp. 790–798, 2011.

[16] M. S. Farash, M. Bayat, and M. A. Attari, "Vulnerability of two multiple-key agreement protocols," *Computers and Electrical Engineering*, vol. 37, no. 2, pp. 199–204, 2011.

[17] N.-Y. Lee, C.-N. Wu, and C.-C. Wang, "Authenticated multiple key exchange protocols based on elliptic curves and bilinear pairings," *Computers and Electrical Engineering*, vol. 34, no. 1, pp. 12–20, 2008.

[18] US National Security Agency, The Case for Elliptic Curve Cryptography, https://www.nsa.gov/business/programs/elliptic_curve.shtml.

[19] A. Jurisic and A. J. Menezes, "Elliptic curves and cryptography," Certicom Whitepaper, 1997.

[20] K. Xue, P. Hong, and C. Ma, "A lightweight dynamic pseudonym identity based authentication and key agreement protocol without verification tables for multi-server architecture," *Journal of Computer and System Sciences*, vol. 80, no. 1, pp. 195–206, 2014.

[21] M. Rivain, "Fast and regular algorithms for scalar multiplication over elliptic curve," The International Association for Cryptologic Research (IACR) Cryptology ePrint Archive 2011/338, https://eprint.iacr.org/2011/338.pdf.

[22] G. D. Sutter, J.-P. Deschamps, and J. L. Imana, "Efficient elliptic curve point multiplication using digit-serial binary field operations," *IEEE Transactions on Industrial Electronics*, vol. 60, no. 1, pp. 217–225, 2013.

[23] R. Sinha, H. K. Srivastava, and S. Gupta, "Performance based comparison study of RSA and elliptic curve cryptography," *International Journal of Scientific & Engineering Research*, vol. 4, no. 5, pp. 720–725, 2013.

[24] D. F. Pigatto, N. B. F. d. Silva, and K. R. L. J. C. Branco, "Performance evaluation and comparison of algorithms for elliptic curve cryptography with el-gamal based on MIRACL and RELIC libraries," *Journal of Applied Computing Research*, vol. 1, no. 2, pp. 95–103, 2012.

4

Fingerprint Quality Evaluation in a Novel Embedded Authentication System for Mobile Users

Giuseppe Vitello,[1] Vincenzo Conti,[1] Salvatore Vitabile,[2] and Filippo Sorbello[3]

[1]*Faculty of Engineering and Architecture, University of Enna Kore, 94100 Enna, Italy*
[2]*Department of Biopathology and Medical Biotechnologies, University of Palermo, 90127 Palermo, Italy*
[3]*Department of Chemical Engineering, Management, Computer Science, and Mechanics, University of Palermo, 90128 Palermo, Italy*

Correspondence should be addressed to Giuseppe Vitello; giuseppe.vitello@unikore.it

Academic Editor: Ilsun You

The way people access resources, data and services, is radically changing using modern mobile technologies. In this scenario, biometry is a good solution for security issues even if its performance is influenced by the acquired data quality. In this paper, a novel embedded automatic fingerprint authentication system (AFAS) for mobile users is described. The goal of the proposed system is to improve the performance of a standard embedded AFAS in order to enable its employment in mobile devices architectures. The system is focused on the quality evaluation of the raw acquired fingerprint, identifying areas of poor quality. Using this approach, no image enhancement process is needed after the fingerprint acquisition phase. The Agility RC2000 board has been used to prototype the embedded device. Due its different image resolution and quality, the experimental tests have been conducted on both PolyU and FVC2002 DB2-B free databases. Experimental results show an interesting trade-off between used resources, authentication time, and accuracy rate. The best achieved false acceptance rate (FAR) and false rejection rate (FRR) indexes are 0% and 6.25%, respectively. The elaboration time is 62.6 ms with a working frequency of 50 MHz.

1. Introduction

The growing number of mobile users has deeply influenced scenarios such as commercial, banking, and government applications. Due to the increasing security requirements, the way people access information resources, data communication and processing, is radically changing [1, 2]. In this field, biometric recognition systems are a good solution for mobile users authentication [3, 4].

Depending on the application context, a biometric recognition system may be used as verification or identification system. A verification system checks the person's identity by comparing the captured biometric characteristic with his/her own biometric template enrolled in the system. It conducts a one-to-one comparison to determine whether the identity claimed by the individual is true. An identification system recognizes the subject by searching the entire template database for a match. It conducts one-to-many comparisons and establishes person's identity or fails if he/she is not enrolled in the system database, without the subject having to claim an identity. A biometric recognition system may be further classified as unimodal, when one or more instances of a single biometric trait (e.g., multiple impressions of a finger) are processed. The system is classified as multimodal, when it uses one or more instances of multiple biometric characteristics (e.g., fingerprint and face images) [5]. Multialgorithmic systems represent a particular multimodal systems class, where the same biometric trait is processed with different algorithms [4].

To reduce the processing time in identification systems, biometric characteristics can be classified in an accurate and consistent way such that the input needs to be matched only with a database subset. Fingerprint classification, for example, can be performed using a wide variety of algorithms, almost all based on one or more of the following features: neural network [6], Gabor filter and support vector machine [7], genetic programming [8], singular points [9], and so forth. Unfortunately, singular points are not always present in a fingerprint image (e.g., in the partially fingerprint image acquisition). In that case, the approach proposed in [10] may be useful, where pseudosingularity points are detected and extracted for fingerprints classification and matching.

Biometric systems are a rapidly evolving technology in mobile devices, with a very strong potential to be widely adopted in a broad range of human scenarios. However, there are many challenges to overcome in designing completely automatic and reliable systems, especially when input data are of poor quality. For example, fingerprint acquisitions not correctly performed, because of skin humidity, impressing pressure, large translation on sensor area, sensing mechanism, and so on, could lead to the following issues [11]:

(i) quite different ridges quality;

(ii) ridges and valleys pattern deformation;

(iii) insufficient contrast;

(iv) small foreground area;

(v) inadequate overlapping area between different images although they are captured from the same finger.

In this paper, a novel embedded automatic fingerprint authentication system (AFAS) for mobile users is described. The goal of the proposed approach is to improve the performance of a standard embedded AFAS, in terms of used resources, execution time, and working frequency, in order to enable its employment in mobile devices architectures. Starting from the work described in [12], focused only on an advanced matching technique for partial fingerprints, the novel embedded AFAS has been prototyped adding the proposed fingerprint image quality evaluation module. This module is designed to find a measure that can characterize the quality of raw fingerprint images, only using the information achieved in the acquisition step. The quality index calculates and merges six different global quality indexes based on image contrast, ridges orientation certainty level, fingerprint's center position, impressing pressure, and fingerprint size over the entire image. It is also specialized in identifying areas of poor quality. If the image overcomes the quality constraints only good areas are processed reducing the potential false minutiae. Otherwise, if the image is rejected, the system suggests to user a set of information about the not correct acquisition step, helping him to follow correct guidelines to obtain a better image quality in the next fingerprint acquisition task (Figure 1).

The proposed AFAS architecture, designed for field programmable gate array (FPGA) devices using pipeline techniques and parallelisms to reduce the execution time, has been prototyped on the Agility RC2000 development board, equipped with a Xilinx Virtex-II xc2v6000 FPGA [13]. To evaluate the effectiveness of the proposed approach, three tests have been conducted starting from two different free databases, chosen for their different characteristics in terms of resolution and quality.

The AFAS described in [14] has been extended with the proposed fingerprint image quality evaluation module. Experimental trials on the FVC2002 DB2-B database [15] show that the accuracy performance has been strongly increased. Then, the matching algorithm has been replaced with the advanced technique for partial fingerprints proposed

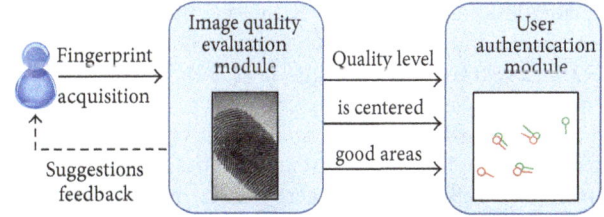

FIGURE 1: Image quality evaluation module classifies the fingerprint image quality and identifies high quality areas. It checks if the fingerprint is centered over the image. If an image is rejected, a suggestions feedback is given, to the user for the next fingerprint acquisition tasks.

in [12]. Experimental results on the PolyU database [16] show an interesting trade-off between required hardware resources, authentication time, and accuracy rate. Finally, the fingerprint image quality evaluation module has been replaced with a preprocessing task to enhance fingerprint images, a Gabor filter, and the system has been tested on the same PolyU database. The obtained experimental results show the validity of the proposed novel AFAS.

The paper is structured as follows. Section 2 reports the main literature works on fingerprint image quality evaluation methods. Section 3 describes the proposed novel fingerprint authentication system. Section 4 outlines the experimental results. Finally, conclusions are reported.

2. Remarks on Fingerprint Image Quality Evaluation Methods

One of the main techniques to test the performance of an automatic fingerprint recognition system relies heavily on the quality analysis of the acquired fingerprint image [17]. In literature many researchers have studied, proposed, and implemented different methods for evaluating the images quality, using, for example, artificial neural networks, micro- and macrofeatures analysis, and texture feature estimates.

In [11] the authors propose a hybrid scheme to measure the quality of fingerprint images by combining both local and global characteristics. It uses not only local texture features but also some global factors such as the standard deviation of Gabor features, the foreground area and central position, the number of minutiae, and the existence of singular points. The authors define seven quality indexes and also two weighting methods, an overlapping area based method and a linear regression method, for computing the correlation between the final quality value and each quality index.

In [18] the authors present a fast fingerprint enhancement algorithm, based on the estimated local ridge orientation and frequency, which can adaptively improve the clarity of ridge and valley structures of input fingerprint images. It models the ridge and valley patterns as a sinusoidal wave and then calculates the amplitude, frequency, and variance of the wave to determine the quality of the fingerprint regions.

In [19] the authors define a method not aimed at selecting images of good visual appearance but aimed at identifying poor quality as well as invalid fingerprints for automatic fingerprint identification systems. It analyzes the image in the spatial domain and uses the orientation certainty to certify the localized texture pattern, while it uses ridge and valley structure to detect invalid images.

In [20] the authors implement an effective quality classification method for fingerprint images based on neural networks. It uses effective area, energy concentration, spatial consistency, and directional contrast as quality indexes. A comparison with individual quality index thresholding and linear weighted sum method, on a private database, shows the higher quality classification accuracy of their method.

In [21] the authors describe a novel method for estimating the quality of fingerprint images using both local and global analyses. They propose a fusion method mixing the information from ridge and valley line resolution, fingerprint area, and gray levels average and variance, using the golden section method to select the relevant weights value.

In [22] the authors propose a novel quality-checking algorithm which considers the condition of the input fingerprints and the orientation estimation errors. First, the 2D gradients of the fingerprint image is separated into two sets of 1D gradients, and then the shape of the probability density functions of these gradients is measured in order to determine the fingerprint quality.

In [23] the authors present an image quality assessment technique for a novel fingerprint multimodal algorithm to provide high accuracy under nonideal conditions. It uses the redundant discrete wavelet transform to assess the image quality, for high resolution fingerprint databases, by determining the presence of noise, smoothness, and edge information in a fingerprint image. Successively, in [24] the authors extend this technique designing a local image quality assessment algorithm. They use it as the first step of a novel algorithm for fast extraction and identification of level-3 features, such as pores, ridge contours, dots, and incipient ridges.

After an exhaustive analysis of the above described methods for fingerprint image quality evaluation and in order to achieve the best trade-off between execution time and used resources for embedded devices, a mixed method has been designed and integrated in the proposed novel embedded AFAS. It is based on a fingerprint image global analysis in the spatial domain and inspired by works described in [11, 19].

3. The Proposed Novel Embedded Fingerprint Authentication System

The proposed minutiae based AFAS is focused on the acquired raw image quality evaluation identifying poor quality areas, such as dry and moist portions, in order to overcome the common problems in wrong acquisitions on mobile devices. The system checks if the distance between image center and the fingerprint center coordinates is lower than an experimental fixed threshold in order to extract the maximum number of corresponding minutiae. If this condition is verified and the image overcomes the quality constraints, only high quality image portions are processed. Otherwise, the image is rejected and the system gives to the user suggestion feedbacks about the wrong acquisition

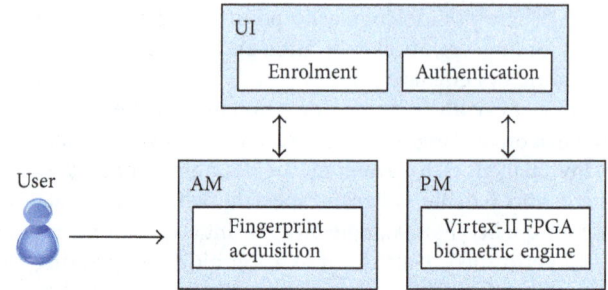

FIGURE 2: System's components: the user interface (UI), the acquisition module (AM), and the processing module (PM).

step, helping him to obtain a better image quality in the next fingerprint acquisition task. In addition, an advanced matching technique for user recognition, based on partial fingerprints, is performed to improve system accuracy [12]. This technique calculates a likelihood ratio by trying every possible overlap of the acquired fingerprint with the enrolled one. The rototranslation parameters computation is based on the similar minutiae pairs identification belonging to both fingerprints.

Considering the functionalities of the proposed system, three main components can be identified: the interface module (IM), which enables the user to interact with the system, the acquisition module (AM), which deals with the fingerprint image acquisition, and the processing module (PM), based on the FPGA processing engine implementing the authentication phase (Figure 2).

Using the proposed PM, no image enhancement after fingerprint acquisition is performed. Therefore, a considerable saving in terms of execution time and hardware resources has been achieved with respect to a standard AFAS implementation. With more details, the proposed AFAS requires an image quality evaluation module, including a binarization module, a thinning module, a feature extraction module, an alignment module, and, finally, a matching module. Despite a standard AFAS implementation no normalization, enhancement, field orientation, filtered orientation and, smoothing tasks are required (Figure 3).

In the following subsections the main submodules of the proposed novel AFAS will be described.

3.1. Image Quality Evaluation Module. This module, inspired by works described in [11, 19], evaluates the fingerprint image quality through a global analysis in the spatial domain. With more details, it analyzes the image by blocks, calculates the fingerprint central position, identifies the dry and moist blocks, and classifies the image quality into two levels.

Figure 4 shows the architecture of the proposed module, while the following subsections describe each submodule.

3.1.1. Blocks_Generator Submodule. This submodule reads a gray levels fingerprint image from the on-board memory, divides it into an ideal grid of $N = N_x * N_y$ nonoverlapping blocks, and sends them, pixel by pixel, to the Fingerprint_Quality_Level_Evaluator submodule. Each block has a fixed size depending on the used database: 30×30 and

Enrolment	Fingerprint image processing	Authentication	Enrolment	Fingerprint image processing	Authentication
Task 0	Acquisition	Task 0	Task 0	Acquisition	Task 0
Task 1	Image quality evaluation	Task 1	Task 1	Segmentation	Task 1
			Task 2	Normalization	Task 2
			Task 3	Enhancement	Task 3
			Task 4	Field orientation	Task 4
			Task 5	Filtered orientation	Task 5
Task 2	Binarization	Task 2	Task 6	Binarization	Task 6
			Task 7	Smoothing	Task 7
Task 3	Thinning	Task 3	Task 8	Thinning	Task 8
Task 4	Features extraction	Task 4	Task 9	Features extraction	Task 9
Task 5	Alignment	Task 5	Task A	Alignment	Task A
Task 6	Matching	Task 6	Task B	Matching	Task B

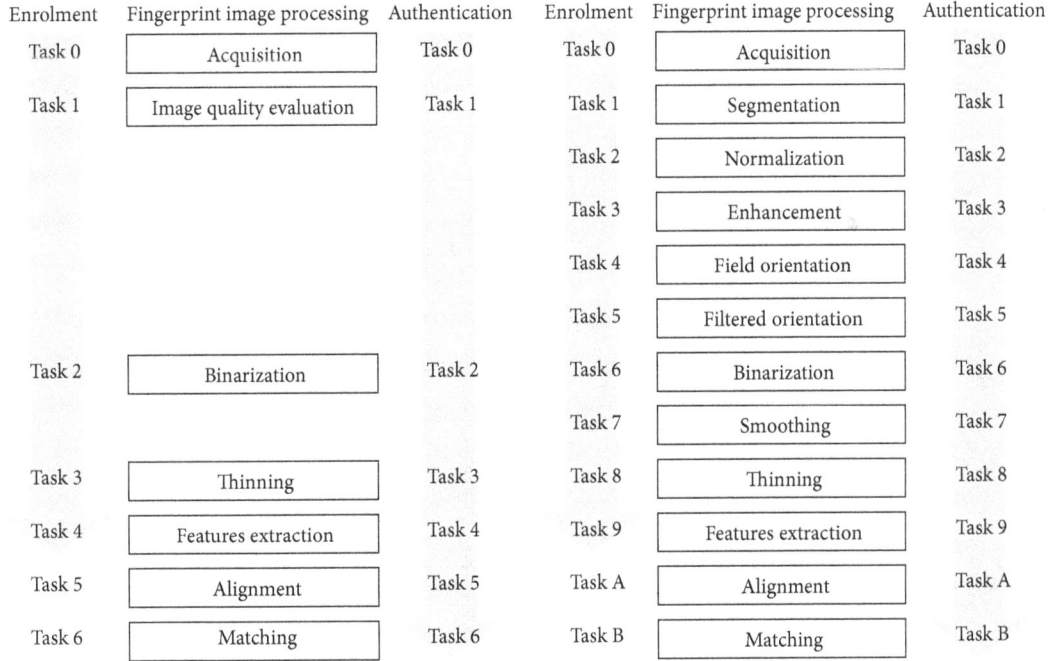

FIGURE 3: Comparison between the proposed AFAS (on the left) and the standard AFAS (on the right).

40×40 pixels for the FVC2002 DB2-B and the PolyU database, respectively. This system also counts the blocks sent and sets the *new_block* signal when the last pixel of the current block is sent.

3.1.2. Fingerprint_Quality_Level_Evaluator Submodule.
This submodule identifies the dry and moist fingerprint portions allowing the subsequent features extraction task to discard them in order to reduce the potential false minutiae number. In concurrent way, the submodule checks if the fingerprint is centered over the image and calculates six indexes, each measuring an image qualitative characteristic. It performs a linear combination of them obtaining the final quality index. Finally, it classifies the image quality into two classes.

In the following subsections, the Fingerprint_Quality_Level_Evaluator submodules are described.

(1) Image_Blocks_Analyzer Submodule. This submodule is able to process block by block the fingerprint image. For each block it calculates, in a concurrent way, the following features:

(i) max and min gray level: these local values are used to calculate the global max and min gray level of the entire image;

(ii) gray levels average and variance: these values are used to classify blocks as foreground/background and as dry/moist/good;

(iii) ridges orientation certainty level (ocl): this value, only for foreground blocks, is added to the ocl_accumulator signal, subsequently used for the calculation of the 2nd index.

After that, in a concurrent way, it identifies the fingerprint high quality areas and calculates the fingerprint central position. The following subsections describe the main submodules of the proposed Image_Blocks_Analyzer submodule.

Orientation_Certainty_Level_Calculator Submodule. A fingerprint image block generally consists of ridges separated by valleys with the same orientation. Ridges and valleys constant structure and regular orientation can be used to evaluate the quality of each considered block. They are analytically calculated through the gradient of the gray levels along the x and y directions of a pixel [19]. The covariance matrix C of the gradient vector for an image block of M points is given by

$$C = E\left\{ \begin{bmatrix} dx \\ dy \end{bmatrix} [dx \ dy] \right\} = \begin{bmatrix} a & c \\ c & b \end{bmatrix}, \qquad (1)$$

where

$$E\{\bullet\} = \frac{1}{M}\sum_M \bullet. \qquad (2)$$

The ridges orientation certainty level (ocl) is calculated as shown in

$$ocl = 100 * \frac{(a+b) - \sqrt{(a-b)^2 + 4c^2}}{(a+b) + \sqrt{(a-b)^2 + 4c^2}}. \qquad (3)$$

With low (high) ocl values, the local structure and orientation of ridges and valleys are very regular (irregular), and therefore the block has good (wrong) quality (Figure 5). With more details, this submodule is further composed of two submodules, implementing a two-stage pipeline (Figure 6). While the first submodule calculates the covariance matrix C of block j, the second submodule calculates the ocl value of j-1 block.

Average_Calculator and Variance_Calculator Submodules. Average and variance are important characteristics for

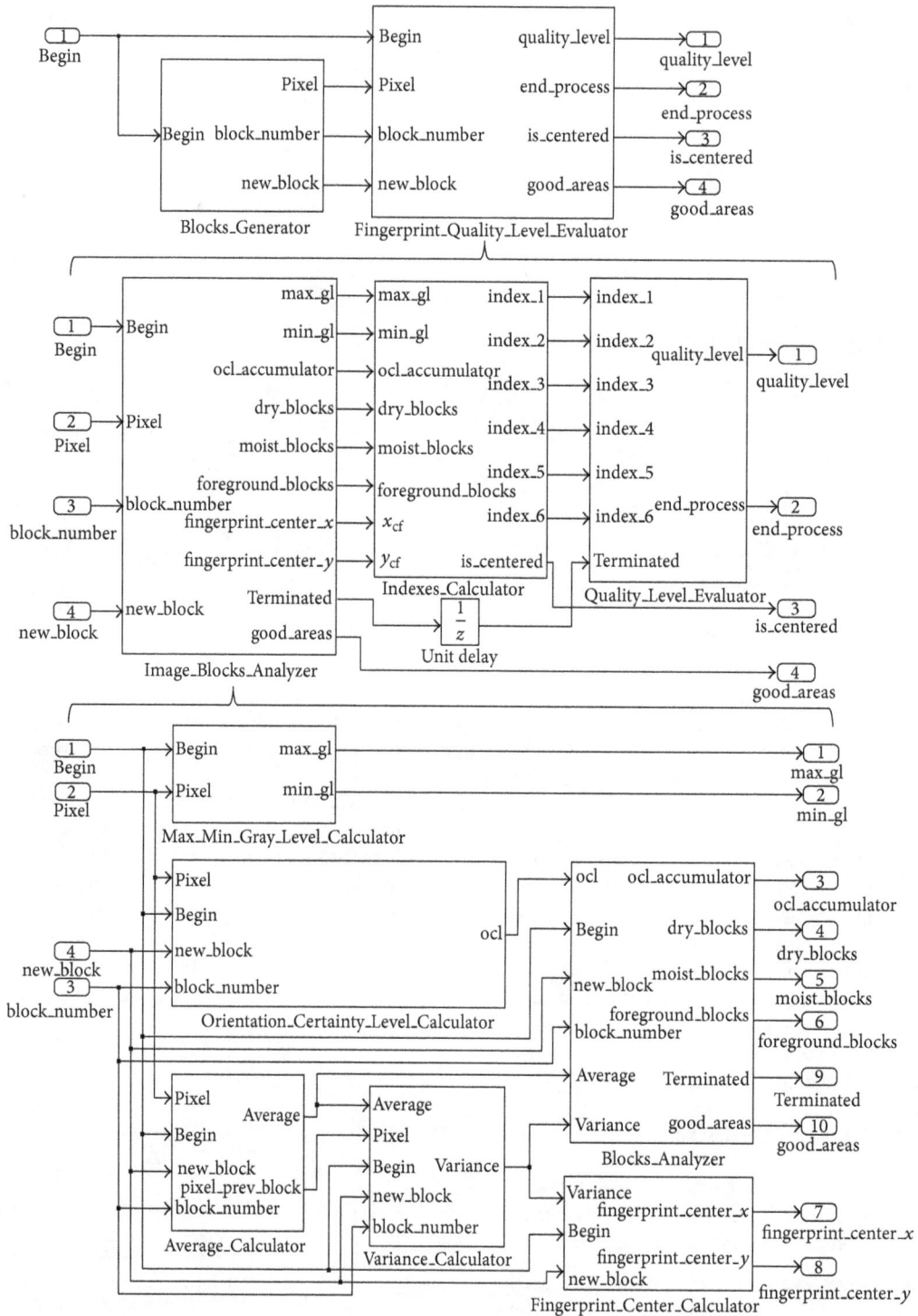

FIGURE 4: The proposed architecture evaluates the fingerprint image quality level. Fingerprint quality evaluator module is composed of image blocks analyzer submodule, indexes calculator submodule, and quality level evaluator submodule. The Image Block Analyzer submodule is composed of Max Min Level Calculator submodule, Orientation Certainty Level Calculator submodule, Average Calculator submodule, Variance Calculator submodule, Block Analyzer submodule, and Fingerprint Center Calculator submodule.

evaluating the block quality: average measures the luminosity, while variance measures the contrast. A low average value is linked to a block prevalently containing ridges (because it is dark), while a low variance value entails that the block does not contain any useful portion of the fingerprint (because it has a low contrast).

The Average_Calculator submodule stores the incoming block pixels on a shift register and sends the pixels of

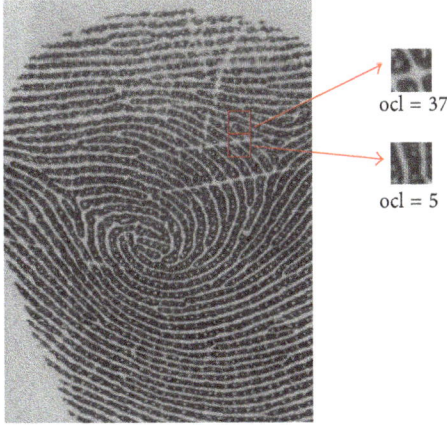

FIGURE 5: Examples of different ocl values.

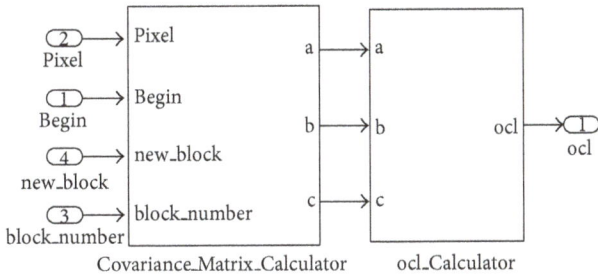

FIGURE 6: Orientation_Certainty_Level_Calculator submodule.

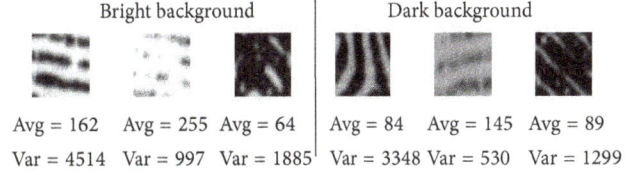

FIGURE 7: Examples of average and variance values with dark and bright background.

the previous block in order to achieve the best trade-off between requested resources and execution time.

Block_Analyzer Submodule. The ocl characteristic is not sufficient to quantify the clearness of the fingerprint ridges and valleys pattern when the skin humidity is also considered. For a moist block the ridges are too thick, since it has low average value. On the other hand, the ridges are too thin for a dry block, since it has a high average value. So, the average value is heavily influenced by the background gray level intensity (Figure 7). In this work, the gray level intensity of the image background is fixed to be the average value of the first image block, since it does not usually contain part of the fingerprint. If the block contains part of the fingerprint (i.e., the fingerprint covers the entire image) the background gray level is assumed as dark.

This submodule compares the average value of the first block with an experimental fixed threshold classifying the background as dark or bright and setting moist and dry thresholds. These values are experimentally fixed and depend on the used database. For example, on the FVC2002 DB2-B, the dry thresholds are 140 and 180 for bright and dark background, respectively, while the moist thresholds are 80 for dark background and 100 otherwise. Successively, it classifies each block as foreground or background using the incoming variance value. The foreground threshold is not influenced by the background gray level and it is experimentally fixed to 190.

Fingerprint_Center_Calculator Submodule. This submodule calculates, in a concurrent way, the fingerprint central position (Figure 8). It checks if the considered block belongs to the column $N_x/2$, $N_x/4$, or $3N_x/4$. If so then, if it is of foreground, a value equal to the block size is added to the relevant column foreground accumulator (an accumulator for each considered column); otherwise only if this accumulator value is zero, the same value is added to the relevant column background accumulator (i.e., the background blocks below the fingerprint are discharged). Concurrently, the same check is performed on the rows $N_y/2$, $N_y/4$, or $3N_y/4$, and, in the same way, the relevant row background or foreground accumulator is increased. Finally, for the last block, the column foreground accumulator with the highest value is selected and the y-coordinate of the fingerprint's center is calculated as the sum of the half value stored in the selected foreground accumulator and the relevant column background accumulator value. Concurrently, the x-coordinate of the fingerprint's center is calculated in the same way.

(2) Indexes_Calculator Submodule. Among common quality indexes present in literature and reported in the related works section, this subsystem concurrently calculates six global indexes, designed in order to realize a module reducing used resources and execution time. To make all indexes compatible, they have normalized in the range of [0, 100]. High index value entails a good image quality.

Index1_Calculator Submodule. The first index measures the contrast between fingerprint and background. This value is calculated as the difference between the maximum and the minimum gray level value of the entire image:

$$\text{index\_1} = 100 * \frac{(\max\_\text{gl} - \min\_\text{gl})}{255}. \tag{4}$$

Index2_Calculator Submodule. The second index extends to the whole image the considerations about the block orientation certainty level estimation, thus globally measuring the clarity and continuity of ridges and valleys orientation. It is calculated by averaging all the ocl values relating to only foreground blocks:

$$\text{index\_2} = 100 - \frac{\text{ocl\_accumulator}}{\text{foreground\_blocks}}. \tag{5}$$

Index3_Calculator Submodule. The third index measures the humidity of the entire image and it is calculated as the ratio

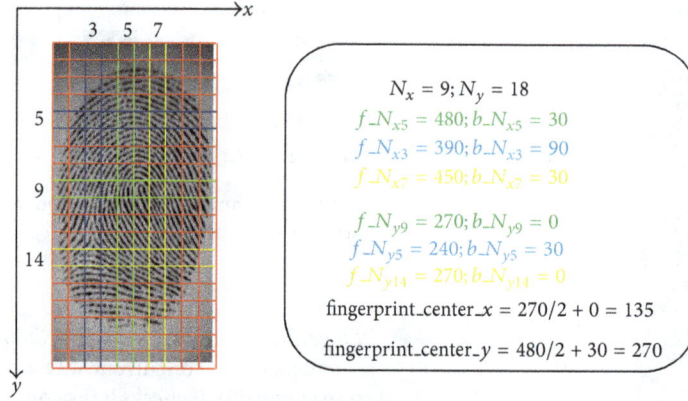

FIGURE 8: Example of a FVC2002 fingerprint's center calculation.

between the number of moist blocks and the number of foreground blocks:

$$\text{index\_3} = 100 - \left(100 * \frac{\text{moist\_blocks}}{\text{foreground\_blocks}} \right). \quad (6)$$

Index4_Calculator Submodule. The fourth index measures the dryness of the entire image and it is calculated as the ratio between the number of dry blocks and the number of foreground blocks:

$$\text{index\_4} = 100 - \left(100 * \frac{\text{dry\_blocks}}{\text{foreground\_blocks}} \right). \quad (7)$$

Index5_Calculator Submodule. The fifth index measures the image area occupied by the foreground blocks. It is an estimate of the fingerprint size over the entire image and it is calculated as the ratio between the number of foreground blocks and the total number of blocks:

$$\text{index\_5} = 100 * \frac{\text{foreground\_blocks}}{N}. \quad (8)$$

Index6_Calculator Submodule. The sixth index measures the position of the fingerprint over the entire image: too large translation caused by human behavior can generate an insufficient overlapping area between images captured from the same finger. It is calculated as the average of two values, $i6_x$ and $i6_y$:

$$\text{index\_6} = \frac{i6_x + i6_y}{2} \quad (9)$$

with

$$i6_x = 100 - \left(100 * \frac{|x_{cf} - x_{ci}|}{x_{ci}} \right),$$

$$\qquad\qquad\qquad\qquad\qquad (10)$$

$$i6_y = 100 - \left(100 * \frac{|y_{cf} - y_{ci}|}{y_{ci}} \right),$$

where x_{cf} and y_{cf} are the coordinates of the fingerprint's center, while x_{ci} and y_{ci} are the coordinates of the image's center.

In addition, this subsystem checks if the distance between the respective coordinates of the image's center and the fingerprint's center is lower than a threshold (experimentally fixed to 100) and then sets the *is_centered* signal.

(3) Quality_Level_Calculator Submodule. First, this subsystem calculates the final fingerprint quality index as linear combination of the previous six indexes. As described in [11], a linear regression method is used for weights calculation. They are experimentally determined by performing tests to observe the behavior of the change in the final quality index while one index is changing and the others are constant. Experimental results show that the most relevant indexes are ocl, fingerprint moisture, and fingerprint dryness. Then, by comparing the final quality index value with a threshold (experimentally fixed to 65), this subsystem classifies the image quality level as *Good* or *Bad*. Finally, the subsequent tasks are performed only if the quality is *Good* and the fingerprint is centered over the image.

3.2. Binarization Module. This module gives out an image where pixels assume a binary value: white as background and black as foreground (Figure 9). Binarization is performed using the local gray range technique described in [25]. In this adaptive technique the threshold is set at the average of the maximum and minimum gray values in a local window of size 9×9.

3.3. Thinning Module. This module reduces the ridge thickness to the unitary value (Figure 10), using the Zhang-Suen algorithm described in [26]. For the realization of the thinning algorithm on FPGA, a 3×3 mask has been used in order to implement a two-stage pipeline.

3.4. Features Extraction Module. For the minutiae extraction, the algorithm proposed in [14] has been optimized and extended: in order to reduce the system execution time and the potential false minutiae, only the good areas, computed by the image quality evaluation module, of a central area of 240×320 pixels, are processed. The proposed approach improves the performance of a standard embedded AFAS, such as

FIGURE 9: Example of fingerprint binarization.

FIGURE 10: Example of fingerprint thinning.

would a Gabor filtering process in order to reconstruct the poor quality areas. Figure 11 shows the minutiae extracted using the Gabor filter, to reconstruct image areas of poor quality, and using the image quality evaluation, to discard those areas. As depicted, the Gabor filter approach introduces two false bifurcations and discards two terminations, while the proposed approach discards two bifurcations and one termination.

3.5. Alignment and Matching Modules. The computation of a likelihood ratio in fingerprint authentication is obtained by trying all the possible overlapping of the acquired fingerprint with the one enrolled in the system [12]. The rototranslation parameters computation is based on the identification of two similar pairs of minutiae belonging to both fingerprints (Figure 12). A threshold (experimentally fixed to 175) based on Euclidean distance is used to generate the minutiae pairs.

First, rototranslation parameters are computed only if the value of Euclidean distance between each minutiae pair of both fingerprints is lower than a threshold (experimentally

fixed to 20). The rotation parameter is based on the differences between the corresponding angles in the selected minutiae pairs. If the gap between each of these differences with respect to the other is lower than a threshold (experimentally fixed to 1.5) the rotation parameter is the average of the calculated differences. In the same way, the translation parameter is based on the differences between the respective Cartesian coordinates in the selected minutiae pairs. If the gap between each coordinate distance is lower than a threshold (experimentally fixed to 30) the translation parameter is the average of the respective calculated differences.

Then, the rototranslation is performed and, for each minutia, differences between respective coordinates x-y (diff_{xy}) and angles ($\text{diff}_{\text{theta}}$) are calculated. Only when these differences are lower than two thresholds ($xy_{\text{threshold}}$ and $\text{theta}_{\text{threshold}}$, experimentally fixed to 15 and 0.785, resp.) a first partial score is obtained and normalized in the range of $[0, 1]$. The complete score is calculated as

$$s_i = 0.75 * \left(1 - \frac{\max\left(\text{diff}_{xy}\right)}{xy_{\text{threshold}}} \right) + 0.25 * \left(1 - \frac{\max\left(\text{diff}_{\text{theta}}\right)}{\text{theta}_{\text{threshold}}} \right), \tag{11}$$

where higher importance has been made to the differences between respective coordinates rather than to angles, due to rounding problems on data.

Finally, among all complete scores, only the greater is considered. Therefore, the final matching score is calculated adding the 12 highest obtained scores. In accordance with the USA guidelines in the forensic field, when two fingerprints have a minimum of 12 corresponding minutiae, these are regarded as coming from the same finger [27].

4. Experimental Results

The proposed approach introduces interesting characteristics for mobile devices. The architectural implementation on FPGA, considering its working frequency (50 MHz), achieves the performance of the highly competitive systems, realizing a good trade-off between accuracy rate, used resources, and execution time. To evaluate the accuracy performances of the proposed authentication system, the well-known false recognition rate (FRR) and false acceptance rate (FAR) indexes have been used and two different free databases with different characteristics in terms of resolution and quality have been used.

The following subsections report the used databases and datasets description, the execution time, the required hardware resources, and the authentication performance of the proposed AFAS.

4.1. Databases Description

4.1.1. FVC2002 DB2-B Database. This free downloadable database has been made available for the second edition of the international fingerprint verification competition [28]. It contains 80 fingerprint images of 296×560 pixels, with

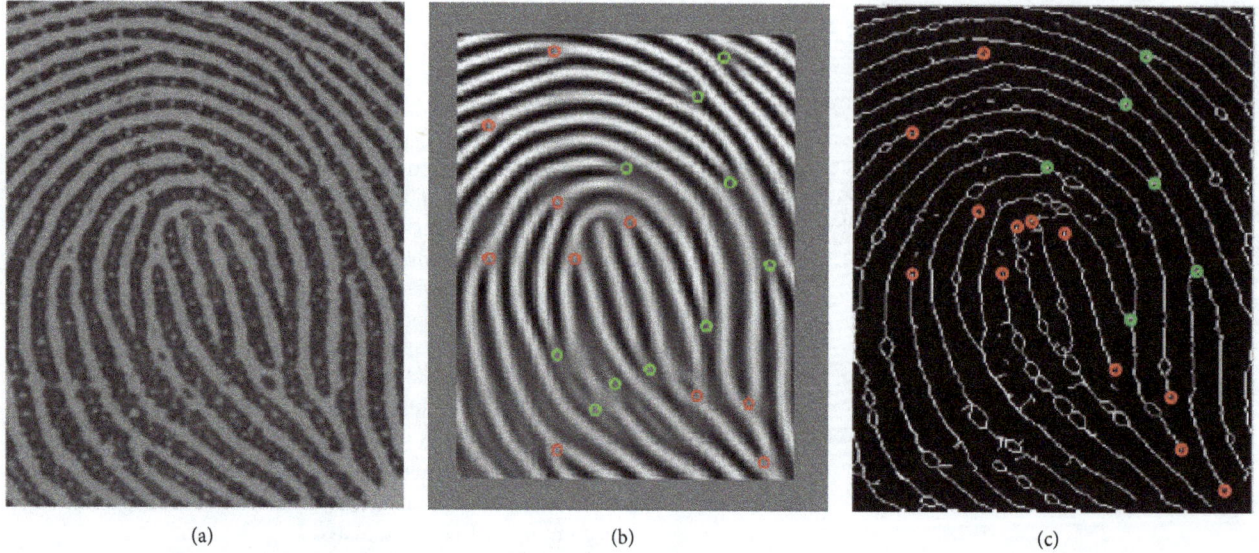

FIGURE 11: (a) Image 2_1_5 from PolyU database; (b) minutiae extracted with a Gabor filter and without the image quality evaluation; (c) minutiae extracted with the image quality evaluation and without a Gabor filter.

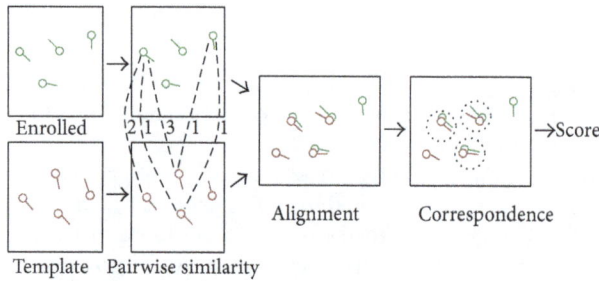

FIGURE 12: Rototranslation parameters computation.

a resolution of 569 dpi. The images has been acquired from 10 users (8 acquisitions for user of the same finger), via the scanner Biometrika FX2000 [29], with a maximum rotation of about 35 degrees between impressions (Figure 13).

4.1.2. PolyU Database.
This free downloadable database has been built at the Hong Kong Polytechnic University [16]. It contains 1480 fingerprint images of 480×640 pixels, with a resolution around 1,200 dpi of 148 users (10 acquisitions for user of two fingers, Figure 14). Each image name has been described using three numbers in the following way: first number represents the user, second number represents the finger, and third number represents the different acquisition.

4.1.3. Datasets Description.
Starting from the above description databases, two different datasets have been built:

(i) the *dataset1* has been generated using the entire FVC2002 DB2-B database (10 users, 8 acquisitions for user);

(ii) the *dataset2* has been generated using a consistent subset of the PolyU database (100 users with 5 acquisitions for user of the same finger).

TABLE 1: FAR and FRR indexes of the three performed tests.

Test number	FAR	FRR
1.	0%	6.25%
2.	0%	8.00%
3.	0%	9.00%

4.2. Authentication Performance.
Starting from the AFAS described in [14] and used as comparison, three different tests have been conducted:

(1) the AFAS has been extended with the proposed fingerprint image quality evaluation module and tested on the *dataset1*;

(2) the AFAS has been extended with the proposed fingerprint image quality evaluation module and, moreover, the matching algorithm has been replaced with the advanced technique, based on partial fingerprints, proposed in [12] and tested on the *dataset2*;

(3) the AFAS has been extended with a preprocessing task, based on the Gabor filter, to enhance fingerprint images and, moreover, the matching algorithm has been replaced with the advanced technique, based on partial fingerprints, proposed in [12] and tested on the *dataset2*.

Table 1 illustrates the authentication performance in terms of FAR and FRR indexes for the three performed tests.

4.3. Execution Time.
The following tables (Tables 2, 3, and 4) and Figure 15 illustrate the elaboration times, for the three performed tests, required by each single task, with a working frequency of 50 MHz.

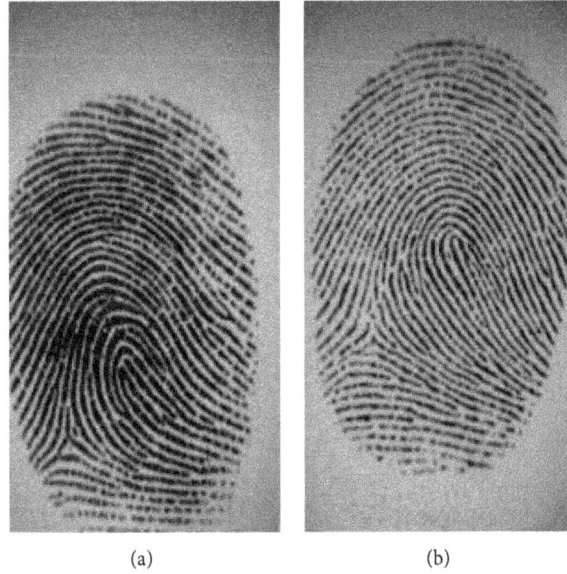

FIGURE 13: Two example images of the FVC2002 DB2-B acquired by Biometrika FX2000 sensor.

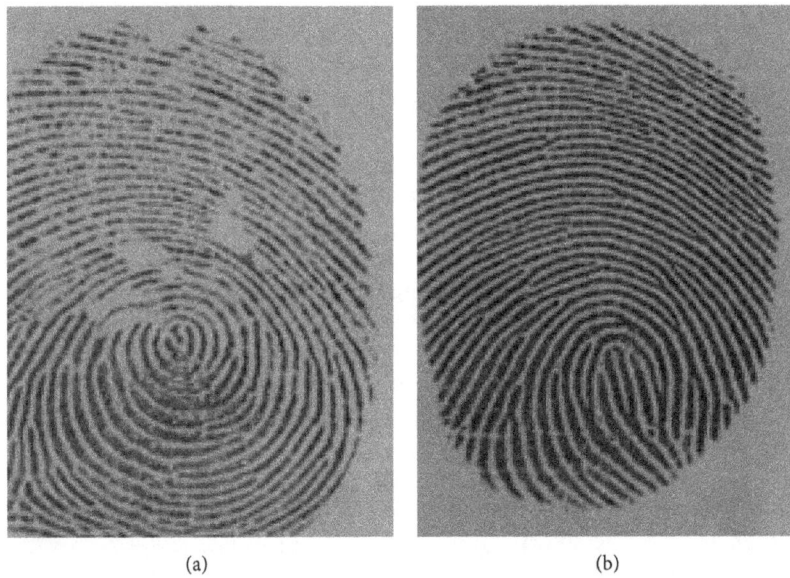

FIGURE 14: Two example images of the Hong Kong Polytechnic University.

TABLE 2: Execution times of test number 1.

Task	Execution time (msec)
Image quality evaluation	3.9
Binarization	2.2
Thinning	39.0
Minutiae extraction	13.7
Matching	3.8
Total	62.6

TABLE 3: Execution times of test number 2.

Task	Execution time (msec)
Image quality evaluation	3.9
Binarization	2.2
Thinning	39.0
Minutiae extraction	13.7
Matching	2.35×10^3
Total	2.4×10^3

4.4. Hardware Resources. The following tables (Tables 5, 6, and 7) depict the required hardware resources, for the three performed tests, used by each single task on the Agility RC2000 development board. Figure 16 illustrates the total used hardware resources for the three performed tests.

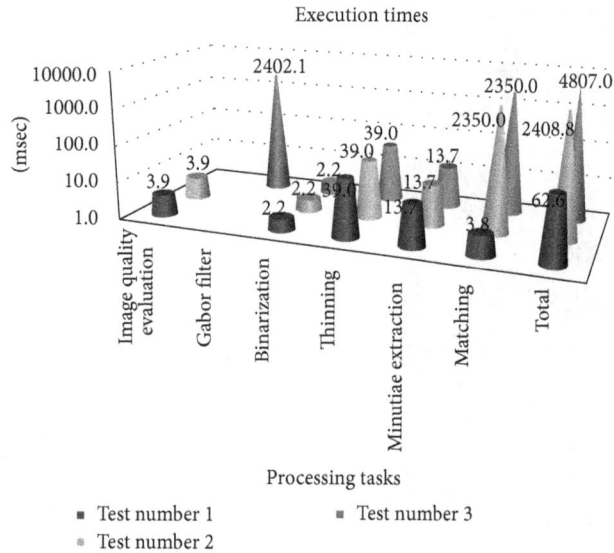

FIGURE 15: Elaboration times required by each processing task for the three performed tests.

TABLE 4: Execution times of test number 3.

Task	Execution time (msec)
Gabor filter	2.4×10^3
Binarization	2.2
Thinning	39.0
Minutiae extraction	13.7
Matching	2.35×10^3
Total	4.8×10^3

4.5. Discussion and Comparisons. User authentication is one of the most challenging issues for system and network security. A robust authentication mechanism is based on the use of biometric access control methods, processing one or more biometrics (such as a fingerprint). There are many approaches to deal with fingerprint verification. In recent literature publications, few findings have been on design and prototyping of an embedded biometric recognizer. For example, in [30] the authors proposed an implementation of a hardware identification system. However, the fingerprint matching phase was not developed and presented, so that no direct comparison with this work can be addressed. The remaining fingerprint processing tasks had been implemented in a FPGA device with a clock frequency of 27.65 MHz and a processing time of 589.6 ms. Compared with this system, the achieved execution times denote high performance levels. In [11] the authors use local texture features as well as some global factors such as the standard deviation of Gabor features, the foreground area and central position, the number of minutiae, and the existence of singular points. They produce a good analysis about equal error rate (EER) for three databases: FVC2002 DB2A, Fujitsu database, and FVC2002 DB4A. In [18] the authors have developed a software fast fingerprint enhancement algorithm which can adaptively improve the clarity of ridge

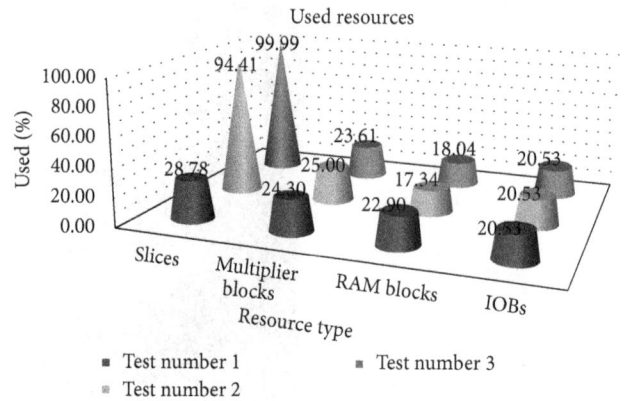

FIGURE 16: Used hardware resources, for the three performed tests.

and valley structures based on the local ridge orientation and ridge frequency. Experimental results show that their enhancement algorithm is capable of improving both the goodness index and the verification performance. The whole execution time of the enhancement algorithm on a Pentium 200MHZ is 2.49 sec, with FAR = 0.01% and FRR = 27% (without enhancement) and FRR = 9% (with enhancement) using the MSU fingerprint database (700 live-scan images; 10 per individual each). In [23, 24] the authors present an image quality assessment software technique for a novel fingerprint multimodal algorithm to provide high accuracy under nonideal conditions. Their study was based on a small number of minutia features. This is likely to be the case with latent fingerprints collected at a crime scene. Specifically, the performance of their fusion algorithm is studied when the number of minutiae is between 5 and 10. Experimental results show that while the performance of existing fusion algorithm decreases if compared to the performance of complete rolled fingerprints, the proposed approach is able to compensate for the limited partial information. The approach shows FRR between 91.35% and 97.98% with FAR = 0.01%, using a comprehensive database with rolled and partial fingerprint images of different quality and arbitrary number of features.

5. Conclusion

In this work a novel embedded AFAS improving the performance in terms of both used resources and execution time has been proposed. It is focused on the raw image quality evaluation of the acquired fingerprint, identifying areas of poor quality. It is designed to find a measure to characterize the quality of raw fingerprint images, using only the information obtained in the acquisition step. In addition, an advanced matching technique for user recognition using partial fingerprints has been developed to increase system accuracy. The best achieved FAR and FRR indexes are 0% and 6.25%, respectively. The required elaboration time is 62.6 ms with a working frequency of 50 MHz.

TABLE 5: Used resources of test number 1.

Resource type	Image quality evaluation	Binarization	Thinning	Minutiae extraction	Matching
Slices	5.83%	0.14%	0.67%	21.80%	0.34%
Multiplier blocks	4.17%	0.00%	0.00%	19.44%	0.69%
RAM blocks	0.69%	0.69%	0.69%	14.58%	6.25%
IOBs	4.98%	4.98%	0.00%	0.00%	10.57%

TABLE 6: Used resources of test number 2.

Resource type	Image quality evaluation	Binarization	Thinning	Minutiae extraction	Matching
Slices	5.83%	0.14%	0.67%	21.80%	65.97%
Multiplier blocks	4.17%	0.00%	0.00%	19.44%	1.39%
RAM blocks	0.69%	0.69%	0.69%	14.58%	0.69%
IOBs	4.98%	4.98%	0.00%	0.00%	10.57%

TABLE 7: Used resources of test number 3.

Resource type	Gabor filter	Binarization	Thinning	Minutiae extraction	Matching
Slices	11.41%	0.14%	0.67%	21.80%	65.97%
Multiplier blocks	2.78%	0.00%	0.00%	19.44%	1.39%
RAM blocks	1.39%	0.69%	0.69%	14.58%	0.69%
IOBs	4.98%	4.98%	0.00%	0.00%	10.57%

The proposed prototype has been implemented on the Agility RC2000 development board, addressing interesting characteristics for security in mobile device applications and enabling its use in commercial, banking, and government scenarios.

Conflict of Interests

The authors declare that there is no conflict of interests regarding the publication of this paper.

References

[1] Apple Inc., http://support.apple.com/kb/HT5883.

[2] Samsung, http://www.samsung.com/it/consumer/mobile-devices/smartphones/smartphones/SM-G900FZKAITV-spec.

[3] C. Militello, V. Conti, F. Sorbello, and S. Vitabile, "An embedded iris recognizer for portable and mobile devices," *International Journal of Computer Systems Science and Engineering*, vol. 25, no. 2, pp. 119–131, 2010.

[4] V. Conti, C. Militello, F. Sorbello, and S. Vitabile, "A multimodal technique for an embedded fingerprint recognizer in mobile payment systems," *International Journal of Mobile Information Systems*, vol. 5, no. 2, pp. 105–124, 2009.

[5] D. Maltoni, D. Maio, A. K. Jain, and S. Prabhakar, *Handbook of Fingerprint Recognition*, Springer, New York, NY, USA, 2003.

[6] V. Conti, C. Militello, F. Sorbello, and S. Vitabile, "An embedded fingerprints classification system based on weightless neural networks," *Frontiers in Artificial Intelligence and Applications*, vol. 193, no. 1, pp. 67–75, 2009.

[7] D. Batra, G. Singhal, and S. Chaudhury, "Gabor filter based fingerprint classification using support vector machines," in *Proceedings of the IEEE 1st India Annual Conference (INDICON '04)*, pp. 256–261, December 2004.

[8] J. Hu and M. Xie, "Fingerprint classification based on genetic programming," in *Proceedings of the 2nd International Conference on Computer Engineering and Technology (ICCET '10)*, vol. 6, pp. 193–196, Chengdu, China, April 2010.

[9] A. Tariq, M. U. Akram, and S. A. Khan, "An automated system for fingerprint classification using singular points for biometric security," in *Proceedings of the International Conference for Internet Technology and Secured Transactions (ICITST '11)*, pp. 170–175, December 2011.

[10] V. Conti, C. Militello, F. Sorbello, and S. Vitabile, "A frequency-based approach for features fusion in fingerprint and iris multimodal biometric identification systems," *IEEE Transactions on Systems, Man and Cybernetics Part C: Applications and Reviews*, vol. 40, no. 4, pp. 384–395, 2010.

[11] J. Qi, D. Abdurrachim, D. Li, and H. Kunieda, "A hybrid method for fingerprint image quality calculation," in *Proceedings of the 4th IEEE Workshop on Automatic Identification Advanced Technologies*, pp. 124–129, October 2005.

[12] V. Conti, G. Vitello, F. Sorbello, and S. Vitabile, "An advanced technique for user identification using partial fingerprint," in *Proceedings of the 7th International Conference on Complex, Intelligent, and Software Intensive Systems (CISIS '13)*, pp. 236–242, July 2013.

[13] Xilinx Inc, http://www.xilinx.com/support/documentation/data_sheets/ds031.pdf.

[14] V. Conti, S. Vitabile, G. Vitello, and F. Sorbello, "An embedded biometric sensor for ubiquitous authentication," in *Proceedings of the AEIT Annual Conference: Innovation and Scientific and Technical Culture for Development*, October 2013.

[15] FVC Databases, http://bias.csr.unibo.it/fvc2002/databases.asp.

[16] PolyU Database, http://www4.comp.polyu.edu.hk/~biometrics/HRF/HRF_old.htm.

[17] E. Tabassi, C. Wilson, and C. Watson, "Fingerprint image quality," NIST Research Report NISTIR7151, 2004.

[18] L. Hong, Y. Wan, and A. Jain, "Fingerprint image enhancement: algorithm and performance evaluation," *IEEE Transactions on Pattern Analysis and Machine Intelligence*, vol. 20, no. 8, pp. 777–789, 1998.

[19] E. Lim, X. D. Jiang, and W. Y. Yau, "Fingerprint quality and validity analysis," in *Proceedings of the International IEEE Conference on Image Processing*, vol. 1, pp. 469–472, September 2002.

[20] X. Yang and Y. Luo, "A classification method of fingerprint quality based on neural network," in *Proceedings of the International Conference on Multimedia Technology (ICMT '11)*, pp. 20–23, IEEE, Hangzhou, China, July 2011.

[21] F.-J. An and X.-P. Cheng, "Approch for estimating the quality of fingerprint Image based on the character of ridge and valley lines," in *Proceedings of the International Conference on Wavelet Active Media Technology and Information Processing (ICWAMTIP '12)*, pp. 113–116, Chengdu, China, December 2012.

[22] S. Lee, H. Choi, K. Choi, and J. Kim, "Fingerprint-quality index using gradient components," *IEEE Transactions on Information Forensics and Security*, vol. 3, no. 4, pp. 792–800, 2008.

[23] M. Vatsa, R. Singh, A. Noore, and M. M. Houck, "Quality-augmented fusion of level-2 and level-3 fingerprint information using DSm theory," *International Journal of Approximate Reasoning*, vol. 50, no. 1, pp. 51–61, 2009.

[24] M. Vatsa, R. Singh, A. Noore, and S. K. Singh, "Quality induced fingerprint identification using extended feature set," in *Proceedings of the 2nd IEEE International Conference on Biometrics: Theory, Applications and Systems*, pp. 1–6, October 2008.

[25] J. Bernsen, "Dynamic thresholding of gray-level images," in *Proceedings of the 8th International Conference on Pattern Recognition*, pp. 1251–1255, Paris, France, 1986.

[26] T. Y. Zhang and C. Y. Suen, "A fast parallel algorithm for thinning digital patterns," *Communications of the ACM*, vol. 27, no. 3, pp. 236–239, 1984.

[27] NSCT, "Fingerprint Recognition," http://www.biometrics.gov/documents/fingerprintrec.pdf.

[28] D. Maio, D. Maltoni, R. Cappelli, J. L. Wayman, and A. K. Jain, "FVC: The Second International Competition for Fingerprint Verification Algorithms," http://bias.csr.unibo.it/fvc2002/.

[29] Biometrika FX2000, http://www.biometrika.it/eng/fx2000.html.

[30] V. Bonato, R. F. Molz, J. C. Furtado, M. F. Ferrão, F. G. Moraes, and M. F. Ferrão, "Propose of a hardware implementation for fingerprint systems," in *Field Programmable Logic and Application: 13th International Conference, FPL 2003, Lisbon, Portugal, September 1–3, 2003 Proceedings*, vol. 2778 of *Lecture Notes in Computer Science*, pp. 1158–1161, 2003.

5

New Construction of PVPKE Scheme and Its Application in Information Systems and Mobile Communication

Minqing Zhang,[1,2] **Xu An Wang,**[2] **Xiaoyuan Yang,**[2] **and Weihua Li**[1]

[1]*School of Computer Science, Northwestern Polytechnical University, Xi'an 710072, China*
[2]*Key Laboratory of Information and Network Security, Engineering University of Chinese Armed Police Force, Xi'an 710086, China*

Correspondence should be addressed to Xu An Wang; wangxazjd@163.com

Academic Editor: David Taniar

In SCN12, Nieto et al. discussed an interesting property of public key encryption with chosen ciphertext security, that is, ciphertexts with public verifiability. Independently, we introduced a new cryptographic primitive, CCA-secure publicly verifiable public key encryption without pairings in the standard model (PVPKE), and discussed its application in proxy reencryption (PRE) and threshold public key encryption (TPKE). In Crypto'09, Hofheiz and Kiltz introduced the group of signed quadratic residues and discussed its application; the most interesting feature of this group is its "gap" property, while the computational problem is as hard as factoring, and the corresponding decisional problem is easy. In this paper, we give new constructions of PVPKE scheme based on signed quadratic residues and analyze their security. We also discuss PVPKE's important application in modern information systems, such as achieving ciphertext checkable in the cloud setting for the mobile laptop, reducing workload by the gateway between the open internet and the trusted private network, and dropping invalid ciphertext by the routers for helping the network to preserve its communication bandwidth.

1. Introduction

In modern information systems such as mobile wireless network, social network, open internet, and cloud computation, security is an important issue [1, 2]. Public key encryption [3] is among the most important basic tools to strengthen the whole system's security. Along with the development of information system, the security notion for public key encryption has been strengthened. The first proposal on public key encryption, RSA, though a great breakthrough in cryptography, only achieves the security notion of one-way security [4]. In 1984, Goldwasser and Micali [5] proposed the notion of semantic security (also known as indistinguishable security (IND-CPA)). This security notion states that the challenge ciphertext needs to contain no more information than a randomly chosen ciphertext. Although it is a reasonable security notion, many applications using public key encryption as a basic tool need stronger security notion, that is, chosen ciphertext security (IND-CCA). Compared with the semantic security notion, this security notion considers that the adversary can get help from the decryption oracle

(the adversary can query the decryption oracle with his chosen ciphertexts, except the challenge ciphertext which cannot be queried). Until now, many CCA-secure PKE schemes have been proposed [6–11].

Active attackers play more and more important role in breaking the security of modern information systems [1, 2]; thus chosen ciphertext security of the encryption scheme is essential for these systems. However, if the validity can only be checked by the decrypter privately with his secret key, the whole system can easily suffer from ciphertext-malleable attack. The active attackers can easily modify the right ciphertext transferred in the network to get numerous malicious ciphertexts and thus cost the precious bandwidth greatly. Although these ciphertexts can be rejected by the decrypter at the last moment, they have already caused great problem in the systems. These problems can affect the users' feeling on using the system. Even more seriously, they cause shutting down the whole system and bring damage to the service providing corporations. If the validity of these ciphertexts can be checked publicly, the problems can be easily solved, the routers or the access infrastructure can drop these maliciously created ciphertexts, and the bandwidth has been effectively preserved [12]. As a concrete example, can you imagine, when using mobile phone for secure instant-message talking like MSN, you always have to deal with nonsense invalid ciphertexts maliciously created by active attackers? But if the access infrastructure equipped with PVPKE can help you to filter these invalid ciphertexts, you certainly will feel better. In one word, PVPKE is an important tool for smoothly running modern information systems if these systems have employed public key encryption as a basic way to achieve security.

However, researchers give little care to the property of public verifiability of the chosen ciphertext-secure ciphertexts. In bilinear map setting or by using the random oracle, public verifiability of ciphertexts coming from an IND-CCA-secure public key encryption can be easily achieved. Thus, in this paper, we care about how to construct publicly verifiable public key encryption without pairing in the standard model. Recently, in [13], we introduced an interesting cryptographic primitive: PVPKE, defined as publicly verifiable chosen ciphertext-secure public key encryption in the standard model without pairing. PVPKE is a very powerful building block to construct some other interesting cryptographic protocols and cloud computation [14, 15]. For example, it can be used to construct chosen ciphertext-(CCA-) secure threshold public key encryption (TPKE) [16–20]. In TPKE, chosen ciphertext security always requires that the distributed decryption server can check the ciphertext's validity before decryption; otherwise some valuable information about decryption will be returned to the adversary and this will help the adversary to break the chosen ciphertext security. For another example, PVPKE can be a core block to construct chosen ciphertext-secure proxy reencryption (PRE) [21–26]. Chosen ciphertext attackers can query the delegator and delegatee's decryption oracle arbitrarily; if invalid ciphertexts forwarded by the proxy to the delegatee have been decrypted by the delegatee, the attackers can get useful information to break CCA security. Since the proxy without secret keys needs to check the validity of the ciphertext for the delegatee before reencryption, thus public verifiability of the ciphertext seems to be an essential requirement for achieving CCA security for proxy reencryption.

In SCN12, Nieto et al. [27] discussed an interesting property of public key encryption with chosen ciphertext security, that is, ciphertexts with public verifiability. They also demonstrated an important application of this new primitive, that is, "nontrivial filtering" of an incoming IND-CCA-secure ciphertext to be an IND-CPA-secure ciphertext with reduced workload by a gateway. They formally defined (nontrivial) public variability of ciphertexts for general encryption schemes, key encapsulation mechanisms, and hybrid encryption schemes, encompassing public key, identity-based, and tag-based encryption and also gave several concrete constructions. But we also note that their constructions cannot simultaneously satisfy the four requirements on "PVPKE": (1) chosen ciphertext-secure; (2) publicly verifiable; (3) in the standard model; (4) without pairing. Thus their work further explores PVPKE's application but does not give concrete construction of PVPKE.

In Crypto'09, Hofheinz and Kiltz [28] introduced the group of signed quadratic residues and discussed its application; the most interesting feature of this group is its "gap" property, while the computational problem is as hard as factoring, and the corresponding decisional problem is easy. Membership in QR_N^+ can be publicly and efficiently verified while it inherits some nice intractability properties of the quadratic residues. For example, computing square roots in QR_N^+ is also equivalent to factoring the modulus N. We therefore have a gap group, in which the corresponding decisional problem (i.e., deciding if an element is a signed square) is easy, whereas the computational problem (i.e., computing a square root) is as hard as factoring. We also can show that, in the group of signed quadratic residues, the Strong Diffie-Hellman problem is implied by the factoring assumption.

1.1. Our Contribution. In [13], based on the core idea of changing the prime modular field to the composite modular field and masking the verifying secret key with secret order of the composite group and making the resulting "pseudosecret key" public, we find it is relatively easy to construct PVPKE scheme based on the Cramer-Shoup encryption and the Hanaoka-Kurosawa CCA-secure public key encryption.

In this paper, we show that, in case of basing some of Nieto et al.'s schemes on signed quadratic residues, the resulting schemes can meet the requirements of PVPKE. The core idea about this construction is that the DDH oracle can be publicly instantiated by bilinear pairing, while DDH oracle cannot be instantiated by discrete logarithm group or RSA group. But, in signed quadratic residues, the DDH oracle can be efficiently publicly instantiated. Based on this observation, we give new constructions of PVPKE scheme based on signed quadratic residues and discuss their security.

Furthermore, we discuss PVPKE's important application in modern information system, such as achieving ciphertext checkable in the cloud setting for the mobile laptop, reducing the workload by the gateway between the open internet and the trusted private network, and dropping the invalid ciphertext by the routers for helping the network to preserve its communication bandwidth effectively.

1.2. Related Works

1.2.1. Chosen Ciphertext Security in the Standard Model. Naor and Yung [29] introduced the notion of CCA security for public key encryption, and this notion was further extended by Rackoff and Simon [30], Dolev et al. [31], and Sahai [32]. *Noninteractive zero-knowledge (NIZK) proofs* are core blocks of these constructions, which is a relatively inefficient paradigm and its efficient realization always relies on bilinear pairing or random orale. In 1993, Bellare and Rogaway [33] introduced a so-called *random oracle* which idealizes the hash function as a perfect random function to devise efficient CCA-secure public key encryption with provable security. However, random oracle model has seen criticism by cryptographers for its unrealistic assumption [34]. More and more cryptographers show interest in constructing efficient

CCA-secure PKE in the standard model. Till now, there are at least four ways to construct efficient CCA-secure PKE in the standard model. The first way is proposed by Cramer and Shoup [8], which was further extended by themselves and other cryptographers [35–37]. The second way to construct CCA-secure PKE is the paradigm of IBE *transformation*, which allows transforming selective-ID CPA-secure identity-based encryption (IBE) into a CCA-secure PKE [38–41]. The third way is based on *verifiable broadcast encryption*, which is proposed by Hanaoka and Kurosawa [9]. The fourth way is by relying on lossy trapdoor function introduced by Peikert and Waters [42] and further extended by Rosen and Segev [43] and many other works. Among the CCA-secure PKE schemes from these four ways, only the ones from the IBE transformation are publicly verifiable. However, most of existing practical IBE are based on the time-consuming pairings.

1.2.2. Without Pairings. The bilinear pairings enable the construction of first practical identity-based encryption by Boneh and Franklin [44]. Since then, many wonderful results can be achieved by using the bilinear pairings, such as fully collusion resistant broadcast encryption [45], efficient practical zero-knowledge proof [46], searchable public key encryption [47, 48], attribute based encryption [49], and predicate encryption [50].

But we note that, on the one hand, bilinear pairing is a very powerful cryptographic tool; on the other hand, the implementation speed of bilinear pairing is still relatively slower. So recently many researchers show interest in construction of schemes without pairings, because, on the one hand, it can clarify to us which cryptographic task inherits the bilinear property of pairings and which does not; on the other hand, it gives us a new view on old cryptographic problems. For example, Baek et al. constructed the first certificateless public key encryption without pairing [51], while the concept of certificateless public key cryptography was first raised by using bilinear pairings [52]. Other examples include Deng et al. and Shao and Cao's CCA-secure proxy reencryption without pairing [53, 54].

1.2.3. Verifiable Public Key Encryption. Another related research area is (private) verifiable public key encryption, such as Camenisch and Shoup's work [55]. However, their work was concerned with only the decryptor's verifiability of the ciphertext instead of *public* verifiability. Kiayias et al. extended their work by introducing some new concepts for constructing group encryption [56]. Owing to bilinear property of pairings, CCA-secure public key encryption with public verifiability can be easily achieved in the bilinear pairing setting. However, the situation is completely different in the "without pairing" setting; constructing PVPKE scheme remains as an open problem left for almost decades.

1.3. Organization. We organize our paper as follows: In Section 2, we give some preliminaries. In Section 3, we give our PVPKE's construction based on signed quadratic residues and analyse its security. In Section 4, we discuss PVPKE's applications. In the last section, we give our conclusion.

2. Preliminaries

2.1. Publicly Verifiable Public Key Encryption. A publicly verifiable public key encryption system consists of the following algorithms.

(i) The randomized key generation algorithm Gen takes as input a security parameter 1^k and outputs a public key (PK) and a secret key (SK). We write (PK, SK) ← Gen(1^k).

(ii) The randomized encryption algorithm $\mathcal{E}$ takes as input a public key (PK) and a message $m \in \{0, 1\}^*$ and outputs a ciphertext C. We write $C \leftarrow \mathcal{E}_{PK}(m)$.

(iii) The verification algorithm $\mathcal{V}$ takes as input a ciphertext C and a public key (PK). It returns valid or invalid to indicate whether the ciphertext is valid or not. Note that the validity of C can be verified publicly.

(iv) The decryption algorithm $\mathcal{D}$ takes as input a ciphertext C and a secret key (SK). It returns a message $m \in \{0.1\}^*$ or the distinguished symbol $\perp$. We write $m \leftarrow \mathcal{D}_{SK}(C)$.

We require that, for all (PK, SK) output by Gen, all $m \in \{0, 1\}^*$, and all C output by $\mathcal{E}_{PK}(m)$, we have $\mathcal{D}_{SK} = m$.

2.2. Chosen Ciphertext Security. We recall the standard definition of security against adaptive chosen ciphertext attack. A publicly verifiable public key encryption (PKE scheme is secure against adaptive chosen ciphertext attacks (i.e., "CCA-secure") if the advantage of any PPT adversary A in the following game is negligible in the security parameter k.

(1) Gen(1^k) outputs (PK, SK). Adversary A is given 1^k and PK.

(2) The adversary may make many polynomial-many queries to a decryption oracle $\mathcal{D}_{SK}(\cdot)$.

(3) The adversary may make many polynomial-many queries to a verification oracle $\mathcal{V}_{PK}(\cdot)$.

(4) At some point, A outputs two messages m_0, m_1 with $|m_0| = |m_1|$. A bit b is randomly chosen and the adversary is given a "challenge ciphertext" $C^* \leftarrow \mathcal{E}_{PK}(m_b)$.

(5) A may continue to query its decryption oracle $\mathcal{D}_{SK}(\cdot)$ except that it may not request the decryption of C^*.

(6) A may continue to make polynomial-many queries to a verification oracle $\mathcal{V}_{PK}(\cdot)$.

(7) Finally, A outputs a guess b'.

We say that A succeeds if $b' = b$ and denote the probability of this event by $\Pr_{A,PKE}[Succ]$. The adversary's advantage is defined as $|\Pr_{A,PKE}[Succ] - 1/2|$.

2.3. The Group of Signed Quadratic Residues

2.3.1. RSA Instance Generator.
Let $0 \leq \delta \leq 1/2$ be a constant and let $n(k)$ be a function. Let RSAgen be an algorithm that generates elements (N, P, Q), such that $N = PQ$ is an n-bit Blum integer ($N = PQ$ (where $P = 3 \bmod 4$ and $Q = 3 \bmod 4$)) and all prime factors of $\phi(N)/4$ are pairwise distinct and at least δn-bit integers.

2.3.2. Factoring Assumption.
The factoring assumption is that computing P, Q from N (generated by RSAgen) is hard. We write

$$\mathrm{Adv}^{fac}_{\mathcal{A}, \mathrm{RSAGen}}$$

$$= \Pr\left[\{P, Q\} \longleftarrow_R \mathcal{A}(N) : (N, P, Q) \longleftarrow_R \mathrm{RSAGen}\left(1^k\right)\right]. \tag{1}$$

The factoring assumption for RSAgen holds if $\mathrm{Adv}^{fac}_{\mathcal{A}, \mathrm{RSAGen}}$ is negligible for all efficient $\mathcal{A}$.

2.3.3. The Group of Signed Quadratic Residues.
Let N be an integer. For $x \in \mathbb{Z}_N$ we define $|x|$ as the absolute value of x, where x is represented as a signed integer in the set $\{-(N-1)/2, \ldots, (N-1)/2\}$. For a subgroup $\mathbb{G}$ of $\mathbb{Z}_N^*$, we define the signed group, $\mathbb{G}^+$, as the group

$$\mathbb{G}^+ = \{|x| : x \in \mathbb{G}\} \tag{2}$$

with the following group operation. Namely, for $g, h \in \mathbb{G}^+$ and an integer x, we define

$$g \circ h = |g \cdot h \bmod N|,$$
$$g^x = g \circ g \circ \cdots \circ g = |g^x \bmod N|. \tag{3}$$

More complicated expressions in the exponents are computed modulo the group order; for example, $g^{1/2} = g^{2^{-1} \bmod \mathrm{ord}(\mathbb{G}^+)}$. Note that taking the absolute value is a surjective homomorphism from $\mathbb{G}$ to $\mathbb{G}^+$ with trivial kernel if -1 does not belong to $\mathbb{G}$ and with kernel $\{-1, 1\}$ if $-1 \in \mathbb{G}$.

Let N be a Blum integer such that -1 does not belong to QR_N. We will mainly be interested in QR_N^+, which we call **signed quadratic residues (modulo N)**. QR_N^+ is a subgroup of $\mathbb{Z}_N^*/\pm 1$, with absolute values as a convenient computational representation. The following basic facts hold.

Theorem 1. *Let N be a Blum integer; then we have the following.*

(1) *$(QR_N^+, \circ)$ is a group of order $\phi(N)/4$.*

(2) *$QR_N^+ = J_N^+$. In particular, QR_N^+ is efficiently recognizable (given only N).*

(3) *If QR_N is cyclic, so is QR_N^+.*

2.3.4. Strong DH Assumption Reduced to Factoring Assumption.
Hofheinz and Kiltz [28] also proved that the strong DH assumption can be reduced to factoring assumption. Here we review the theorem and its proof.

Theorem 2. *If the factoring assumption holds then the strong DH assumption holds relative to RSAgen. In particular, for every strong DH adversary $\mathcal{A}$, there exists a factoring adversary $\mathcal{B}$ (with roughly the same complexity as $\mathcal{A}$) such that*

$$\mathrm{Adv}^{SDH}_{\mathcal{A}, \mathrm{RSAgen}}(k) \leq \mathrm{Adv}^{fac}_{\mathcal{B}, \mathrm{RSAgen}}(k) + O\left(2^{-\delta n(k)}\right). \tag{4}$$

Proof. We construct $\mathcal{B}$ from given $\mathcal{A}$. Concretely, $\mathcal{B}$ receives a challenge $N = PQ$, chooses uniformly $u \longleftarrow_R (\mathbb{Z}_N^*)^+ \setminus QR_N^+$, and sets $h = u^2$. Note that, by definition of N, we have $\langle h \rangle = QR_N^+$ except with probability $O(2^{-\delta n(k)})$. Then $\mathcal{B}$ chooses $a, b \in [N/4]$ and sets

$$g := h^2, \quad := h \circ g^a, \quad := h \circ g^b \tag{5}$$

(here we omit mod$N$ operation, and hereafter we continue to omit mod$N$ for typical exponential modular operation). This implicitly defines

$$\begin{aligned} d\log_g^X &= a + \frac{1}{2} \bmod \mathrm{ord}\left(QR_N^+\right), \\ d\log_g^Y &= b + \frac{1}{2} \bmod \mathrm{ord}\left(QR_N^+\right), \end{aligned} \tag{6}$$

where the discrete logarithms are of course considered in $(QR_N^+, \circ)$. Again, by definition of N, the statistical distance between these (g, X, Y) and the input of $\mathcal{A}$ in the strong DH experiment is bounded by $O(2^{-\delta n(k)})$. So $\mathcal{B}$ runs $\mathcal{A}$ on input (g, X, Y) and answers $\mathcal{A}$'s oracle queries $(\widehat{Y}, \widehat{Z})$ as follows. First, we may assume that $(\widehat{Y}, \widehat{Z}) \in QR_N^+$ since $QR_N^+ = J_N^+$ is efficiently recognizable. Next, since N is a Blum integer, the group order $\mathrm{ord}(QR_N^+) = (P-1)(Q-1)/4$ is odd, and hence

$$\begin{aligned} \widehat{Y}^{d\log_g^X} &= \widehat{Z} \\ \Longleftrightarrow \widehat{Y}^{2d\log_g^X} &= \widehat{Z}^2 \\ \Longleftrightarrow \widehat{Y}^{2a+1} &= \widehat{Z}^2. \end{aligned} \tag{7}$$

Thus, $\mathcal{B}$ can implement the strong DH oracle by checking whether $\widehat{Y}^{2a+1} = \widehat{Z}^2$ hold.

Consequently, with probability $\mathrm{Adv}^{SDH}_{\mathcal{A}, \mathrm{RSAgen}}(k) - O(2^{-\delta n(k)})$, $\mathcal{A}$ will finally output

$$\begin{aligned} Z &= g^{(d\log_g^X)(d\log_g^Y)} = g^{(a+1/2)(b+1/2)} \\ &= h^{2ab+a+b+1/2} \in QR_N^+ \end{aligned} \tag{8}$$

from which $\mathcal{B}$ can extract $v := h^{1/2} \in QR_N^+$ (using its knowledge about a and b). Since u is not in QR_N^+ and $v \in QR_N^+$ are two nontrivially different square roots of h, $\mathcal{B}$ can factor N by computing $gcd(u - v, N)$. $\square$

3. CCA-Secure Publicly Verifiable Public Key Encryption in the Standard Model Based on Signed Quadratic Residues

3.1. Review of Nieto et al.'s Publicly Verifiable PKE Scheme.
Their construction is inspired by the IND-CCA public key

KEM of Kiltz [57]; the PG(ParamGen) algorithm is similar to [57] except that it uses gap groups: $PG(1^k)$ outputs public parameters par $= (\mathbb{G}, p, g, \text{DDH}, H)$, where $\mathbb{G} = g$ is a multiplicative cyclic group of prime order p, $2^k \le p \le 2^{k+1}$, DDH is an efficient algorithm such that $\text{DDH}(g^a, g^b, g^c) = 1 \leftrightarrow c = ab(p)$, and $H : \mathbb{G} \to \{0, 1\}^{l_1(k)}$ is a cryptographic hash function such that $l_1(k)$ is a polynomial in k. We also use a strong one-time signature scheme OTS = (KG, Sign, Vrfy) with verification key space $\{0, 1\}^{l_2(k)}$ such that $l_2(k)$ is a polynomial in k and a target collision resistant hash function $\text{TCR} : \mathbb{G} \times \{0, 1\}^{l_2(k)} \to Z_p$. The message space is MsgSp = $\{0, 1\}^{l_1(k)}$. The scheme works as follows.

(i) PKE.KG(par)

$$x \longleftarrow Z_p^*$$

$$u \longleftarrow g^x, \quad v \longleftarrow \mathbb{G}$$

$$ek \longleftarrow (u, v), \quad dk \longleftarrow x \tag{9}$$

$$\text{Return } (ek, dk)$$

(ii) PKE.Enc(par, ek, M)

$$(u, v) \longleftarrow ek$$

$$(vk, sig k) \longleftarrow \text{OTS.KG}\left(1^k\right)$$

$$r \longleftarrow_R Z_p^*, \quad c_1 \longleftarrow g^r$$

$$t \longleftarrow \text{TCR}\left(c_1, vk\right), \quad \pi \longleftarrow \left(u^t v\right)^r \tag{10}$$

$$K \longleftarrow H\left(u^r\right), \quad c_2 \longleftarrow M \oplus K$$

$$c \longleftarrow \left(c_1, c_2, \pi\right)$$

$$\delta \longleftarrow \text{OTS.Sign}\left(sig k, c\right)$$

$$\text{Return } C = (c, \delta, vk)$$

(iii) PKE.Ver(par, ek, C)

$$(u, v) \longleftarrow ek$$

$$(c, \delta, vk) \longleftarrow C$$

$$(c_1, c_2, \phi) \longleftarrow c$$

$$t \longleftarrow \text{TCR}\left(c_1, vk\right) \tag{11}$$

$$\text{If DDH}\left(c_1, u^t v, \pi\right) \neq \quad \text{Or}$$

$$\text{OTS.Vrfy}\left(c, \delta, vk\right) = \bot, \text{ return } \bot$$

$$\text{Return } C' = \left(c_1, c_2\right)$$

(iv) PKE.Dec$'$(par, ek, dk, C')

$$(c_1, c_2) \longleftarrow C'$$

$$x \longleftarrow dk$$

$$K \longleftarrow H\left(c_1^x\right), \quad M \longleftarrow c_2 \oplus K \tag{12}$$

$$\text{Return } M.$$

3.2. Our Proposed PVPKE Scheme Based on Signed Quadratic Residues. First we give the core idea behind our construction. We observe that Nieto et al.'s PKE scheme actually is a PVPKE scheme, but the only issue is that they use an abstract DDH oracle. They instantiate this oracle by bilinear pairings, but we require that PVPKE scheme cannot rely on bilinear pairings. We also observe that signed quadratic residues can also instantiate the abstract DDH oracle, so we modify Nieto et al.'s scheme to be based on signed quadratic residues group, which now give a natural new PVPKE scheme. Notation: we omit the $\text{mod} N$ operation and every modular exponentiation in signed quadratic residues such as the fact that $h = u^{\underline{2}}$ is represented as $h = u^2$, which implies all the modular exponentiation and other operations obey the rules defined in [28] instead of obeying the normal group rules. The following is the concrete scheme.

(i) PVPKE.PG(1^k) is as follows.

  (a) Here we focus on QR_N^+ group; we first generate an RSA modulus $N = PQ$ with $\text{RSAgen}(1^k)$ [28], then choose uniformly $u \longleftarrow_R (Z_N^*)^+ \setminus QR_N^+$, and set $h = u^2$. Note that, by definition of N, we have $G = \langle h \rangle = QR_N^+$ except with probability $O(2^{-\delta n(k)})$.

  (b) $H : \mathbb{G} \to \{0, 1\}^{l_1(k)}$ is a cryptographic hash function such that $l_1(k)$ is a polynomial in k.

  (c) We also use a strong one-time signature scheme OTS = (KG, Sign, Vrfy) with verification key space $\{0, 1\}^{l_2(k)}$ such that $l_2(k)$ is a polynomial in k and a target collision resistant hash function $\text{TCR} : \mathbb{G} \times \{0, 1\}^{l_2(k)} \to Z_p$. The message space is MsgSp = $\{0, 1\}^{l_1(k)}$.

  (d) DDH is an efficient algorithm such that $\text{DDH}(g^a, g^b, g^c) = 1 \leftrightarrow c = ab \mod p$. For the scheme relying on QR_N^+ group, we can easily decide the DDH tuple; concretely, we do the following.

    (1) Choose $a, b \in [N/4]$ and $m, n \in \text{ord}(QR_N^+)$ satisfying $2^m(a + 1/2) > n \times \text{ord}(QR_N^+)$, $2^m(b + 1/2) > n \times \text{ord}(QR_N^+)$, and m is not very little. Then set

$$g := h^2, \quad X := h \circ g^a, \quad Y := h \circ g^b. \tag{13}$$

    (2) Publish $a' = 2^m(a + 1/2) \mod n \times \text{ord}(QR_N^+)$, $b' = 2^m(b + 1/2) \mod n \times \text{ord}(QR_N^+)$ as the parameters for public verifying.

(3) The $\text{DDHParams} = (g, X, Y, a', b', 2^m)$.

(e) $\text{PG}(1^k)$ outputs public parameters $\text{par} = (\mathbb{G}, N, \text{DDHParams}, H, \text{OTS}) = (\mathbb{G}, N, g, X, Y, a', b', 2^m, H, \text{OTS})$.

(ii) PVPKE.KG(par)

$$x \longleftarrow Z_N^*$$

$$u \longleftarrow g^x, \quad X = h \circ g^a, \quad Y = h \circ g^b \tag{14}$$

$$ek \longleftarrow (u, X, Y), \quad dk \longleftarrow x$$

Return (ek, dk)

(iii) PVPKE.Enc(par, ek, M)

$$(u, X, Y) \longleftarrow ek$$

$$(vk, \text{sig}\,k) \longleftarrow \text{OTS.KG}\left(1^k\right)$$

$$r \longleftarrow_R Z_N^*, \quad c_1 \longleftarrow g^r$$

$$t \longleftarrow \text{TCR}\left(c_1, vk\right), \quad \pi \longleftarrow \left(X^t Y\right)^r \tag{15}$$

$$K \longleftarrow H\left(u^r\right), \quad c_2 \longleftarrow M \oplus K$$

$$c \longleftarrow (c_1, c_2, \pi)$$

$$\delta \longleftarrow \text{OTS.Sign}\left(\text{sig}\,k, c\right)$$

Return $C = (c, \delta, vk)$

(iv) PVPKE.Ver(par, ek, C)

$$(u, X, Y) \longleftarrow ek$$

$$(c, \delta, vk) \longleftarrow C$$

$$(c_1, c_2, \pi) \longleftarrow c$$

$$t \longleftarrow \text{TCR}\left(c_1, vk\right) \tag{16}$$

If $c_1^{a't+b'} \neq (\pi)^{2^m}$ Or

$\text{OTS.Vrfy}\left(c, \delta, vk\right) = \perp, \; \text{return } \perp$

Return $C' = (c_1, c_2)$

(v) PVPKE.Dec$'$(par, ek, dk, C$'$)

$$(c_1, c_2) \longleftarrow C'$$

$$x \longleftarrow dk$$

$$K \longleftarrow H\left(c_1^x\right), \quad M \longleftarrow c_2 \oplus K \tag{17}$$

Return M.

3.3. Security Analysis. Based on Nieto et al.'s security result and the property of signed quadratic residues, we can give the following theorem.

Theorem 3. *Assume that* TCR *is a target collision resistant hash function and* OTS *is a strongly unforgeable one-time signature scheme. Under a variant of hashed Diffie-Hellman assumption for* $\mathbb{G}$ *(signed quadratic residues group) and* H, *the factoring assumption of* RSAGen *(which implies the strong Diffie-Hellman assumption in signed quadratic residues group proved in [28]), our PVPKE scheme based on signed quadratic residues is IND-CCA-secure.*

Proof. In the following we give our scheme's security proof roughly.

(1) We observe that, in Nieto et al.'s PKE scheme, u plays two roles: one used to be deriving the DEM message mask key and the other used to be as part of the DDH test. But many research results show that it is secure to split these two roles separately [8]; thus we introduce X as the role of part of the DDH test, while maintaining u as the source of deriving DEM message mask key, which is the reason why we use $(X^t Y)$ instead of $(u^t v)$ in our scheme.

(2) In our scheme, we adopt Hofheinz and Kiltz's technique of reducing SDH assumption to the factoring assumption; concretely, we set X, Y, g, h, a, and b the same as theirs, but we make $a' = 2^m(a + 1/2) \bmod n \times \text{ord}(QR_N^+)$ and $b' = 2^m(b+1/2) \bmod n \times \text{ord}(QR_N^+)$ public, which is used for public verifying. The verifying equation $(g^r)^{a't+b'} = ((X^t Y)^r)^{2^m}$ can also be used for deciding the DDH relationship of $(g, X^t Y, g^r, (X^t Y)^r)$, but an attacker cannot figure out $\pi = (X^t Y)^r$ through finding $1/2^m$ root of $(g^r)^{a't+b'}$, for we know finding square root in QR_N is as hard as factoring and this also holds in QR_N^+.

(3) We require $2^m(a+1/2) > n \times \text{ord}(QR_N^+)$, $2^m(b+1/2) > n \times \text{ord}(QR_N^+)$ for avoiding the trivial attack of computing $a + 1/2 = a'/2^m$ and $b + 1/2 = b'/2^m$ without any modular operation, and thus trivial computing $\pi = (X^t Y)^r = (g^r)^{(a+1/2)t+(b+1/2)}$. Obviously this attack can easily forge a valid π and thus a valid ciphertext and break the IND-CCA property. We also require that m is not too little to resist the brute force attack on finding a from a'.

(4) Generally speaking, our scheme is almost identical to Nieto et al.'s scheme; thus the security proof is almost the same as theirs. Below are the details.

Let (c^*, δ^*, vk^*) be the challenge ciphertext. The proposed PKE without the CHK transform can be seen as a KEM/DEM combination, which is at least IND-CPA-secure due to Herranz et al. [58]. As for the KEM, a variant of the hashed Diffie-Hellman (HDH) assumption [48] can be used to prove the IND-CPA security of the resulting PKE. Note that the message does not depend on vk^* and is just the signature on c^*. Therefore c^* being an output of the IND-CPA-secure scheme hides the value of the chosen b from the adversary.

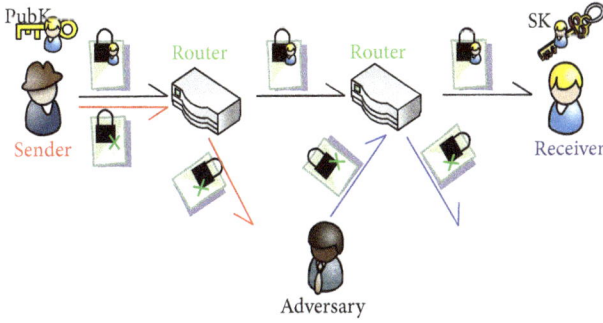

FIGURE 1: Routers drop the invalid ciphertexts via PVPKE.

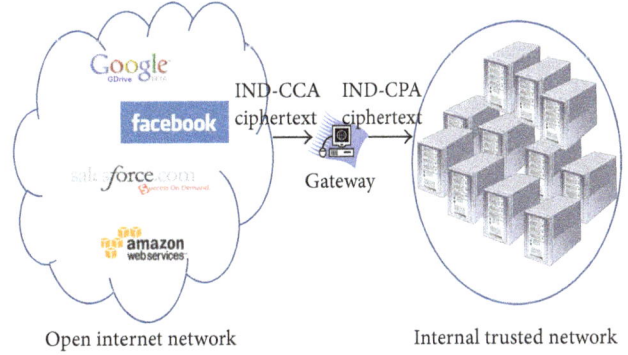

FIGURE 2: Gateways reduce the workload via PVPKE.

Below we prove that the IND-CCA adversary $\mathcal{A}$ may access decryption oracle and will gain no help in guessing the value of b. Suppose the adversary submits a ciphertext $(c', \delta', vk') \neq (c^*, \delta^*, vk^*)$ to the decryption oracle. Now there are two cases.

(i) When $vk' = vk^*$, the decryption oracle will output $\bot$ as the adversary fails to break the underlying strongly unforgeable one-time signature scheme with respect to vk'.

(ii) When $vk' \neq vk$, the attacker $\mathcal{B}$ against the variant of HDH problem can set the public keys as seen in the IND-CCA security proof for the KEM by Kiltz [57] such that (1) $\mathcal{B}$ can answer except for the challenge ciphertext all decryption queries from $\mathcal{A}$ even without the knowledge of the secret key and (2) $\mathcal{B}$ solves HDH if $\mathcal{A}$ wins. Note in Nieto et al.'s scheme u, v is the public key while in our scheme u, X, Y is the public key, but we observe v is randomly chosen from G, while in our scheme X, Y are set as $h \circ g^a, h \circ g^b$ which are also random because a, b are random. Thus our scheme roughly shares the same security proof outline as in [57] except that our scheme is in signed quadratic residues. □

4. Applications

4.1. Application 1: The Routers Drop the Invalid Ciphertexts via PVPKE. As shown in Figure 1, PVPKE can be used in the open internet network to help the routers to filter the invalid ciphertexts, while traditional IND-CCA-secure public key encryption does not have this function. First a sender (encrypter) wants to encrypt his message to a receiver (decrypter) by using public key encryption, and the ciphertexts in many cases have to be sent through open networks, which are not equipped with security guards to resist malicious attack; thus the sender should better choose an IND-CCA-secure public key encryption to encrypt his message. When an error or a data loss occurs in the ciphertexts through the transferring, the PVPKE can help the routers drop invalid ciphertexts by using the algorithm of public verifying. Note here the routers need not any secret,

which will greatly reduce the cost of resetup of the old system. Also, if there exists malicious attacker modifying the ciphertexts, the invalid ciphertexts will also be dropped. This will greatly help the network to preserve its communication band only to effective data blocks and help the routers and the receiver to reduce the workload for they now only need to do the necessary computation. However, PVPKE cannot resist the following case: an attacker generates a ciphertext following the right encryption algorithm and this ciphertext will certainly pass through the algorithm of public verifying. We think this time the attacker is indeed an encrypter, which will be a trivial case, and any verifying algorithm cannot avoid it.

4.2. Application 2: The Gateways Reduce the Workload via PVPKE. The following scenarios are always existing: ciphertexts need to be transferred from a public open network like internet to an internal network like the government's network. As shown in Figure 2, PVPKE can be used to help the gateways reduce the workload: transforming an IND-CCA ciphertext to be an IND-CPA ciphertext. When an IND-CCA ciphertext was captured by the gateway, the gateway first verifies its validity by using the publicly verifying algorithm. If it has passed, then the gateway can drop one part of the ciphertext: the part which is used to authenticate the ciphertext, like (δ, vk) in our PVPKE and Nieto et al.'s PKE scheme (here we do not claim that any PVPKE scheme has this separate authentication part, for there exist PVPKE schemes in which the authentication part has been integrated in the other parts of the ciphertext as a whole). Thus the remaining ciphertext will be IND-CPA-secure and will be shorter compared with the original ciphertext. Because the government's network usually will be protected well with many security mechanisms, IND-CPA security is enough to assure the security of the ciphertext. This will also reduce the workload of the employees who work on the internal network of the government.

4.3. Application 3: Achieving Ciphertext Checkable in the Clouds via PVPKE. Today more and more people prefer to upload their personal data contents to the clouds, but they do not want the cloud to know what the data contents

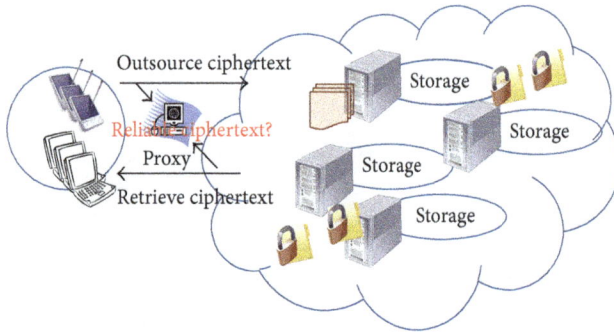

FIGURE 3: Achieving ciphertext checkable in the clouds via PVPKE.

are. Thus they need to encrypt the personal data contents before uploading them to the clouds. PVPKE can be used to achieve ciphertext checkable in this case, which can be seen in Figure 3. When the data owner uploads the ciphertexts to the cloud, there may exist incident things, like data loss or malicious attacker modifying the ciphertexts; in these cases, a proxy can be used to check the ciphertext's validity by using PVPKE. When the data owner or data user needs to retrieve the content, the clouds return the corresponding ciphertext to them. Also this time the proxy can be used to check the ciphertext's validity by using PVPKE. Note here that the proxy needs only to be semitrusted; it can perform the check without any secret; this will greatly benefit reducing the system management. For example, the proxy can be the access infrastructure in the wireless network setting. Note here that we do not claim that every ciphertext needs to be checked, which will be too heavy. This check must be run probabilistically with randomly chosen ciphertext.

5. Conclusions

PVPKE is a very powerful block to construct other cryptographic primitives or protocols, and its construction remains open for almost decades. In [13], we give several constructions and analyze their security. In this paper, by using the fact that the DDH oracle can be instantiated in signed quadratic residues, we give new PVPKE construction and roughly prove its security. The future work will be further exploring our idea and prove our proposal's security strictly.

Disclosure

This paper is a revised and expanded version of a paper titled "New Construction of PVPKE Scheme Based on Signed Quadratic Residues" presented at the Incos 2013 Conference [59]. The second author is the corresponding author.

Conflict of Interests

The authors declare that there is no conflict of interests regarding the publication of this paper.

Acknowledgments

The authors would like to express their gratitude to the editors for many helpful comments. This work is supported by the National Natural Science Foundation of China under Contracts nos. 61103230, 61272492, 61103231, and 61202492.

References

[1] A. J. Jara, S. Varakliotis, A. F. Skarmeta, and P. Kirstein, "Extending the Internet of things to the future internet through IPv6 support," *Mobile Information Systems*, vol. 10, no. 1, pp. 3–17, 2014.

[2] A. J. Jara, D. Fernandez, P. Lopez, M. A. Zamora, and A. F. Skarmeta, "Lightweight MIPv6 with IPSec support," *Mobile Information Systems*, vol. 10, no. 1, pp. 37–77, 2014.

[3] W. Diffie and M. E. Hellman, "New directions in cryptography," *IEEE Transactions on Information Theory*, vol. 22, no. 6, pp. 644–654, 1976.

[4] R. L. Rivest, A. Shamir, and L. Adleman, "A method for obtaining digital signatures and public-key cryptosystems," *Communications of the Association for Computing Machinery*, vol. 21, no. 2, pp. 120–126, 1978.

[5] S. Goldwasser and S. Micali, "Probabilistic encryption," *Journal of Computer and System Sciences*, vol. 28, no. 2, pp. 270–299, 1984.

[6] M. Abe, E. Kiltz, and T. Okamoto, "Chosen ciphertext security with optimal ciphertext overhead," in *Advances in Cryptology—ASIACRYP*, vol. 5350 of *Lecture Notes in Computer Science*, pp. 355–371, Springer, Berlin, Germany, 2008.

[7] M. Bellare and P. Rogaway, "Optimal asymmetric encryption: how to encrypt with RSA," in *Advances in Cryptology—EUROCRYPT'94*, vol. 950 of *Lecture Notes in Computer Science*, pp. 92–111, 1994.

[8] R. Cramer and V. Shoup, "A practical public key cryptosystem provably secure against adaptive chosen ciphertext attack," in *Advances in Cryptology—CRYPTO '98*, vol. 1462 of *Lecture Notes in Computer Science*, pp. 13–25, Springer, Berlin, Germany, 1998.

[9] G. Hanaoka and K. Kurosawa, "Efficient chosen ciphertext secure public key encryption under the computational Diffie-Hellman assumption," in *Advances in Cryptology—ASIACRYPT 2008*, vol. 5350 of *Lecture Notes in Computer Science*, pp. 308–325, Springer, Berlin, Germany, 2008.

[10] D. Hofheinz, E. Kiltz, and V. Shoup, "Practical chosen ciphertext secure encryption from factoring," *Journal of Cryptology*, vol. 26, no. 1, pp. 102–118, 2013.

[11] Y. Lindell, "A simpler construction of cca2-secure public-key encryption under general assumptions," in *Advances in Cryptology—EUROCRYPT 2003*, vol. 2656 of *Lecture Notes in Computer Science*, pp. 241–254, Springer, Berlin, Germany, 2003.

[12] K. Goto, Y. Sasaki, T. Hara, and S. Nishio, "Data gathering using mobile agents for reducing traffic in dense mobile wireless sensor networks," *Mobile Information Systems*, vol. 9, no. 4, pp. 295–314, 2013.

[13] M. Zhang, X. A. Wang, W. Li, and X. Yang, "CCA secure publicly verifiable public key encryption without pairings nor random oracle and its applications," *Journal of Computers*, vol. 8, no. 8, pp. 1987–1994, 2013.

[14] X. Chen, J. Li, and W. Susilo, "Efficient fair conditional payments for outsourcing computations," *IEEE Transactions on Information Forensics and Security*, vol. 7, no. 6, pp. 1687–1694, 2012.

[15] X. Chen, J. Li, J. Ma, Q. Tang, and W. Lou, "New algorithms for secure outsourcing of modular exponentiations," in *Computer Security—ESORICS 2012*, vol. 7459 of *Lecture Notes in Computer Science*, pp. 541–556, Springer, Berlin, Germany, 2012.

[16] R. Canetti and S. Goldwasser, "An efficient *threshold* public key cryptosystem secure against adaptive chosen ciphertext attack," in *Advances in Cryptology—EUROCRYPT'99*, vol. 1592 of *Lecture Notes in Computer Science*, pp. 90–106, Springer, Berlin, Germany, 1999.

[17] J. Baek and Y. Zheng, "Identity-based threshold decryption," in *Public Key Cryptography—PKC 2004*, vol. 2947 of *Lecture Notes in Computer Science*, pp. 262–276, Springer, Berlin, Germany, 2004.

[18] D. Boneh, X. Boyen, and S. Halevi, "Chosen ciphertext secure public key threshold encryption without random oracles," in *Topics in Cryptology—CT-RSA 2006*, vol. 3860 of *Lecture Notes in Computer Science*, pp. 226–243, 2006.

[19] V. Shoup and R. Gennaro, "Securing threshold cryptosystems against chosen ciphertext attack," *Journal of Cryptology*, vol. 15, no. 2, pp. 75–96, 2002.

[20] C. Delerablée and D. Pointcheval, "Dynamic threshold public-key encryption," in *Advances in Cryptology—CRYPTO*, vol. 5157 of *Lecture Notes in Computer Science*, pp. 317–334, Springer, Berlin, Germany, 2008.

[21] G. Ateniese, K. Fu, M. Green, and S. Hohenberger, "Improved proxy re-encryption schemes with applications to secure distributed storage," in *Proceedings of the 12th Annual Network and Distributed System Security Symposium (NDSS '05)*, pp. 29–43, San Diego, Calif, USA, 2005.

[22] G. Ateniese, K. Fu, M. Green, and S. Hohenberger, "Improved proxy re-encryption schemes with applications to secure distributed storage," *ACM Transactions on Information and System Security*, vol. 9, no. 1, pp. 1–30, 2006.

[23] B. Libert and D. Vergnaud, "Unidirectional chosen-ciphertext secure proxy re-encryption," in *Public Key Cryptography—PKC 2008*, vol. 4939 of *Lecture Notes in Computer Science*, pp. 360–379, Springer, Berlin, Germany, 2008.

[24] R. Canetti and S. Hohenberger, "Chosen ciphertext secure proxy re-encryption," in *Proceedings of the 14th ACM Conference on Computer and Communications Security (CCS '07)*, pp. 185–194, ACM, 2007.

[25] J. Zhang and X. A. Wang, "On the security of a multi-use CCA-secure proxy re-encryption scheme," in *Proceedings of the 4th International Conference on Intelligent Networking and Collaborative Systems (INCoS '12)*, pp. 571–576, Bucharest, Romania, September 2012.

[26] J. Zhang and X. Wang, "Security analysis of a multi-use identity based CCA-secure proxy reencryption scheme," in *Proceedings of the 4th International Conference on Intelligent Networking and Collaborative Systems (INCoS '12)*, pp. 581–586, September 2012.

[27] J. Nieto, M. Manulis, B. Poettering, J. Rangasamy, and D. Stebila, "Publicly verifiable ciphertexts," in *Proceedings of the 8th International Conference on Security and Cryptography for Networks (SCN '12)*, vol. 7485 of *Lecture Notes in Computer Science*, pp. 393–410, Amalfi, Italy, 2012.

[28] D. Hofheinz and E. Kiltz, "The group of signed quadratic residues and applications," in *Advances in Cryptology—CRYPTO 2009*, vol. 5677 of *Lecture Notes in Computer Science*, pp. 637–653, Springer, Berlin, Germany, 2009.

[29] M. Naor and M. Yung, "Public-key cryptosystems provably secure against chosen ciphertext attacks," in *Proceedings of the 22nd Annual ACM Symposium on Theory of Computing (STOC '90)*, pp. 427–437, May 1990.

[30] C. Rackoff and D. R. Simon, "Non-interactive zero-knowledge proof of knowledge and chosen ciphertext attack," in *Advances in Cryptology—CRYPTO '91*, vol. 576 of *Lecture Notes in Computer Science*, pp. 433–444, Springer, Berlin, Germany, 1992.

[31] D. Dolev, C. Dwork, and M. Naor, "Non-malleable cryptography," in *Proceedings of the 23rd Annual ACM Symposium on Theory of Computing (STOC '91)*, pp. 542–552, May 1991.

[32] A. Sahai, "Non-malleable non-interactive zero knowledge and adaptive chosen-ciphertext security," in *Proceedings of the 40th Annual Symposium on Foundations of Computer Science (IEEE FOCS '99)*, pp. 543–553, New York, NY, USA, October 1999.

[33] M. Bellare and P. Rogaway, "Random oracles are practical: a paradigm for designing efficient protocols," in *Proceedings of the 1st ACM Conference on Computer and Communications Security (CCS '93)*, pp. 62–73, November 1993.

[34] R. Canetti, O. Goldreich, and S. Halevi, "Random oracle methodology, revisited," in *Proceedings of the 30th Annual ACM Symposium on Theory of Computing (STOC '98)*, pp. 209–218, May 1998.

[35] R. Cramer and V. Shoup, "Design and analysis of practical public-key encryption schemes secure against adaptive chosen ciphertext attack," *SIAM Journal on Computing*, vol. 33, no. 1, pp. 167–226, 2003.

[36] K. Kurosawa and Y. Desmedt, "A new paradigm of hybrid encryption scheme," in *Advances in Cryptology—CRYPTO 2004*, vol. 3152 of *Lecture Notes in Computer Science*, pp. 426–442, 2004.

[37] M. Abe, R. Gennaro, K. Kurosawa, and V. Shoup, "Tag-kem/dem: a new framework for hybrid encryption and a new analysis of kurosawa-desmedt kem," in *Advances in Cryptology—EUROCRYPT 2005*, vol. 3494 of *Lecture Notes in Computer Science*, pp. 128–146, Springer, Berlin, Germany, 2005.

[38] R. Canetti, S. Halevi, and J. Katz, "Chosen-ciphertext security from identity-based encryption," in *Advances in Cryptology—EUROCRYPT 2004*, vol. 3027 of *Lecture Notes in Computer Science*, pp. 207–222, Springer, Berlin, Germany, 2004.

[39] D. Boneh and J. Katz, "Improved efficiency for CCA-secure cryptosystems built using identity-based encryption," in *Topics in Cryptology—CT-RSA 2005*, vol. 3376 of *Lecture Notes in Computer Science*, pp. 87–103, Springer, Berlin, Germany, 2005.

[40] X. Boyen, Q. Mei, and B. Waters, "Direct chosen ciphertext security from identity-based techniques," in *Proceedings of the 12th ACM Conference on Computer and Communications Security (CCS '05)*, pp. 320–329, November 2005.

[41] E. Kiltz, "Chosen-ciphertext security from tag-based encryption," in *Theory of Cryptography*, vol. 3876 of *Lecture Notes in Computer Science*, pp. 581–600, Springer, Berlin, Germany, 2006.

[42] C. Peikert and B. Waters, "Lossy trapdoor functions and their applications," in *Proceedings of the 40th Annual ACM Symposium on Theory of Computing (STOC '08)*, pp. 187–196, 2008.

[43] A. Rosen and G. Segev, "Chosen-ciphertext security via correlated products," in *Theory of Cryptography*, vol. 5444, pp. 419–436, Springer, Berlin, Germany, 2009.

[44] D. Boneh and M. Franklin, "Identity-based encryption from the Weil pairing," in *Advances in Cryptology—CRYPTO 2001: Proceedings of the 21st Annual International Cryptology Conference, Santa Barbara, California, USA, August 19–23, 2001*, vol. 2139 of

Lecture Notes in Computer Science, pp. 213–229, Springer, Berlin, Germany, 2001.

[45] D. Boneh, C. Gentry, and B. Waters, "Collusion resistant broadcast encryption with short ciphertexts and private keys," in *Proceedings of the 25th Annual International Cryptology Conference (CRYPTO '05)*, vol. 3621 of *Lecture Notes in Computer Science*, pp. 258–275, Santa Barbara, Calif, USA, 2005.

[46] J. Groth and A. Sahai, "Efficient non-interactive proof systems for bilinear groups," in *Advances in Cryptology—EUROCRYPT 2008*, vol. 4965 of *Lecture Notes in Computer Science*, pp. 415–432, Springer, Berlin, Germany, 2008.

[47] D. Boneh, G. D. Crescenzo, R. Ostrovsky, and G. Persiano, "Public key encryption with keyword search," in *Computational Science and Its Applications—ICCSA 2008*, vol. 3089 of *Lecture Notes in Computer Science*, pp. 31–45, Springer, Berlin, Germany, 2004.

[48] M. Abdalla, M. Bellare, D. Catalano et al., "Searchable encryption revisited: consistency properties, relation to anonymous IBE, and extensions," in *Advances in Cryptology—CRYPTO 2005*, vol. 3621 of *Lecture Notes in Computer Science*, pp. 205–222, Springer, Berlin, Germany, 2005.

[49] V. Goyal, O. Pandey, A. Sahai, and B. Waters, "Attribute-based encryption for fine-grained access control of encrypted data," in *Proceedings of the 13th ACM Conference on Computer and Communications Security (CCS '06)*, pp. 89–98, November 2006.

[50] J. Katz, A. Sahai, and B. Waters, "Predicate encryption supporting disjunctions, polynomial equations, and inner products," in *Advances in Cryptology—EUROCRYPT 2008*, vol. 4965 of *Lecture Notes in Computer Science*, pp. 146–162, Springer, Berlin, Germany, 2008.

[51] J. Baek, R. Safavi-Naini, and W. Susilo, "Certificateless public key encryption without pairing," in *Information Security*, vol. 3650 of *Lecture Notes in Computer Science*, pp. 134–148, Springer, Berlin, Germany, 2005.

[52] S. Al Riyami and K. Paterson, "Certificateless public key cryptography," in *Advances in Cryptology—ASIACRYPT 2003*, vol. 2894 of *Lecture Notes in Computer Science*, pp. 452–473, Springer, 2003.

[53] R. Deng, J. Weng, S. Liu, and K. Chen, "Chosen ciphertext secure proxy re-encryption without pairings," in *Cryptology and Network Security*, vol. 5339 of *Lecture Notes in Computer Science*, pp. 1–17, Springer, Berlin, Germany, 2008, http://eprint.iacr.org/2008/509.

[54] J. Shao and Z. Cao, "CCA-secure proxy re-encryption without pairings," in *Public Key Cryptography—PKC 2009*, vol. 5443 of *Lecture Notes in Computer Science*, pp. 357–376, Springer, Berlin, Germany, 2009.

[55] J. Camenisch and V. Shoup, "Practical verifiable encryption and decryption of discrete logarithms," in *Advances in Cryptology—CRYPTO 2003*, vol. 2729 of *Lecture Notes in Computer Science*, pp. 126–144, Springer, Berlin, Germany, 2003.

[56] A. Kiayias, Y. Tsiounis, and M. Yung, *Group Encryption*, Cryptology ePrint Archive, 2007, http://eprint.iacr.org/2007/015.pdf.

[57] E. Kiltz, "Chosen-ciphertext secure key-encapsulation based on gap hashed Diffie-Hellman," in *Public Key Cryptography—PKC*, vol. 4450 of *Lecture Notes in Computer Science*, pp. 282–297, Springer, Berlin, Germany, 2007.

[58] J. Herranz, D. Hofheinz, and E. Kiltz, "KEM/DEM: necessary and sucffcient conditions for secure hybrid encryption," in *IACR Cryptology ePrint Archive*, Report 2006/256, IACR, 2006.

[59] J. Zhang and X. Wang, "New construction of PVPKE scheme based on signed quadratic residues," in *Proceedings of the 5th International Conference on Intelligent Networking and Collaborative Systems (INCoS '13)*, pp. 434–437, September 2013.

6

Verifiable Rational Secret Sharing Scheme in Mobile Networks

En Zhang,[1,2,3] **Peiyan Yuan,**[1,3] **and Jiao Du**[4]

[1]*College of Computer and Information Engineering, Henan Normal University, Xinxiang 453007, China*
[2]*State Key Laboratory of Information Security, Institute of Information Engineering, Chinese Academy of Sciences, Beijing 100093, China*
[3]*Engineering Lab of Intelligence Business & Internet of Things, Xinxiang, Henan 453007, China*
[4]*College of Mathematics and Information Science, Henan Normal University, Xinxiang 453007, China*

Correspondence should be addressed to En Zhang; zhangenzdrj@163.com

Academic Editor: Francesco Gringoli

With the development of mobile network, lots of people now have access to mobile phones and the mobile networks give users ubiquitous connectivity. However, smart phones and tablets are poor in computational resources such as memory size, processor speed, and disk capacity. So far, all existing rational secret sharing schemes cannot be suitable for mobile networks. In this paper, we propose a verifiable rational secret sharing scheme in mobile networks. The scheme provides a noninteractively verifiable proof for the correctness of participants' share and handshake protocol is not necessary; there is no need for certificate generation, propagation, and storage in the scheme, which is more suitable for devices with limited size and processing power; in the scheme, every participant uses her encryption on number of each round as the secret share and the dealer does not have to distribute any secret share; every participant cannot gain more by deviating the protocol, so rational participant has an incentive to abide by the protocol; finally, every participant can obtain the secret fairly (means that either everyone receives the secret, or else no one does) in mobile networks. The scheme is coalition-resilient and the security of our scheme relies on a computational assumption.

1. Introduction

1.1. Background. Secret sharing is playing a more and more important role in modern cryptography. In classical (t, n) secret sharing schemes [1, 2], a secret can be shared among n participants. At least t or more participants can reconstruct the secret, but $t - 1$ or fewer participants cannot obtain anything about the secret. Recently, a series of secret sharing schemes were proposed in [3–6]. However, the works in [1–6] cannot prevent the dealer or players from cheating. For example, in Shamir's scheme, we assume that one party does not broadcast his share, while exactly $t-1$ other players reveal their shares. He can still reconstruct the secret although his cheating can be detected by the scheme [7–9].

Motivated by the desire to develop more realistic models, the cryptographic community has significant interest in exploring protocols for rational secret sharing. Halpern and Teague [10] firstly introduced the notion of rational secret sharing. They pointed out that there exist many Nash equilibriums which, in some sense, are unreasonable. Therefore, they focus on one particular refinement of Nash equilibrium that is determined by iterated deletion of weakly dominated strategies. However, their protocols cannot work for 2 out of 2 secret sharing and require the online dealer. Later, a series of rational secret sharing schemes [11–20] were proposed. However, none of them are fully satisfactory. The works in [11–13] rely on secure multiparty computation which is strong. Kol and Naor's scheme [14] has information theoretic security. However, their scheme fails to resist against coalitions. The works in [15, 16] require the involvement of some trusted external parties during the reconstruction phase which is difficult to find. The solution in [17] constructs a rational scheme based on repeated games. However, every player has high probability to learn the secret in his last round. The works of Lepinski et al. [19, 20] and Izmalkov et al. [15, 18] can guarantee fairness, prevent coalitions, and eliminate side information. However, their solutions rely on physical assumption such as secure envelopes and ballot boxes. The works in [10–14, 17, 21–25] assume the existence of broadcast channel which is not realistic. The works in [11–13, 19–27] need to exchange public keys associated with certificate management, including revocation, storage and distribution, and the computational cost of certificate verification. Nowadays, with the development of mobile network, a large percent of the world's population now has access to mobile phones and incredibly fast mobile networks give users ubiquitous connectivity. New devices like smart phones and tablets are providing users with a lot of applications and services and have fundamentally changed our lives. However, smart phones and tablets are poor in computational resources such as processor speed, memory size, and disk capacity. A drawback of public key infrastructure (PKI) is that they are computationally very intensive, which makes them less suitable mobile phones. From the discussion above, it seems clear that all of above schemes cannot work in a mobile system.

1.2. Our Results. In this paper, we propose a verifiable rational secret sharing scheme in mobile networks. The major contribution of this work is as follows. We present a new verifiable random function for multiparty case, which provides a noninteractively verifiable proof for the correctness of participants' share and handshake protocol is not necessary; there is no need for certificate generation, propagation, and storage in the scheme, which is more suitable for devices with limited size and processing power; the public key in our approach is based on each participant's identity (e.g., telephone number or email address), which can be very much shorter as compared to the 1024 bits public key in RSA cryptosystem; in the scheme, every participant uses her/his encryption on number of each round as the secret share and the dealer does not have to distribute any secret share, which reduce the computational consumption and communicational overhead; the participants do not know whether the current round is a test round or not, and every participant cannot gain more by cheating. Finally, every player can obtain the secret fairly (means that either everyone receives the secret, or else no one does) in mobile networks. To the best of our knowledge, we propose the first rational secret sharing scheme over mobile networks.

1.3. Overview. The rest of this paper is organized as follows. In Section 2, the preliminary of game theory and cryptography for rational secret sharing are introduced. Section 3 introduces the rational secret scheme in mobile networks. In Section 4, we analyze the new scheme. Finally, we present our conclusions in Section 5.

2. Preliminaries

2.1. Basics of Game Theory. We begin by introducing some basic terminology of game theory in this section. For more details, please refer to [28].

Game theory aims to help us understand situations in which decision-makers interact. A strategic game consists of three components: (a) a set of players; (b) a set of actions for each player; (c) for each player, preferences over the set of action profiles.

Let $a = (a_1, \ldots, a_n)$ be profile of players, a_i denote the strategy employed by player P_i, a_{-i} be a strategy profile of all players except for the player P_i, and $(a_i', a_{-i}) = (a_1, \ldots, a_{i-1}, a_i', a_{i+1}, \ldots, a_n)$ denote the strategy vector a with P_i's strategy changed to a_i'; $u_i(a)$ represents P_i's preferences, which rational players wish to maximize.

Definition 1 (Nash equilibrium). Let $\Gamma = (\{A_i\}, \{u_i\}_{i=1}^n)$ be a game presented in normal form. A strategy profile $a = (a_1, \ldots, a_n) \in A$ is Nash equilibrium if, for all i and every $a_i' \in A_i$, it holds that

$$u_i(a_i', a_{-i}) \le u_i(a). \tag{1}$$

Generally speaking, Nash equilibrium holds the idea that no rational party has an incentive to deviate from the protocol. Everyone is playing a best response to everyone else and no individual can do strictly better by moving away. The definition of Nash equilibrium is designed to model a steady state among experienced players. In a steady state, no player wishes to change her behavior, considering the other players' behavior.

In a traditional secret sharing scheme, a player is thought as either honest or malicious. However, in a rational secret sharing scheme, it may make more sense to view the players, not as good or bad, but as rational individuals trying to maximize their own utility [10]. For any player P_i, assume that any rational player prefers to get the secret rather than miss it. And secondarily, prefer that as few as possible of the other players get it.

Now, let we introduce the definition of computational C-immune [13] in which utility functions take the security parameter k as input.

Definition 2 (computational C-immune). Let σ be an efficient protocol for a computing game and $\mathbb{C}$ be a set of coalitions (subsets of players). Let R^t be the set of sequences of random tapes for the first t iterations that do not cause σ to end. A sequence $r \in R^t$ is of the form $r = (r^1, \ldots, r^t)$ where $r^s = (r_1^s, \ldots, r_n^s)$ and r_j^s is the random tape used by player j in iteration s.

The protocol σ is computational C-immune if, for every coalition $C \in \mathbb{C}$ and every sequence of tapes $r_0 = (r_0^1, \ldots, r_0^t) \in R^t$ used by the players in the first t round, there exists a negligible function $\varepsilon(k)$ such that, for every player $i \in C$, every efficient (deviating) joint strategy σ_C' for players in C, and every efficient joint strategy τ_{-C} for players in N/C implementing σ_{-C}, it holds that

$$E\left[u_i\left(\tau_{-C}(k), \sigma_C(k)\right)\right] + \varepsilon(k)$$
$$\ge E\left[u_i\left(\tau_{-C}(k), \sigma_C'(k)\right)\right]. \tag{2}$$

2.2. Cryptographic Terminology

Definition 3 (bilinear pairing). Let G_1 and G_2 be multiplicative groups of prime order p. g is the generator of G_1.

A bilinear pairings is a map $e : G_1 \times G_1 \rightarrow G_2$ with the following properties.

(1) Bilinear: for all $u, v \in G_1$ and all $a, b \in Z$, one has $e(u^a, v^b) = e(u, v)^{ab}$.

(2) Nondegenerate: $e(g, g) \neq 1$.

(3) Computable: there is an efficient algorithm to compute $e(u, v)$ for all u and $v \in G_1$.

We describe decisional bilinear Diffie-Hellman inversion assumption below.

Given $(g, g^x, \ldots, g^{(x^q)})$ as input, to distinguish $e(g, g)^{1/x}$ from random. An algorithm A has advantage ε in solving the q-DBDHI problem if

$$
\begin{aligned}
&\left| \Pr \left[A \left(g, g^x, \ldots, g^{(x^q)}, e(g, g)^{1/x} \right) = 1 \right] \right. \\
&\left. - \Pr \left[A \left(g^x, \ldots, g^{(x^q)}, \Gamma \right) = 1 \right] \right| \leq \varepsilon,
\end{aligned}
\tag{3}
$$

where $x \in Z_p^*$ and $\Gamma \in G_2$.

We say that the (t, q, ε)-DBDHI assumption holds in G_1, if no t-time algorithm A has advantage at least ε in solving the q-DBDHI problem in G_1.

2.3. Verifiable Random Function from Identity-Based Key Encapsulation (IB-KEM).
Verifiable random function (VRF) was firstly introduced by Micali et al. [29]. A VRF is a pseudo-random function that provides a noninteractively verifiable proof for the correctness of its output, and the VRF has many useful applications. References [29–32], respectively, constructed a VRF. Next we briefly recall the VRF from a VRF-suitable IB-KEM [32].

The IB-KEM Scheme. An identity-based key encapsulation mechanism (IB-KEM) scheme allows a sender and a receiver to agree on a random session key K. And it is defined by four algorithms: Setup(1^k) takes a security parameter as input and outputs a master key pairs (mpk, msk); KeyDer(msk, ID) uses the master secret key to compute sk_{ID} for identity ID; Encap(mpk, ID) computes a random session key K and a ciphertext C; Decap(C, sk_{ID}) allows the receiver to decapsulate C to get back a session key K. An VRF-suitable IB-KEM scheme [33] is defined by the following algorithms.

(i) Setup(1^k) is a probabilistic algorithm that takes in input a security parameter k and outputs a master public key mpk and a master secret key msk. Let G_1, G_2 be bilinear groups of prime order q. Additionally, let $e: G_1 \times G_1 \rightarrow G_2$ denote the bilinear map. The description of G_1 contains a generator $g \in G_1$. Then the algorithm picks a random $s \leftarrow Z_p^*$, sets $h = g^s$, and outputs a master key pairs (mpk $= (g, h)$, msk $= s$).

(ii) KeyDer(msk, ID): the key derivation algorithm uses the master secret key to compute a secret key $sk_{ID} = g^{1/(s+ID)}$ for identity ID.

(iii) Encap(mpk, ID): the encapsulation algorithm picks a random $t \leftarrow Z_q$ and computes a random session key $K = e(g, g)^t$ using (mpk, ID). Moreover it uses (mpk, ID) to computes a ciphertext $C = (g^s g^{ID})^t$ encrypted under the identity ID.

(iv) Decap(C, sk_{ID}) allows the possessor of sk_{ID} to compute a session key K from a ciphertext C as follows: $K = e(C, sk_{ID})$.

The VRF (Gen, Func, and Ver) Construction Is as follows

(i) Gen(1^k) runs (mpk, msk) $\leftarrow$ Setup(1^k), chooses an arbitrary identity $ID_0 \in ID$, where ID is the identity space, and computes $C_0 \leftarrow$ Encap(mpk, ID_0). Then it sets vpk $= (mpk, C_0)$ and vsk $= msk$.

(ii) Func$_{vsk}(x)$ computes $\pi_x = (sk_x, aux_x) =$ KeyDer(msk, x) and $y =$ Decap(C_0, π_x). It returns (y, π_x) where y is the output and π_x is the proof.

(iii) Ver(vpk, x, y, π_x) first checks if π_x is a valid proof for x by computing $(C, K) =$ Encap(mpk, x, aux$_x$) and checking if $K =$ Decap(C, π_x). Then it checks the validity of y by testing if Decap(C_0, π_x) $= y$. If both the tests are true, then the algorithm returns 1, otherwise it returns 0.

With a modification, we extend the VRF from a VRF-suitable IB-KEM [32] to multiparty case, and this can be used in our rational secret sharing schemes. Let $p_1, \ldots, p_n$ be n participants, $ID_i \in ID$ ($i = 1, \ldots, n$) be the identity of p_i, where ID is the identity space, and d_i be the private key of p_i.

(i) Gen(1^k) takes a security parameter k, returns mpk$_i$, msk$_i$ and computes $C_0^i \leftarrow$ Encap(mpk$_i$, ID_i). Then it sets vpk$_i = (mpk_i, C_0^i)$ and $d_i = msk_i$.

(ii) Func$_{d_i}(x)$ computes $\pi_{d_i}(x) = (sk_x^i, aux_x^i) =$ KeyDer(msk$_i$, x) and $E_{d_i}(x) =$ Decap(C_i, π_x). It returns $(E_{d_i}(x), \pi_{d_i}(x))$ where the VRF output is $E_{d_i}(x)$ and $\pi_{d_i}(x)$ is the proof.

(iii) VER(vpk$_i$, $x, E_{d_i}(x), \pi_{d_i}(x)$) checks if $\pi_{d_i}(x)$ is a valid proof by computing $(C_i, K_i) =$ Encap(mpk$_i$, x, aux$_x^i$) and checking if $K_i =$ Decap($C_i, \pi_{d_i}(x)$). Then it checks the validity of y by testing if Decap(C_0^i, $\pi_{d_i}(x)$) $= E_{d_i}(x)$. If both the tests are true, then the algorithm returns 1, otherwise it returns 0.

2.4. The Model of Security

Init. The adversary declares the identity set $\varphi = (ID_1, ID_2, \ldots, ID_n)$ that he wants to be challenged.

Setup. The challenger runs the setup phase of the algorithm and tells the adversary the public parameter.

Phase 1. The adversary is allowed to issue queries for private keys for many identities χ_i, where $|\chi_i \cap \varphi| < t$.

Challenge. The adversary output a message x^*. The challenger flips a random coin b and obtains a session key K_b. If $b = 0$, then K_b is a correct form, otherwise K_b is random. Finally, it sends K_b to the adversary.

Phase 2. This goes exactly as phase 1.

Guess. The adversary outputs a guess b' of b. The adversary wins if $b' = b$.

We define the advantage of an adversary in this game as $\Pr[b' = b] - 1/2$.

3. The Rational Secret Sharing Scheme

3.1. System Parameters. Let $p_1, \ldots, p_n$ be n participants and l be the secret. Assume $\mathrm{ID}_i \in \mathrm{ID}$ $(i = 1, \ldots, n)$ is the identity of p_i, where $\mathrm{ID} \in Z_p^*$ is the identity space and $h : \{0,1\}^* \to Z_p^*$ is a collision resistance hash function. Let d_i be the private key of p_i.

3.2. Protocol for Sharing Phase

Step 1. The dealer chooses an integer $r^{\mathrm{real}} \in Z_p^*$ according to a geometric distribution with parameter λ. We discuss how to set λ below. The dealer computes $\mathrm{Gen}(1^k)$ and obtains d_i.

Step 2. Choose a prime p and construct two $(n-1)$ degree polynomials. One is $W(x)$ with the knowledge of n pairs of $(\mathrm{ID}_1 \parallel r^{\mathrm{real}}, E_{d_1}(r^{\mathrm{real}})), \ldots, (\mathrm{ID}_n \parallel r^{\mathrm{real}}, E_{d_n}(r^{\mathrm{real}}))$ as (4). The other is $W'(x)$ with the knowledge of n pairs of $(\mathrm{ID}_1 \parallel (r^{\mathrm{real}} + 1), E_{d_1}(r^{\mathrm{real}} + 1)), \ldots, (\mathrm{ID}_n \parallel (r^{\mathrm{real}} + 1), E_{d_n}(r^{\mathrm{real}} + 1))$ as (5):

$$W(x) = \sum_{i=1}^{n} E_{d_i}\left(r^{\mathrm{real}}\right)$$
$$\cdot \prod_{j=1, j \neq i}^{n} \frac{x - h\left(\mathrm{ID}_j \parallel r^{\mathrm{real}}\right)}{h\left(\mathrm{ID}_i \parallel r^{\mathrm{real}}\right) - h\left(\mathrm{ID}_j \parallel r^{\mathrm{real}}\right)} \bmod p, \tag{4}$$

$$W'(x) = \sum_{i=1}^{n} E_{d_i}\left(r^{\mathrm{real}} + 1\right)$$
$$\cdot \prod_{j=1, j \neq i}^{n} \frac{x - h\left(\mathrm{ID}_j \parallel \left(r^{\mathrm{real}} + 1\right)\right)}{h\left(\mathrm{ID}_i \parallel \left(r^{\mathrm{real}} + 1\right)\right) - h\left(\mathrm{ID}_j \parallel \left(r^{\mathrm{real}} + 1\right)\right)} \bmod p \tag{5}$$
$$= a_0^{r^{\mathrm{real}}+1} + a_1^{r^{\mathrm{real}}+1} x + a_2^{r^{\mathrm{real}}+1} x^2 + \cdots + a_{n-1}^{r^{\mathrm{real}}+1} x^{n-1},$$

$$\text{let } M^{r^{\mathrm{real}}} = W(0), \tag{6}$$

$$\text{value} = l \oplus M^{r^{\mathrm{real}}}. \tag{7}$$

Step 3. The dealer chooses the $n - t$ minimum integers $m_1, \ldots, m_{n-t}$ from $[p, q-1] - (\mathrm{ID}_i \parallel r)$ for $r = 1, 2, \ldots, r^{\mathrm{real}}$ and computes $W(m_k)$ and $W'(m_k)$ for $k = 1, 2, \ldots, n-t$.

Step 4. The dealer publishes the values $((m_k, W(m_k), (m_k, W'(m_k))$ for $k = 1, 2, \ldots, n-t$, value and $h(a_j^{r^{\mathrm{real}}+1})$ for $j = 0, 1, \ldots, n-1)$, and sends d_i to p_i.

3.3. Protocols for Reconstruction Phase. Let $T = \{p_{a_1}, p_{a_2}, \ldots, p_{a_t}\}$ be the set of the t active participants and $(E_{d_{a_i}}(r), \pi_{d_{a_i}}(r))$

be the share of p_{a_i} $(1 \leq i \leq t)$. In each iteration $(r = 0, 1, \ldots)$ the players execute the following steps.

Step 1. When $r \equiv i \pmod{t}$, each of the t active participants sends her share in the order $p_{a_{i+1}}, p_{a_{i+2}}, \ldots, p_{a_t}, p_{a_1}, p_{a_2}, \ldots, p_{a_i}$ for $0 \leq i \leq t - 1$.

Step 2. $p_{a_j} \in T$ receives the share from $p_{a_i} \in T$. If $\mathrm{VER}(\mathrm{vpk}_i, r, E_{d_{a_i}}(r), \pi_{d_{a_i}}(r)) = 0$, $\pi_{d_{a_i}}(r)$ is an invalid proof of $E_{d_{a_i}}(r)$, then, with the knowledge of t pairs of $(\mathrm{ID}_{a_i} \parallel (r-1), E_{d_{a_i}}(r-1)), \ldots, (\mathrm{ID}_{a_t} \parallel (r-1), E_{d_{a_t}}(r-1))$ and $n - t$ pairs of $(m_1, W(m_1)), \ldots, (m_{n-t}, W(m_{n-t}))$, the $(n-1)$ degree polynomial $B(x)$ can be uniquely determined as follows:

$$B(x) = \sum_{i=1}^{t} E_{d_{a_i}}(r-1)$$

$$\cdot \prod_{j=1, j \neq i}^{t} \frac{x - h\left(\mathrm{ID}_{a_i} \parallel (r-1)\right)}{h\left(\mathrm{ID}_{a_i} \parallel (r-1)\right) - h\left(\mathrm{ID}_{a_j} \parallel (r-1)\right)}$$

$$\cdot \prod_{j=1}^{n-t} \frac{x - m_i}{h\left(\mathrm{ID}_{a_i} \parallel (r-1)\right) - m_i} + \sum_{i=1}^{n-t} W(m_i) \tag{8}$$

$$\cdot \prod_{j=1, j \neq i}^{n-t} \frac{x - m_j}{m_i - m_j} \prod_{j=1}^{t} \frac{x - h\left(\mathrm{ID}_{a_j} \parallel (r-1)\right)}{m_i - h\left(\mathrm{ID}_{a_j} \parallel (r-1)\right)} \bmod p.$$

We let $M^{r-1} = B(0)$. The secret can be obtained as $l' = \text{value} \oplus M^{r-1}$ and then output l' and terminate the protocols. If $\mathrm{VER}(\mathrm{vpk}_i, r, E_{d_{a_i}}(r), \pi_{d_{a_i}}(r)) = 1$, $\pi_{d_{a_i}}(r)$ is a valid proof of $E_{d_{a_i}}(r)$, then the protocol continues.

Step 3. With the knowledge of t pairs of $((\mathrm{ID}_{a_1} \parallel r), E_{d_{a_1}}(r)), \ldots, ((\mathrm{ID}_{a_t} \parallel r), E_{d_{a_t}}(r))$ and $n - t$ pairs of $(m_1, W'(m_1)), \ldots, (m_{n-t}, W'(m_{n-t}))$, the $(n-1)$ degree polynomial $B'(x)$ can be uniquely determined as follows:

$$B'(x) = \sum_{i=1}^{t} E_{d_{a_i}}(r) \prod_{j=1, j \neq i}^{t} \frac{x - h\left(\mathrm{ID}_{a_j} \parallel r\right)}{h\left(\mathrm{ID}_{a_i} \parallel r\right) - h\left(\mathrm{ID}_{a_j} \parallel r\right)}$$

$$\cdot \prod_{j=1}^{n-t} \frac{x - m_i}{h\left(\mathrm{ID}_{a_i} \parallel r\right) - m_i} + \sum_{i=1}^{n-t} W'(m_i) \tag{9}$$

$$\cdot \prod_{j=1, j \neq i}^{n-t} \frac{x - m_j}{m_i - m_j} \prod_{j=1}^{t} \frac{x - h\left(\mathrm{ID}_{a_j} \parallel r\right)}{d_{a_i} - h\left(\mathrm{ID}_{a_j} \parallel r\right)} \bmod p = b_0^r$$

$$+ b_1^r x + b_2^r x^2 + \cdots + b_{n-1}^r x^{n-1}.$$

Step 4. If $h(b_j^r) \neq h(a_j^{r^{\mathrm{real}}+1})$ for $j = 0, 1, \ldots, n-1$ then the protocol goes to next iteration, else if $h(b_j^r) = h(a_j^{r^{\mathrm{real}}+1})$, then $r = r^{\mathrm{real}} + 1$, with the knowledge of t pairs of $((\mathrm{ID}_{a_1} \parallel r^{\mathrm{real}}), E_{d_{a_1}}(r^{\mathrm{real}})), \ldots, ((\mathrm{ID}_{a_t} \parallel r^{\mathrm{real}}), E_{d_{a_t}}(r^{\mathrm{real}}))$ and $n - t$

pairs of $(m_1, W(m_1)), \ldots, (m_{n-t}, W(m_{n-t}))$, the $(n-1)$ degree polynomial $B^{\text{real}}(x)$ can be uniquely determined as follows:

$$B^{\text{real}}(x) = \sum_{i=1}^{t} E_{d_{a_i}}\left(r^{\text{real}}\right)$$

$$\cdot \prod_{j=1, j \neq i}^{t} \frac{x - h\left(\text{ID}_{a_j} \parallel r^{\text{real}}\right)}{h\left(\text{ID}_{a_i} \parallel r^{\text{real}}\right) - h\left(\text{ID}_{a_j} \parallel r^{\text{real}}\right)}$$

$$\cdot \prod_{j=1}^{n-t} \frac{x - m_i}{h\left(\text{ID}_{a_i} \parallel r^{\text{real}}\right) - m_i} + \sum_{i=1}^{n-t} W(m_i) \tag{10}$$

$$\cdot \prod_{j=1, j \neq i}^{n-t} \frac{x - m_j}{m_i - m_j} \prod_{j=1}^{t} \frac{x - h\left(\text{ID}_{a_j} \parallel r^{\text{real}}\right)}{d_{a_i} - h\left(\text{ID}_{a_j} \parallel r^{\text{real}}\right)} \bmod p.$$

Let $M^{r^{\text{real}}} = B^{\text{real}}(0)$. The secret can be obtained as $l = \text{value} \oplus M^{r^{\text{real}}}$. Then output l and terminate the protocols.

4. Proof of Security

In this section, the poof of the security is discussed.

Theorem 4. *If an adversary can break our scheme, then one can build a simulator to solve the q-DBDHI assumption with a nonnegligible advantage.*

Proof. We assume there exists an adversary A that has nonnegligible advantage $\varepsilon(k)$ into breaking the protocol. Then we can build a simulator B which is able to break the q-DBDHI assumption with nonnegligible advantage.

Input to the Reduction. Algorithm B receives a tuple $(g, g^{\alpha}, \ldots, g^{(\alpha^q)}, \Gamma) \in G_1^{q+1} \times G_2$, and output 1 if $\Gamma = e(g, g)^{1/\alpha}$, or 0 otherwise.

Key Generation. Assume that A tries to guess the challenge message $x_0 \in Z_p^*$. Let $s = \alpha - x_0$. Using the binomial theorem, it computes $(g, g^s, \ldots, g^{(s^q)})$. Then B define $f(z) = \prod_{w \neq x_0}^{w \in z_q}(z + w) = \sum_{i=0}^{q-1} c_i z^i$ and compute the new base $g' = g^{f(s)} = \prod_{i=0}^{q-1} g^{s^i c_i}$. Finally it computes $h = (g')^s = \prod_{i=1}^{q-1} g^{s^i c_{i-1}}$, picks a random t, and sets $C_0 = (g')^t$. Then it gives g', h, C_0 as the public key to A.

Phase 1. The adversary A is allowed to issue queries for private keys for many identities χ_i, where $|\chi_i \cap \varphi| < t$ and $\varphi = (\text{ID}_1, \text{ID}_2, \ldots, \text{ID}_n)$. Consider the ith query ($1 \leq i < q$) on message x_i. If $x_i = x_0$, then B fails. Otherwise B can compute the secret key as follows. Firstly it defines $f_i(z) = f(z)/(z + x_i) = \sum_{i=0}^{q-2} z^i v_i$. Then it computes $\text{sk}_{x_i} = (g')^{1/(s+x_i)} = g^{f_i(s)} = \sum_{i=0}^{q-2} g^{s^i v_i}$ and returns it to A as the private key of χ_i. With the knowledge of t pairs of $((\text{ID}_{a_1} \parallel r), E_{d_{a_1}}(r)), \ldots, ((\text{ID}_{a_t} \parallel r), E_{d_{a_t}}(r))$ and $n - t$ pairs of $(m_1, W'(m_1)), \ldots, (m_{n-t}, W'(m_{n-t}))$, the simulator can construct the $(n-1)$ degree polynomial $B'(x)$ by using the

Lagrange interpolation polynomial. However, the coefficient of the $B'(x)$ is identical to that of the original scheme.

Challenge. The adversary A output a message x^*. If $x^* \neq x_0$, then B fails. Otherwise, the challenger can compute a session key K_b in the following way. Let $f'(z) = f(z)/(z + x_0) - \gamma/(z + x_0) = \sum_{i=0}^{q-2} z^i \gamma_i$ and compute $Z_0 = (\prod_{i=0}^{q-1} \prod_{j=0}^{q-2} e(g^{s^i}, g^{s^j}))(\prod_{m=0}^{q-2} e(g, g^{s^m})^{\gamma \gamma_m}) = e(g, g)^{(f(s) - \gamma^2)/\alpha}$. The simulator flips a random coin, b, and sets a session key $K_b = (Z^{\gamma^2} \cdot Z_0)^t$, if $b = 0$, then $Z = e(g, g)^{1/\alpha}$ and $K_b = e(g', g')^{t/(s+x_i)}$ is a correct form. Otherwise Z is a random, and so is K_b. Finally, it sends K_b to the adversary.

Phase 2. This goes exactly as phase 1.

Guess. The adversary A outputs a guess b' of b. B returns b' as its guess as well.

For the sake of contradiction, suppose there exists a probabilistic polynomial time attacker A can break the protocol with probability $1/2 + \varepsilon(k)$. Then we can build a simulator B which is able to break the q-DBDHI assumption with probability $1/2 + \varepsilon(k)$. (The output of B is the same as the output of A.) Because the q-DBDHI assumption is hard to be solved, there is no any adversary A that has nonnegligible advantage $\varepsilon(k)$ into breaking the protocol. This completes the proof. □

Theorem 5. *The above rational secret sharing scheme is computational C-immune, and rational participant has an incentive to abide by the protocol.*

Proof. Given the $n - t$ public values $W(m_k), W'(m_k)$, the two $(n - 1)$ degree polynomials $W(x), W'(x)$ cannot be constructed by anyone. So, an adversary can learn nothing about the secret. Any $t - 1$ or fewer participants cannot obtain the secret too. In the scheme, any rational participant can detect and determine who is cheating. Suppose that $p_{a_j} \in T$ receives the share from $p_{a_i} \in T$. If $\text{VER}(\text{vpk}_i, r, E_{d_{a_i}}(r), \pi_{d_{a_i}}(r)) = 0$, $\pi_{d_{a_i}}(r)$ is an invalid proof of $E_{d_{a_i}}(r)$, and p_{a_j} terminates the protocols. If $\text{VER}(\text{vpk}_i, r, E_{d_{a_i}}(r), \pi_{d_{a_i}}(r)) = 1$, $\pi_{d_{a_i}}(r)$ is a valid proof of $E_{d_{a_i}}(r)$, and p_{a_j} continues the protocols. Assume that P_i who is the member of the collusion C does not know which round is r^{real}. He can only guess the secret and get U_i^+ with probability β, if the collusion C does not participate in the scheme. On the contrary, he can guess a wrong secret and get U_i^- with probability $1 - \beta$. So, when the collusion C does not participate in the protocols, the expected utility of P_i is as in

$$E(U_i^{\text{guess}}) = \beta * U_i^+ + (1 - \beta) * U_i^-. \tag{11}$$

The participant P_i will get utility U_i^+, if the collusion C participates in the protocols and aborts in real round with probability λ. Otherwise, the participant P_i's utility is $E(U_i^{\text{guess}})$. Therefore, when the collusion C deviates, the expected utility of P_i is at most

$$\lambda * U_i^+ + (1 - \lambda) * E(U_i^{\text{guess}}). \tag{12}$$

When the collusion C abides by the protocol, the utility of the participant P_i is U_i. So, rational collusion C has an inventive not to deviate from the protocol if the protocol satisfies

$$U_i > \lambda * U_i^+ + (1 - \lambda) * E\left(U_i^{\text{guess}}\right). \tag{13}$$

We denote λ' the probability that players in C can only have a negligible advantage over λ. There exists a negligible function $\xi(k)$ such that for every k it holds that

$$\lambda' \leq \lambda + \xi(k). \tag{14}$$

We let $U^{*'}$ denote the utility when allowing for the computationally secure. Then

$$\begin{aligned} U^{*'} &= \lambda' U_i^+ + \left(1 - \lambda'\right) * E\left(U_i^{\text{guess}}\right) \\ &= \lambda'\left(U_i^+ - E\left(U_i^{\text{guess}}\right)\right) + E\left(U_i^{\text{guess}}\right) \\ &\leq (\lambda + \xi(k))\left(U_i^+ - E\left(U_i^{\text{guess}}\right)\right) + E\left(U_i^{\text{guess}}\right) \\ &\leq \lambda * U_i^+ + (1 - \lambda) * E\left(U_i^{\text{guess}}\right) \\ &\quad + \xi(k)\left(U_i^+ - E\left(U_i^{\text{guess}}\right)\right) < U_i + \varepsilon(k). \end{aligned} \tag{15}$$

That is for every iteration and for all $C \subset [n]$ with $|C| \leq t - 1$, all $i \in C$, and any $a_C' \in \Delta(A_C)$, no information about the secret is revealed. So, the scheme is computational C-immune and rational player has an incentive to abide by the protocol. $\square$

5. Comparison

We compare the efficiency and security with previous rational secret sharing scheme as follows.

The work of Halpern and Teague [10] assumes the existence of simultaneous broadcast channels (SBC). Their schemes fail to resist against coalitions and have expected round complexity $O(5/\alpha^3)$. The works in [11–13] rely on secure multiparty computation which are inefficient. The works of Kol and Naor [14] have shown how to avoid simultaneous broadcast, at the cost of increasing the round complexity. In addition, the scheme is not collusion-free, and the round complexity is $O(n/\beta)$ and the works in [15, 16] require the involvement of some trusted external parties during the reconstruction phase which is difficult to find. The round complexity of Maleka et al. [17] is $O(n^2)$. The works of Izmalkov et al. [18] and Lepinski et al. [19, 20] rely on a physical assumption such as secure envelopes and ballot boxes. The works in [10–14, 17, 21–25] assume the existence of broadcast channel which is not realistic. The works in [11–13, 19–27] need handshake protocol and exchange public keys associated with certificate management, including distribution, storage, revocation, and the computational cost of certificate verification, which are relatively expensive and limit their practical application to mobile networks. In contrast with prior schemes, the round complexity is $O(1/\lambda)$ (the value of α, β, and λ is roughly the same) in our scheme, and we do not assume multiparty computations, physical assumption,

or trust party, which is more practical; the scheme provides a noninteractively verifiable proof for the correctness of participants' share and handshake protocol is not necessary; there is no need for certificate generation, propagation, and storage in the scheme, which is more suitable for devices with limited size and processing power; the public key in our approach is based on each participant's identity which can be very much shorter as compared to the 1024 bits public key in RSA cryptosystem; in the scheme, every participant uses her encryption on number of each round as the secret share and the dealer does not have to distribute any secret share, which reduce the computational consumption and communicational overhead; the scheme can withstand the conspiracy attack and no player of the coalition C can do better, even if the whole coalition C cheats.

6. Conclusions

We propose a rational secret sharing scheme in mobile networks. The scheme, without needing to resort to broadcast channel, eliminates the online certificate authority and simplifies key management, which is more practical for devices of limited size and processing power, such as mobile phones. In addition, the scheme assumes neither the availability of a trusted party nor multiparty computations in the reconstruction phase. Moreover, the scheme can withstand the conspiracy attack and no player of the coalition C can do better, even if the whole coalition C cheats. So, rational players have no incentive to cheat in the scheme, and, finally, every player can obtain the secret fairly in mobile networks.

Conflict of Interests

The authors declare that there is no conflict of interests regarding the publication of this paper.

Acknowledgments

The authors would like to thank the anonymous referees for their suggestions. This work was supported by the National Natural Science Foundation of China (nos. 61170221, 11471104, U1204606, U1404601, and U1404602) and the Key Project of Education Department of Henan Province (no. 14A520032).

References

[1] A. Shamir, "How to share a secret," *Communications of the ACM*, vol. 22, no. 11, pp. 612–613, 1979.

[2] G. R. Blakeley, "Safeguarding cryptographic keys," in *Proceedings of the National Computer Conference*, pp. 313–317, AFIPS Press, New York, NY, USA, 1979.

[3] Y.-C. Hou, Z.-Y. Quan, C.-F. Tsai, and A.-Y. Tseng, "Block-based progressive visual secret sharing," *Information Sciences*, vol. 233, no. 4, pp. 290–304, 2013.

[4] J. Herranz, A. Ruiz, and G. Sáez, "New results and applications for multi-secret sharing schemes," *Designs, Codes, and Cryptography*, vol. 73, no. 3, pp. 841–864, 2014.

[5] J. Shao, "Efficient verifiable multi-secret sharing scheme based on hash function," *Information Sciences*, vol. 278, pp. 104–109, 2014.

[6] M. Fatemi, R. Ghasemi, and T. Eghlidos, "Efficient multistage secret sharing scheme using bilinear map," *IET Information Security*, vol. 8, no. 4, pp. 224–229, 2014.

[7] B. Chor, S. Goldwasser, S. Micali, and B. Awerbuch, "Verifiable secret sharing and achieving simultaneity in the presence of faults," in *Proceedings of the 26th Annual Symposium on Foundations of Computer Science*, pp. 383–395, IEEE Computer Society, Portland, Ore, USA, October 1985.

[8] P. Feldman, "A practical scheme for non-interactive verifiable secret sharing," in *Proceedings of the 28th Annual Symposium on Foundations of Computer Science*, pp. 427–437, IEEE, Los Angeles, Calif, USA, 1987.

[9] T. P. Pedersen, "Distributed provers with applications to undeniable signatures," in *Advances in Cryptology—EUROCRYPT '91*, vol. 547 of *Lecture Notes in Computer Science*, pp. 221–242, Springer, Berlin, Germany, 1991.

[10] J. Halpern and V. Teague, "Rational secret sharing and multiparty computation: extended abstract," in *Proceedings of the 36th Annual ACM Symposium on Theory of Computing*, pp. 623–632, ACM Press, New York, NY, USA, June 2004.

[11] S. D. Gordon and J. Katz, "Rational secret sharing, revisited," in *Security and Cryptography for Networks*, vol. 4116 of *Lecture Notes in Computer Science*, pp. 229–241, Springer, Berlin, Germany, 2006.

[12] E. Zhang and Y. Q. Cai, "A new rational secret sharing scheme," *China Communications*, vol. 7, no. 40, pp. 18–22, 2010.

[13] G. Kol and M. Naor, "Cryptography and game theory: designing protocols for exchanging information," in *Theory of Cryptography: Fifth Theory of Cryptography Conference, TCC 2008, New York, USA, March 19–21, 2008. Proceedings*, vol. 4948 of *Lecture Notes in Computer Science*, pp. 320–339, Springer, Berlin, Germany, 2008.

[14] G. Kol and M. Naor, "Games for exchanging information," in *Proceedings of the 40th Annual ACM Symposium on Theory of Computing (STOC '08)*, pp. 423–432, ACM Press, May 2008.

[15] S. Izmalkov, M. Lepinski, and S. Micali, "Verifiably secure devices," in *Theory of Cryptography: Proceedings of the 5th Theory of Cryptography Conference, TCC 2008, New York, USA, March 19–21, 2008*, vol. 4948 of *Lecture Notes in Computer Science*, pp. 273–301, Springer, Berlin, Germany, 2008.

[16] S. Micali and A. Shelat, "Purely rational secret sharing (extended abstract)," in *Theory of Cryptography: 6th Theory of Cryptography Conference, TCC 2009, San Francisco, CA, USA, March 15–17, 2009. Proceedings*, vol. 5444, pp. 54–71, Springer, Berlin, Germany, 2009.

[17] S. Maleka, A. Shareef, and C. P. Rangan, "The deterministic protocol for rational secret sharing," in *Proceedings of the 22nd IEEE International Parallel and Distributed Processing Symposium (IPDPS '08)*, pp. 1–7, IEEE, Miami, Fla, USA, April 2008.

[18] S. Izmalkov, M. Lepinski, and S. Micali, "Rational secure computation and ideal mechanism design," in *Proceedings of the 46th Annual IEEE Symposium on Foundations of Computer Science (FOCS '05)*, pp. 623–632, IEEE Press, New York, NY, USA, October 2005.

[19] M. Lepinski, S. Micali, and A. Shelat, "Collusion-free protocols," in *Proceedings of the 37th Annual ACM Symposium on Theory of Computing (STOC '05)*, pp. 543–552, ACM, Baltimore, Md, USA, May 2005.

[20] M. Lepinski, S. Micali, C. Peikert, and A. Shelat, "Completely fair SFE and coalition-safe cheap talk," in *Proceedings of the 23rd ACM Symposium on Principles of Distributed Computing (PODC '04)*, pp. 1–10, July 2004.

[21] T. Isshiki, K. Wada, and K. Tanaka, "A rational secret-sharing scheme based on RSA-OAEP," *IEICE Transactions on Fundamentals of Electronics, Communications and Computer Sciences*, vol. 93, no. 1, pp. 42–49, 2010.

[22] Z. Zhang and M. Liu, "Rational secret sharing as extensive games," *Science China Information Sciences*, vol. 56, no. 3, pp. 1–13, 2013.

[23] E. Zhang and Y. Q. Cai, "Collusion-free rational secure sum protocol," *Chinese Journal of Electronics*, vol. 22, no. 3, pp. 563–566, 2013.

[24] Y. Yang and Z. F. Zhou, "An efficient rational secret sharing protocol resisting against malicious adversaries over synchronous channels," in *Information Security and Cryptology*, vol. 7763 of *Lecture Notes in Computer Science*, pp. 69–89, Springer, Berlin, Germany, 2013.

[25] Y. Tian, J. Ma, C. Peng, and Q. Jiang, "Fair (t, n) threshold secret sharing scheme," *IET Information Security*, vol. 7, no. 2, pp. 106–112, 2013.

[26] G. Fuchsbauer, J. Katz, and D. Naccache, "Efficient rational secret sharing in standard communication networks," in *Theory of Cryptography: 7th Theory of Cryptography Conference, TCC 2010, Zurich, Switzerland, February 9–11, 2010. Proceedings*, vol. 5978 of *Lecture Notes in Computer Science*, pp. 419–436, Springer, Berlin, Germany, 2010.

[27] E. Zhang and Y. Q. Cai, "Rational multi-secret sharing scheme in standard point-to-point communication networks," *International Journal of Foundations of Computer Science*, vol. 24, no. 6, pp. 879–897, 2013.

[28] J. Katz, "Bridging game theory and cryptography: recent results and future directions," in *Theory of Cryptography: Fifth Theory of Cryptography Conference, TCC 2008, New York, USA, March 19–21, 2008. Proceedings*, vol. 4948 of *Lecture Notes in Computer Science*, pp. 251–272, Springer, Berlin, Germany, 2008.

[29] S. Micali, M. Rabin, and S. Vadhan, "Verifiable random functions," in *Proceedings of the 40th IEEE Symposium on Foundations of Computer Science*, pp. 120–130, IEEE Press, New York, NY, USA, 1999.

[30] Y. Dodis, "Efficient construction of (distributed) verifiable random functions," in *Public Key Cryptography—PKC 2003*, Y. G. Desmedt, Ed., vol. 2567 of *Lecture Notes in Computer Science*, pp. 1–17, Springer, Berlin, Germany, 2002.

[31] Y. Dodis and A. Yampolskiy, "A verifiable random function with short proof and keys," in *Public Key Cryptography—PKC 2005*, vol. 3386 of *Lecture Notes in Computer Science*, pp. 416–431, Springer, Berlin, Germany, 2005.

[32] M. Abdalla, D. Catalano, and D. Fiore, "Verifiable random functions from identity-based key encapsulation," in *Advances in Cryptology—EUROCRYPT 2009*, vol. 5479 of *Lecture Notes in Computer Science*, pp. 554–571, Springer, Berlin, Germany, 2009.

[33] R. Sakai and M. Kasahara, "ID based cryptosystems with pairing on elliptic curve," in *Proceedings of the Symposium on Cryptography and Information Security*, Report 2003/054, Cryptology ePrint Archive, 2003.

7

Server-Aided Verification Signature with Privacy for Mobile Computing

Lingling Xu,[1] Jin Li,[2] Shaohua Tang,[1] and Joonsang Baek[3]

[1]*School of Computer Science and Engineering, South China University of Technology, Guangzhou 510006, China*
[2]*School of Computer Science and Educational Software, Guangzhou University, Guangzhou 510006, China*
[3]*Khalifa University of Science, Technology and Research, P.O. Box 127788, Abu Dhabi, UAE*

Correspondence should be addressed to Shaohua Tang; csshtang@scut.edu.cn

Academic Editor: David Taniar

With the development of wireless technology, much data communication and processing has been conducted in mobile devices with wireless connection. As we know that the mobile devices will always be resource-poor relative to static ones though they will improve in absolute ability, therefore, they cannot process some expensive computational tasks due to the constrained computational resources. According to this problem, server-aided computing has been studied in which the power-constrained mobile devices can outsource some expensive computation to a server with powerful resources in order to reduce their computational load. However, in existing server-aided verification signature schemes, the server can learn some information about the message-signature pair to be verified, which is undesirable especially when the message includes some secret information. In this paper, we mainly study the server-aided verification signatures with privacy in which the message-signature pair to be verified can be protected from the server. Two definitions of privacy for server-aided verification signatures are presented under collusion attacks between the server and the signer. Then based on existing signatures, two concrete server-aided verification signature schemes with privacy are proposed which are both proved secure.

1. Introduction

Recent advances in wireless technology have led to mobile computing [1, 2] which is a technology that enables access to digital resources at any time, from any location. In mobile computing, much data communication and processing is conducted in mobile devices with wireless connection such as cell-phones, security access-cards, and sensors. Therefore, mobile computing represents the elimination of time-and-place restrictions imposed by desktop computers and wired networks. As we know mobile devices must be light and small to be easily carried around. Such considerations, in conjunction with a given cost and level of technology, will exact a penalty in computational resources of mobile devices such as processor speed. While mobile devices will improve in absolute ability, they will always be computationally weak in relation to static ones. As a consequence there are tasks, which potentially could enlarge a device's range of application, which are beyond its reach. A natural solution is to outsource computations that are too expensive for one device, to other devices which are more powerful or numerous and connected to the device. For example, consider a sensor that is presented with an access-card, sends it a random challenge, and receives a digital signature of the random challenge. The computation is required to verify the signature involves public-key operations which are too expensive in both time and space for the sensor to run. Instead, it could outsource the verification to a powerful device in order to reduce its computational load. Recently, with the development of cloud computing, server-aided computation has received widespread attention which enables power-constrained devices to outsource expensive computational tasks to a server. The related works such as server-aided delegated computation [3–8] and server-aided verification signatures [9–16] have been widely studied. Delegated computation is a protocol between two polynomial-time parties, a client, and a server, to collaborate on the computation of a function F. Concretely, the client wants the server to compute $F(x)$ for any input instance x by the delegated computation protocol and verify the correctness of the results that is returned by the server. A key requirement is that the amount of work performed by the client to generate and verify work instances must be substantially cheaper than performing the computation on its own.

A server-aided verification signature scheme consists of a digital signature scheme and a server-aided verification protocol. Signatures can be verified by executing the server-aided verification protocol with the server, where the verification requires less computation than the original verification algorithm of the digital signature. Different to delegated computation, the existing server-aided verification signature schemes can achieve the soundness of the server-aided verification protocol under their security definitions, namely, a trusted server cannot convince the verifier that an invalid signature is valid, and the verifier cannot directly verify the results computed by the server. The notion of server-aided verification signature was first introduced by Quisquater and de Soete [10] for speeding up RSA verification with a small exponent. Then, Lim and Lee [11] extended this idea into discrete-logarithm based schemes, by proposing efficient protocols for speeding up the verification of discrete-logarithm based identity proofs and signatures. Girault and Quisquater [13] introduced a different approach for server-aided verification signature which does not require precomputation or randomization. Its security remains computational, based on the hardness of a subproblem (viz. factorization) of the initial underlying problem (viz. composite discrete logarithm). Hohenberger and Lysyanskaya [17] addressed the situation in which the server is made of two untrusted softwares, which are assumed not to communicate with each other. Girault and Lefranc [14] presented a generic server-aided verification protocol for digital signatures from bilinear maps which has been used to construct many digital signature schemes such as [18–23].

As to the security of server-aided verification signature, many efforts have been devoted to defining strong security models for it. The schemes [10, 11, 13, 14] considered the security property based on the assumption that the malicious server does not have any valid signatures on the message when it tries to prove an invalid signature of that message to be valid. Among them, the scheme [13] is computationally secure based on the hardness of a subproblem of the underlying complexity problem in the original signature scheme. To give stronger definition of this property, Wu et al. [15] formally defined this security assuming that the malicious server may collude with the signer and obtain the secret key of the signer. They first introduced and defined the existential unforgeability of server-aided verification signatures and considered collusion between a signer and a server, who collaboratively prove an invalid signature to be valid. In addition, under their security models, they introduced the server-aided verification for the Waters signature [21] and the BLS signature [18], respectively.

Though the existing server-aided verification signature schemes above have been devoted many efforts to their security models, they only considered the soundness to protect the malicious server who may try to prove an invalid signature of a message to be valid. However, in some applications where the message-signature to be verified contains some sensitive information, for example, the message contains important business secrets or is related to medical information, the verifier does not want the server learn anything about the message and/or the signature to protect its privacy. So, the message privacy of the server-aided verification protocol is also desired besides the soundness. Though in Wu et al. [15], based on Waters Signature [21] and BLS signature [18], two SA verification signature schemes (see Section 4 in [15]) were presented in which the message to be verified is not revealed to the server, the schemes cannot achieve the soundness under collusion and adaptively chosen message attacks.

In this paper, we will present two privacy definitions for server-aided verification signature under collusion by the server and the signer and adaptive chosen message attacks. A server-aided verification signature scheme with privacy also consists of a digital signature scheme and a server-aided verification protocol.

(1) The first privacy definition for the server-aided verification signature is about message privacy; namely, the server cannot learn anything about the message to be verified during the server-aided verification protocol even if it possesses the secret key of the signer. Generally, when the verifier wants the server to verify a message-signature pair, it will "blind" this message at the beginning of the server-aided verification protocol so that the server cannot obtain any information about this message, while it can verify the validity of the message-signature pair by using the server's responses.

(2) The second privacy definition for the server-aided verification signature is about message-signature privacy which is stronger than the first one, and in this definition, the server can learn nothing about the message-signature pair to be verified even if it colludes with the signer. To achieve this privacy, similarly, the verifier will "blind" the message-signature pair at the beginning of the server-aided verification protocol so that the server cannot obtain any information about the message or the signature; however it can verify the validity of the message-signature pair after the server responds.

For the two privacy notions, we present detailed and strict security models. Then, under the security models, we present two concrete constructions for server-aided verification signature based on Waters signature [21] and BLS signature [18] which, respectively, achieve message privacy and message-signature privacy. The soundness of the two constructions is proved under the strong definition of [15] assuming that the malicious server may collude with the signer and obtain the secret key of the signer. In addition, the efficiency analysis of the server-aided verification protocols shows that our two concrete server-aided verification signature schemes are both computation saving. Computation saving is probably the most obvious property that can distinguish a server-aided verification signature scheme SAV-Σ from an ordinary signature scheme Σ. This property enables the verifier in SAV-Σ

to check the validity of signatures in a more computationally efficient way than that in Σ.

Organization. This paper is organized as follows. In Section 2, we will review some fundamental backgrounds, the definition of server-aided verification signatures and the security notions defined in [15] including existential unforgeability and soundness against collusion and adaptive chosen message attacks. In Section 3, we will present the message privacy of server-aided verification signatures, give a concrete construction based on Waters signature scheme, and prove its security under our security model for message privacy. In Section 4, a stronger privacy of server-aided verification signatures named message-signature privacy will be defined and a provably secure concrete construction will be presented based on BLS signature scheme. Finally we conclude in Section 5.

2. Preliminaries

2.1. Syntax. Throughout the paper, if A is a randomized algorithm, then $y \leftarrow A(x)$ denotes the assignment to y of the output of A on input x. Unless noted, all algorithms are probabilistic polynomial-time (PPT) and we implicitly assume that they take an extra parameter κ in their input, where κ is a security parameter.

2.2. Bilinear Maps. Let $\mathbb{G}_1$, $\mathbb{G}_T$ be two (multiplicative) cyclic groups such that $|\mathbb{G}_1| = |\mathbb{G}_T| = p$, where p is a large prime. Let g be a generator of $\mathbb{G}_1$, and e be an admissible bilinear map: $\mathbb{G}_1 \times \mathbb{G}_1 \rightarrow \mathbb{G}_T$, satisfying (1) for all $a, b \in \mathbb{Z}_p$; it holds that $e(g^a, g^b) = e(g, g)^{ab}$; (2) $e(g, g) \neq 1$; and (3) it is efficiently computable.

We say that $(\mathbb{G}_1, \mathbb{G}_T)$ are bilinear groups if there exists the bilinear map $e : \mathbb{G}_1 \times \mathbb{G}_1 \rightarrow \mathbb{G}_T$ as above, and the group action in $\mathbb{G}_1$ and $\mathbb{G}_T$ can be computed efficiently. Such groups can be built from Weil pairing or Tate pairing on elliptic curves.

2.3. Server-Aided Verification Signature. A server-aided verification signature scheme SAV-Σ consists of six algorithms: ParamGen, KeyGen, Sign, Verify, SA-Verifier-Setup, and SA-Verify. The first four algorithms are the same as those in an ordinary signature scheme Σ. SAV-Σ contains three parties, respectively, a signer, a verifier, and a server.

(i) $param \leftarrow$ ParamGen. This algorithm takes a security parameter κ and returns a string $param$ as input, which denotes the common scheme parameters, including the description of the message space $\mathbb{M}$ and the signature space Ω.

(ii) $(pk, sk) \leftarrow$ KeyGen($param$). This algorithm takes $param$ as input and outputs a key pair (pk, sk), where sk is the signing key and pk is the verification key.

(iii) $\sigma \leftarrow$ Sign($param, m, sk, pk$). The signer takes a message $m \in \mathbb{M}$, the system parameter $param$ and the key pair (sk, pk) as inputs, outputs a signature σ.

(iv) {Valid, Invalid} $\leftarrow$ Verify($param, m, \sigma, pk$). The verifier takes the parameter $param$, a message-signature pair (m, σ) and the public key pk, outputs Valid/Invalid to indicate that σ is a valid/invalid signature on m under pk.

(v) VString $\leftarrow$ SA-Verifier-Setup($param$). The verifier takes as input the system parameter $param$ and outputs a string VString which contains the information which can be precomputed by it.

(vi) {Valid, Invalid} $\leftarrow$ SA-Verify(Server$^{(param)}$, Verifier$^{(m, \sigma, pk, \text{VString})}$). This is an interactive protocol between the server and the verifier where the server takes $param$ as input and the verifier takes $(m, \sigma, pk, \text{VString})$ as inputs. Finally, the verifier outputs Valid if the server can convince it that σ is a valid signature on m. Otherwise, the verifier outputs Invalid.

In a SA verification signature scheme, we assume that the verifier has a limited computational ability and is not able to perform all computations in Verify alone. So, a SA verification signature scheme must satisfy an important property called computation saving property, which requires that the computations performed by the verifier in SA-Verify must be less than those performed in Verify.

2.4. Security Model for Server-Aided Verification Signature. In the following, we will first present the security model for SAV-Σ with message privacy. As for the existential unforgeability of SAV-Σ, we will adopt existential unforgeability of SAV-Σ defined in [15], including the existential unforgeability against adaptive chosen message attacks of Σ defined in [24] and the soundness against collusion and adaptive chosen message attacks of SA-Verify. In the following, we will present the existential unforgeability of SAV-Σ as [15]. It requires that the adversary should not be (computationally) capable of producing a signature of a new message which can be proved as valid by SA-Verify, even if the adversary acts as a server.

Definition 1 (existential unforgeability against adaptive chosen message attacks of Σ). The adversary $\mathcal{A}$ and the challenger $\mathcal{C}$ play the following game.

(i) *Setup.* The challenger $\mathcal{C}$ runs the algorithms ParamGen and KeyGen to obtain system parameter $param$ and one key pair (sk, pk). The adversary $\mathcal{A}$ is given $param$ and pk.

(ii) *Queries.* The adversary $\mathcal{A}$ is allowed to make at most q_s sign queries. For each sign query $m_i \in \{m_1, \ldots, m_{q_s}\}$, the challenger $\mathcal{C}$ returns $\sigma_i = $ Sign($param, m_i, sk, pk$) as the response.

(iii) *Output.* Eventually, the adversary $\mathcal{A}$ outputs a pair (m^*, σ^*) and wins the game if:

    (1) $m^* \notin \{m_1, \ldots, m_{q_s}\}$;
    (2) Verify($param, m^*, \sigma^*, pk$) = Valid.

An adversary $\mathcal{A}$ is said to (t, q_s, ε)-break a signature scheme Σ if $\mathcal{A}$ runs in time at most t and makes at most q_s signature queries and the success probability $\Sigma - \mathrm{Adv}_{\mathcal{A}}$ to win the game above is at most ε.

We say that Σ is existentially unforgeable against adaptive chosen message attacks if there exists an adversary that (t, q_s, ε)-breaks it.

In the following, we will present the soundness against collusion and adaptive chosen message attacks of SA-Verify which means that the server cannot prove an invalid signature to be valid even if it colludes with the signer.

Definition 2 (soundness against collusion and adaptive chosen message attacks of SA-Verify). The adversary $\mathcal{A}$ and the challenger $\mathcal{C}$ play the following game.

(i) *Setup.* The challenger $\mathcal{C}$ runs the algorithms ParamGen, KeyGen and SA-Verifier-Setup to obtain the system parameter *param*, one key pair (sk, pk) and VString. The adversary $\mathcal{A}$ is given *param* and (sk, pk).

(ii) *Queries.* Proceeding adaptively, the adversary $\mathcal{A}$ is allowed to make at most q_v server-aided verification queries. The challenger $\mathcal{C}$ responds by executing SA-Verify with the adversary $\mathcal{A}$, where the adversary $\mathcal{A}$ acts as the server and the challenger $\mathcal{C}$ acts as the verifier. At the end of each execution, the challenger returns the output of SA-Verify to the adversary $\mathcal{A}$.

(iii) *Output.* Eventually, the adversary $\mathcal{A}$ outputs a message m^*. The challenger $\mathcal{C}$ chooses a random invalid signature σ^* on the message m^*. Namely, it chooses a random element σ^* in $\Omega \setminus \Omega_{m^*}$, where Ω and Ω_{m^*} are, respectively, the signature space and the set of valid signatures of m^*. We say that $\mathcal{A}$ wins the game if

$$\text{SA-Verify}\left(\mathcal{A}, \mathcal{C}^{(m^*, \sigma^*, pk, \text{VString})}\right) = \text{Valid.} \qquad (1)$$

An adversary $\mathcal{A}$ is said to (t, q_v, ε)-break SA-Verify's soundness against collusion and chosen message attacks if $\mathcal{A}$ runs in time at most t, makes at most q_v server-aided verification queries and the success probability $\mathrm{Adv}_{\mathcal{A}}^{EU}$ to win the game above is at least ε.

We say that SA-Verify is (t, q_v, ε)-sound against collusion and chosen message attacks if there exists no adversary that (t, q_v, ε)-breaks it.

3. Server-Aided Verification Signature with Message Privacy

In this section, we will present the definition of message privacy for SA-Verify, and then, based on Waters signature scheme [21], present a concrete server-aided verification scheme with this privacy property. This privacy property is called *message privacy against collusion and adaptive chosen message attacks*. In this definition, the server is allowed to collude with the signer. Concretely, the server can obtain

the key pair (pk, sk) of the signer and therefore can create the signature on any message. In addition, we will assume that the server cannot obtain the message-signature pairs that have been created by the signer before, alternatively, the signer will not store any message-signature pair that it has created. (Actually, this can be achieved by performing blind signature scheme presented in [25] between the signer and the verifier instead of performing the ordinary signature scheme. After the blind signature scheme, the verifier can obtain the ordinary message-signature pair without the signer learning anything about this pair. Then the verifier lets the server to verify the message-signature pair by performing SA-Verify. In this sense, even if the server colludes with the signer, it cannot obtain more information about the signed messages from the signer than it can obtain on its own. To clarify our privacy definition below more clearly, we simply assume that the server cannot obtain any message-signature pair which the signer has created for the verifier before.)

3.1. Definition of Message Privacy. A server-aided verification signature scheme with message privacy SAV-Σ also consists of six algorithms: ParamGen, KeyGen, Sign, Verify, SA-Verifier-Setup, and SA-Verify. The following is the definition of message privacy for the server-aided verification protocol under the collusion and adaptive chosen message attacks. In this definition, the server cannot obtain any information about the message to be verified under the collusion and adaptive chosen message attacks.

Definition 3 (message privacy of SA-Verify). We say that SA-Verify satisfies (t, q_v, ε)-message privacy against collusion and adaptive chosen message attacks if there exists no adversary $\mathcal{A}$ who runs in time at most t, makes at most q_v server-aided verification queries, and succeeds with probability at least ε in the following game with the challenger $\mathcal{C}$. The game is defined as follows.

(i) *Setup.* The challenger $\mathcal{C}$ runs the algorithms ParamGen, KeyGen and SA-Verifier-Setup to obtain system parameter *param*, one key pair (sk, pk), and VString. The adversary $\mathcal{A}$ is given *param* and (sk, pk). Note that $\mathcal{A}$ can generate any message-signature pair with the secret-public key pair (sk, pk); however as we assumed, it cannot obtain any message-signature pair that has been created by the signer before.

(ii) *Queries.* Proceeding adaptively, the adversary $\mathcal{A}$ is allowed to make at most q_v server-aided verification queries. The challenger $\mathcal{C}$ responds by executing SA-Verify with the adversary $\mathcal{A}$, where the adversary $\mathcal{A}$ acts as the server and the challenger $\mathcal{C}$ acts as the verifier. At the end of each execution, the challenger returns the output of SA-Verify to the adversary $\mathcal{A}$.

(iii) *Challenge.* $\mathcal{A}$ outputs two messages m_0, m_1, and sends them to the challenger $\mathcal{C}$. $\mathcal{C}$ chooses a bit $b \in \{0, 1\}$ at random and also chooses an element σ either randomly from Ω_{m_0} or randomly from Ω_{m_1}, where Ω_{m_0} and Ω_{m_1} are, respectively, the signature space of m_0 and m_1. Then $\mathscr{C}$ and $\mathscr{A}$ interact with each other by running SA-Verify$(\mathscr{A}, \mathscr{C}^{(m_b, \sigma, pk, \text{VString})})$, where $\mathscr{A}$ plays as a server and $\mathscr{C}$ plays as a verifier. After the interaction, $\mathscr{C}$ sends the output of SA-Verify$(\mathscr{A}, \mathscr{C}^{(m_b, \sigma, pk, \text{VString})})$ to $\mathscr{A}$.

(iv) *Output.* Finally, $\mathscr{A}$ outputs a bit $b' \in \{0, 1\}$. We say that $\mathscr{A}$ wins the game with probability ε if

$$\Pr\left[b = b'\right] \geq \varepsilon. \tag{2}$$

Similar to Wu et al. [15], in the protocol Setup of the game above, VString is not provided to the adversary who now is acting as a server since VString might contain some private information of the verifier, which must be kept secret in server-aided verification signatures. In the definition, adversary $\mathscr{A}$ acts as the server and the challenger $\mathscr{C}$ acts as the verifier which will help $\mathscr{A}$ to extract some information from VString.

3.2. Concrete SA Verification Signature with Message Privacy. In the following, we will first present a concrete SA verification signature scheme with message privacy based on Waters signature [21]. The SA verification signature scheme with message privacy SAV-Σ consists of six algorithms: ParamGen, KeyGen, Sign, Verify, SA-Verifier-Setup, and SA-Verify. The first four algorithms are the same as those in Waters signature scheme [21]. As we know that, due to the elegant properties of pairing computation on elliptic curves, pairing has been widely employed as a building block for lots of cryptographic schemes, in particular in the construction of digital signatures. However, performing a pairing on an elliptic curve requires much more computational cost than executing both an exponentiation and a multiplication [16, 26–30], and for a power-constrained verifier who must execute multiple pairing computations during the verification of a message-signature pair, reducing the computational load of it is a meaningful task. In Waters signature [21], the verifier has to compute two pairings; however in SAV-Σ, its computational load is reduced and it will not compute any pairing. The concrete SA verification signature with message privacy based on Waters signature is described in detail as follows.

(i) $param \leftarrow$ ParamGen. Let κ be a security parameter, $(\mathbb{G}_1, \mathbb{G}_T)$ be bilinear groups where $|\mathbb{G}_1| = |\mathbb{G}_T| = p$ for some prime number $p \geq 2^\kappa$ and g be a generator of $\mathbb{G}_1$. $e : \mathbb{G}_1 \times \mathbb{G}_1 \rightarrow \mathbb{G}_T$ is a bilinear mapping. The system parameters are $param = (\kappa, \mathbb{G}_1, \mathbb{G}_T, p, g, e)$ and the message space is $\mathbb{M} = \{0, 1\}^n$.

(ii) $(pk, sk) \leftarrow$ KeyGen$(param)$. Given the system parameters $(\kappa, \mathbb{G}_1, \mathbb{G}_T, p, g, e)$, the signer chooses a random element $x \in \mathbb{Z}_p$, generates the public key pk as $(\vec{V}, PK)$ and $sk = x$ where $\vec{V}$ is a vector consisting of $n + 1$ elements $V_0, V_1, \ldots, V_n$ randomly selected in $\mathbb{G}_1$ and $PK = e(g, g)^x$.

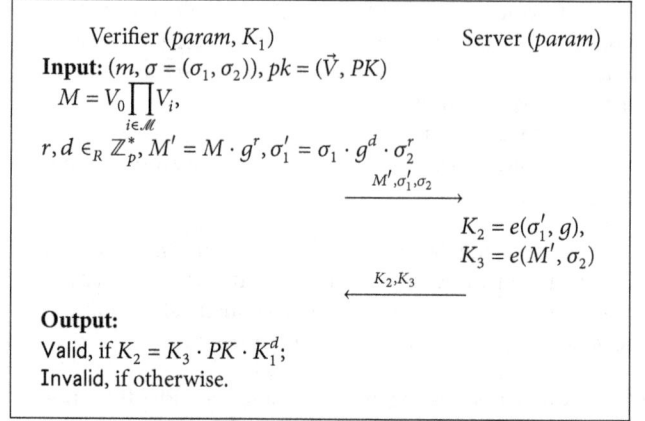

Verifier $(param, K_1)$ Server $(param)$

Input: $(m, \sigma = (\sigma_1, \sigma_2)), pk = (\vec{V}, PK)$

$M = V_0 \prod_{i \in \mathcal{M}} V_i,$

$r, d \in_R \mathbb{Z}_p^*, M' = M \cdot g^r, \sigma_1' = \sigma_1 \cdot g^d \cdot \sigma_2^r$

$\xrightarrow{M', \sigma_1', \sigma_2}$

$K_2 = e(\sigma_1', g),$
$K_3 = e(M', \sigma_2)$

$\xleftarrow{K_2, K_3}$

Output:
Valid, if $K_2 = K_3 \cdot PK \cdot K_1^d$;
Invalid, if otherwise.

ALGORITHM 1: SA-Verify with message privacy based on Waters signature.

(iii) $\sigma \leftarrow$ Sign$(param, m, sk, pk)$. For an n-bit message $m \in \{0, 1\}^n$, let $\mathcal{M} \subseteq \{1, 2, \ldots, n\}$ be the set of all i for which the ith bit of m is 1. The signer selects a random element $t \in \mathbb{Z}_p$ and generates the signature σ as $(\sigma_1, \sigma_2) = (g^x(V_0 \prod_{i \in \mathcal{M}} V_i)^t, g^t)$.

(iv) $\{\text{Valid}, \text{Invalid}\} \leftarrow$ Verify$(param, m, \sigma, pk)$. The verifier takes as input a claimed message-signature pair (m, σ), and outputs Valid if and only if $e(\sigma_1, g) = PK \cdot e(V_0 \prod_{i \in \mathcal{M}} V_i, \sigma_2)$. Otherwise it outputs Invalid.

(v) VString $\leftarrow$ SA-Verify-Setup$(param)$. The verifier takes as inputs the system parameters $(\kappa, \mathbb{G}_1, \mathbb{G}_T, p, g, e)$ and computes $K_1 = e(g, g)$ as VString.

(vi) $\{\text{Valid}, \text{Invalid}\} \leftarrow$ SA-Verify$(\text{Server}^{(param)},$ Verifier$^{(m, \sigma, pk, \text{VString})})$. This is an interactive protocol between the server and the verifier which is shown in Algorithm 1.

   (1) Verifier, for a message-signature pair $(m, \sigma = (\sigma_1, \sigma_2))$ to be verified, first computes $M = V_0 \prod_{i \in \mathcal{M}} V_i$; then selects randomly $r, d \in \mathbb{Z}_p^*$, and blinds the message by computing $M' = M \cdot g^r$, $\sigma_1' = \sigma_1 \cdot g^d \cdot \sigma_2^r$; finally sends $(M', \sigma_1', \sigma_2)$ to the verifier.

   (2) Server computes $K_2 = e(\sigma_1', g)$, $K_3 = e(M', \sigma_2)$ and returns K_2, K_3 to the verifier.

   (3) Verifier checks the equation $K_2 = K_3 \cdot PK \cdot K_1^d$, and outputs Valid if it holds, and otherwise outputs Invalid.

Correctness of SA-Verify. For a claimed message-signature pair (m, σ), when the verifier and the server are both honest, namely, the verifier correctly computes M' and σ_1' and the server correctly computes K_2 and K_3, then the verification equation $K_2 = PK \cdot K_3 \cdot K_1^d$ holds if the message-signature pair

(m, σ) is valid and otherwise does not hold. In the following, we denote $(\sigma_1, \sigma_2) = (g^x(V_0 \prod_{i \in \mathcal{M}} V_i)^t, g^t)$:

$$K_2 = e\left(\sigma_1', g\right)$$
$$= e\left(\sigma_1, g\right) \cdot e\left(g, g\right)^d \cdot e\left(\sigma_2^r, g\right)$$
$$= e\left(M, g\right)^t \cdot PK \cdot K_1^d \cdot e\left(g^{tr}, g\right)$$
$$= e\left(M \cdot g^r, g\right)^t \cdot PK \cdot K_1^d \tag{3}$$
$$= e\left(M', \sigma_2\right) \cdot PK \cdot K_1^d$$
$$= K_3 \cdot PK \cdot K_1^d.$$

In the following, by Theorems 4 and 5, we will show that our SA verification signature scheme above is secure under our security model; namely, the SA verification protocol described in Algorithm 1 is sound against collusion and adaptive chosen message attacks and also satisfies message privacy against collusion and adaptive chosen message attacks.

Since Waters signature scheme has been proved existentially unforgeable against adaptive chosen message attacks, in order to prove that our SA verification signature scheme above is secure, we need only to prove that the SA verification protocol described in Algorithm 1 satisfies soundness against collusion and adaptive chosen message attacks defined in Definition 2 and message privacy against collusion and adaptive chosen message attacks.

Theorem 4. *The SA verification protocol described in Algorithm 1 satisfies $(t, q_v, 1/(p-1))$-soundness against collusion and adaptive chosen message attacks.*

Proof. In order to prove that the SA verification protocol in Algorithm 1 is $(t, q_v, 1/(p-1))$-sound against collusion and adaptive chosen message attacks, we will show that the adversary can only prove an invalid signature as valid with at most probability $1/(p-1)$. The challenger and the adversary play the following game.

(i) *Setup.* The challenger generates the system parameters $(\kappa, \mathbb{G}_1, \mathbb{G}_T, p, g, e)$, chooses a random element $x \in \mathbb{Z}_p$, and sets $sk = x$ and $pk = (\vec{V}, PK)$ where $\vec{V}$ is a vector consisting of $n+1$ elements $V_0, V_1, \ldots, V_n$ randomly selected in $\mathbb{G}_1$ and $PK = e(g, g)^x$. Then it also computes $K_1 = e(g, g)$. Finally the challenger sends $(\kappa, \mathbb{G}_1, \mathbb{G}_T, p, g, e, sk, pk)$ to the adversary.

(ii) *Queries.* The adversary $\mathscr{A}$ is allowed to make at most q_v server-aided verification queries. The challenger $\mathscr{C}$ responds by executing SA-Verify with the adversary $\mathscr{A}$, where the adversary $\mathscr{A}$ acts as the server and the challenger $\mathscr{C}$ acts as the verifier. At the end of each execution, the challenger returns the output of SA-Verify to the adversary $\mathscr{A}$.

(iii) *Output.* Eventually, the adversary $\mathscr{A}$ outputs a message m^*. The challenger $\mathscr{C}$ chooses a random element $\sigma^* = (\sigma_1, \sigma_2)$ in $\mathbb{G}_1^2 \setminus \Omega_{m^*}$, where Ω and Ω_{m^*} are, respectively, the signature space and the set of valid

signatures of m^*. Then they interact with each other as described in the SA-Verify protocol. Concretely, the challenger chooses two random elements $r^*, d^* \in \mathbb{Z}_p^*$, computes $M = V_0 \prod_{i \in \mathcal{M}} V_i$, $M' = M \cdot g^{r^*}$ and $\sigma_1' = \sigma_1 \cdot g^{d^*} \cdot \sigma_2^{r^*}$, and sends $(M', \sigma_1', \sigma_2)$ to the adversary. Then the adversary returns K_2^* and K_3^* to the challenger.

In the following, we will show that $K_2^* = K_3^* \cdot PK \cdot K_1^{d^*}$ happens with probability $1/(p-1)$.

The challenger sends $(M', \sigma_1', \sigma_2)$ to $\mathscr{A}$ such that

$$M' = M \cdot g^{r^*},$$
$$\sigma_1' = \sigma_1 \cdot g^{d^*} \cdot \sigma_2^{r^*} \tag{4}$$
$$\Downarrow$$
$$DL_g M' = DL_g M + r^*,$$
$$DL_g \sigma_1' = DL_g \sigma_1 + d^* + DL_g \sigma_2 \cdot r^*. \tag{5}$$

Since the adversary can only obtain $(M', \sigma_1', \sigma_2)$ from the challenger, from the equation set (5), we can see that the adversary can only obtain d^* with probability $1/(p-1)$. Furthermore, from (6) below, we can directly deduce (7) as follows:

$$K_2^* = K_3^* \cdot PK \cdot K_1^{d^*} \tag{6}$$
$$\Downarrow$$
$$DL_{e(g,g)}\left(\frac{K_2^*}{K_3^*}\right) = DL_{e(g,g)} PK + d^*. \tag{7}$$

Since the adversary can only guess d^* with probability $1/(p-1)$, and (K_2^*/K_3^*) is uniquely determined by d^*, the adversary can only give out a pair (K_2^*, K_3^*) satisfying (6) with probability $1/(p-1)$. This completes the proof of Theorem 4. □

Theorem 5. *The SA verification protocol in Algorithm 1 satisfies $(t, q_v, 1/2)$-message privacy against collusion and adaptive chosen message attacks.*

Proof. In order to prove that the SA verification protocol in Algorithm 1 satisfies $(t, q_v, 1/2)$-message privacy against collusion and adaptive chosen message attacks, we will show that the adversary can only succeed with at most the probability $1/2$ in the game with the challenger described as follows.

(i) *Setup.* The challenger generates the system parameter $(\kappa, \mathbb{G}_1, \mathbb{G}_T, p, g, e)$, and the secret key $sk = x$ and public key $pk = (\vec{V}, PK)$, where $\vec{V}$ is a vector consisting of $n+1$ elements $V_0, V_1, \ldots, V_n$ randomly selected in $\mathbb{G}_1$ and $PK = e(g, g)^x$. Then it also computes $K_1 = e(g, g)$. Finally the challenger sends $(\kappa, \mathbb{G}_1, \mathbb{G}_T, p, g, e, sk, pk)$ to the adversary.

(ii) *Queries*. The adversary $\mathscr{A}$ is allowed to make at most q_v server-aided verification queries. The challenger $\mathscr{C}$ responds by executing SA-Verify with the adversary $\mathscr{A}$, and, at the end of each execution, returns the output of SA-Verify to the adversary $\mathscr{A}$.

(iii) *Challenge*. $\mathscr{A}$ outputs two messages m_0 and m_1 and sends them to the challenger $\mathscr{C}$. $\mathscr{C}$ chooses a bit $b \in \{0, 1\}$ at random and also chooses $\sigma = (\sigma_1, \sigma_2)$ either randomly from Ω_{m_0} or randomly from Ω_{m_1}, where Ω_{m_0} and Ω_{m_1} are, respectively, the signature spaces of m_0 and m_1. Then $\mathscr{C}$ and $\mathscr{A}$ interact with each other by running SA-Verify$(\mathscr{A}, \mathscr{C}^{(m_b, \sigma, pk, \text{VString})})$, where $\mathscr{A}$ plays as a server and $\mathscr{C}$ plays as a verifier. Concretely, the challenger chooses two random elements $r^*, d^* \in \mathbb{Z}_p^*$, computes $M = V_0 \prod_{i \in \mathscr{M}} V_i$, $M' = M \cdot g^{r^*}$, and $\sigma_1' = \sigma_1 \cdot g^{d^*} \cdot \sigma_2^{r^*}$, and sends $(M', \sigma_1', \sigma_2)$ to the adversary. Then the adversary returns K_2^* and K_3^* to the challenger. After the interaction, $\mathscr{C}$ sends the output of SA-Verify$(\mathscr{A}, \mathscr{C}^{(m_b, \sigma, pk, \text{VString})})$ to $\mathscr{A}$.

(iv) *Output*. Finally, $\mathscr{A}$ outputs a bit $b' \in \{0, 1\}$. We will show that $\mathscr{A}$ can only succeed with probability $1/2$.

From the equation set (5) in the proof of Theorem 4, we can see that there exist two pairs (M_0, r_0^*) and (M_1, r_1^*) satisfying $M' = M_b \cdot g^{r_b^*}$ for $b = 0, 1$, and for each r_b^*, there exist $p - 1$ pairs (σ_1, d^*) satisfying $\sigma_1' = \sigma_1 \cdot g^{d^*} \cdot \sigma_2^{r_b^*}$ for $b = 0, 1$. So the adversary cannot obtain anything about (M, σ_1). Though the adversary learns σ_2 which may correspond to m_0 or m_1, it only guess b correctly with probability $1/2$. This completes the proof of Theorem 4. □

Efficiency Analysis. The SA verification signature with privacy based on Waters signature above is computation saving and efficient. In the following, we will analyze the efficiency of SA-Verify algorithm by comparing that of Waters signature scheme. In Waters signature scheme [21], to verify a message-signature pair $(m, \sigma = (\sigma_1, \sigma_2))$, the verifier needs to compute $M = V_0 \prod_{i \in \mathscr{M}} V_i$ which takes n multiplications in $\mathbb{G}_1$, $PK \cdot e(m, \sigma_2)$ which takes 1 multiplication in $\mathbb{G}_T$, and two pairings $e(m, \sigma_2)$ and $e(\sigma_1, g)$. However, in our server-aided verification signature scheme, the verifier can first precompute a pairing $e(g, g)$ which can be used by multiple SA-Verify protocols. Then in a SA-Verify protocol, we can see that the verifier needs to compute totally 3 exponentiations in $\mathbb{G}_1$ and 1 exponentiation in $\mathbb{G}_T$ as well as $3 + n$ multiplications in $\mathbb{G}_1$ and 3 multiplications in $\mathbb{G}_T$. As we know that, performing a pairing on an elliptic curve requires much more computational cost than executing both an exponentiation and a multiplication. So our SA verification signature scheme based on Waters signature is computation saving and efficient.

The concrete computation cost comparison of the verifier in the verification of Waters Signature [21] and our SA-Verify in Algorithm 1 is shown in Table 1.

TABLE 1: Efficiency of the SA verification signature scheme.

Verification	Pairing	Exponentiation	Multiplication
Waters [21]	2	0	$n(\mathbb{G}_1)$
Our Scheme	0	$3(\mathbb{G}_1) + 1(\mathbb{G}_T)$	$(3 + n)(\mathbb{G}_1) + 3(\mathbb{G}_T)$

4. Server-Aided Verification Signature with Message-Signature Privacy

In this section, we will present a stronger definition of privacy, namely, message-signature privacy against collusion and adaptive chosen message attacks. Then based on BLS signature scheme [18], a concrete server-aided verification scheme with this privacy property will be presented. We assume that the server can obtain the key pair (pk, sk) of the signer and cannot obtain the message-signature pairs that have been created by the signer. Under this assumption, the server cannot obtain anything about the message-signature pair.

4.1. Definition of Message-Signature Privacy. A server-aided verification signature scheme with message-signature privacy SAV-Σ also consists of six algorithms: ParamGen, KeyGen, Sign, Verify, SA-Verifier-Setup, and SA-Verify.

Definition 6 (message-signature privacy of SA-Verify). We say that SA-Verify satisfies (t, q_v, ε)-message-signature privacy against collusion and adaptive chosen message attacks if there exists no adversary $\mathscr{A}$ who runs in time at most t, makes at most q_v server-aided verification queries, and succeeds with probability at least ε in the following game with the challenger $\mathscr{C}$. The game is defined as follows.

(i) *Setup*. The challenger $\mathscr{C}$ runs the algorithms ParamGen, KeyGen and SA-Verifier-Setup to obtain the system parameter *param*, one key pair (sk, pk), and VString. The adversary $\mathscr{A}$ is given *param* and (sk, pk). Similar to the definition of message privacy for SA verification signature, $\mathscr{A}$ can generate any message-signature pair with the key pair (sk, pk); however as we assumed, it cannot obtain any message-signature pair that has been created by the signer before.

(ii) *Queries*. Proceeding adaptively, the adversary $\mathscr{A}$ is allowed to make at most q_v server-aided verification queries. The challenger $\mathscr{C}$ responds by executing SA-Verify with the adversary $\mathscr{A}$, where the adversary $\mathscr{A}$ acts as the server and the challenger $\mathscr{C}$ acts as the verifier. At the end of each execution, the challenger returns the output of SA-Verify to the adversary $\mathscr{A}$.

(iii) *Challenge*. $\mathscr{A}$ outputs two pairs (M_0, σ_0) and (M_1, σ_1), where σ_i is a valid signature on M_i for $i = 1, 2$. Then $\mathscr{A}$ sends them to the challenger $\mathscr{C}$. $\mathscr{C}$ chooses a bit $b \in \{0, 1\}$ at random and interacts with $\mathscr{A}$ by running SA-Verify$(\mathscr{A}, \mathscr{C}^{(M_b, \sigma_b, pk, \text{VString})})$ where $\mathscr{A}$ plays as a server and $\mathscr{C}$ plays as a verifier. After the interaction, $\mathscr{C}$ sends the output of SA-Verify$(\mathscr{A}, \mathscr{C}^{(M_b, \sigma_b, pk, \text{VString})})$ to $\mathscr{A}$.

Verifier $(param, K_1)$ Server $(param)$

Input: $(m, \sigma), pk$

$r, t \in_R \mathbb{Z}_p^*, m' = H(m) \cdot g^t,$

$\sigma' = \sigma \cdot g^r; K_2 = e(g, pk)$

$$\xrightarrow{\quad m', \sigma' \quad}$$

$K_3 = e(\sigma', g);$

$K_4 = e(m', pk)$

$$\xleftarrow{\quad K_3, K_4 \quad}$$

Output:

Valid, if $K_3 = K_1^r \cdot K_2^{-t} \cdot K_4$;

Invalid, if otherwise.

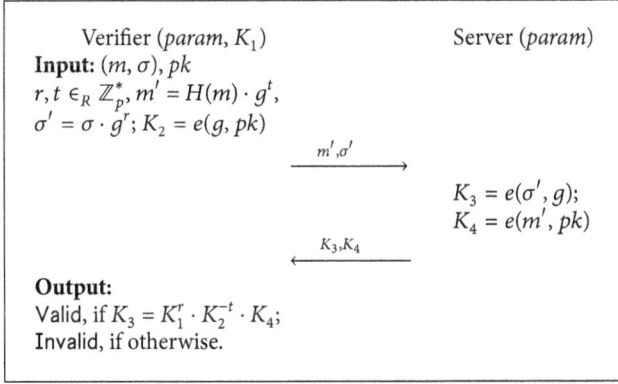

ALGORITHM 2: SA verification Signature with message-signature privacy based on BLS signature.

(iv) *Output.* Finally, $\mathscr{A}$ outputs a bit $b' \in \{0, 1\}$. We say that $\mathscr{A}$ wins the game with probability ε if

$$\Pr\left[b = b'\right] \geq \varepsilon. \tag{8}$$

4.2. Concrete SA Verification Signature with Message-Signature Privacy. In this section, we will present a SA verification signature scheme which satisfies message-signature privacy against collusion and adaptive chosen message attacks. This scheme is constructed based on BLS signature [18], which also consists of six algorithms: ParamGen, KeyGen, Sign, Verify, SA-Verifier-Setup, and SA-Verify. The first four algorithms are the same as those in BLS signature scheme [18]. By executing the SA-Verifier-Setup and SA-Verify algorithms, the computational load of the verifier can be reduced. In BLS signature [18], the verifier has to compute two pairings; however in the following SAV-Σ, it needs only to compute a pairing.

(i) $param \leftarrow$ ParamGen. Let κ be a security parameter, $(\mathbb{G}_1, \mathbb{G}_T)$ be bilinear groups, where $|\mathbb{G}_1| = |\mathbb{G}_T| = p$ for some prime number $p \geq 2^\kappa$, and g be a generator of $\mathbb{G}_1$. $e : \mathbb{G}_1 \times \mathbb{G}_1 \rightarrow \mathbb{G}_T$ is a bilinear mapping. $H : \{0, 1\}^* \rightarrow \mathbb{G}_1$ is a hash function. The system parameter $param = (\kappa, \mathbb{G}_1, \mathbb{G}_T, p, g, e, H)$.

(ii) $(pk, sk) \leftarrow$ KeyGen$(param)$. Given the system parameters $(\kappa, \mathbb{G}_1, \mathbb{G}_T, p, g, e)$, the signer chooses a random element $x \in \mathbb{Z}_p$, and sets the public key $pk = g^x$ and the secret key $sk = x$.

(iii) $\sigma \leftarrow$ Sign$(param, m, sk, pk)$. For a message, the signer generates the signature $\sigma = H(m)^x$.

(iv) $\{$Valid, Invalid$\} \leftarrow$ Verify$(param, m, \sigma, pk)$. The verifier takes as inputs a claimed message-signature pair (m, σ), and outputs Valid if and only if $e(\sigma, g) = e(H(m), pk)$, and otherwise outputs Invalid.

(v) VString $\leftarrow$ SA-Verify-Setup$(param)$. The verifier takes as inputs the system parameter $(\kappa, \mathbb{G}_1, \mathbb{G}_T, p, g, e)$ and computes $K_1 = e(g, g)$ as VString.

(vi) $\{$Valid, Invalid$\} \leftarrow$ SA-Verify(Server$^{(param)}$, and Verifier$^{(m, \sigma, pk, \text{VString})}$). This is an interactive protocol between the server and the verifier which is shown in Algorithm 2.

(1) Verifier, for a message-signature pair (m, σ), blinds the pair by selecting randomly $r, t \in \mathbb{Z}_p^*$ and computing $m' = H(m) \cdot g^t, \sigma' = \sigma \cdot g^r$; $K_2 = e(g, pk)$, and sends (m', σ') to the verifier.

(2) Server computes $K_3 = e(\sigma', g), K_4 = e(m', pk)$ and returns K_3, K_4 to the verifier.

(3) Verifier checks the equation $K_3 = K_1^r \cdot K_2^{-t} \cdot K_4$, and outputs Valid if it holds, and otherwise outputs Invalid.

Correctness of SA-Verify. For a claimed message-signature pair (m, σ), when the verifier and the server are both honest, namely, the verifier correctly computes m', σ' and K_2, and the server correctly computes K_3 and K_4; then the verification equation $K_3 = K_1^r \cdot K_2^{-t} \cdot K_4$ holds if the message-signature pair (m, σ) is valid and otherwise does not hold. Consider

$$
\begin{aligned}
K_3 &= e\left(\sigma', g\right) \\
&= e\left(\sigma, g\right) \cdot e\left(g^r, g\right) \\
&= e\left(H(m)^x, g\right) \cdot e\left(g, g\right)^r \\
&= e\left(H(m), pk\right) \cdot K_1^r \\
&= e\left(m' \cdot g^{-t}, pk\right) \cdot K_1^r \\
&= e\left(m', pk\right) \cdot e\left(g, pk\right)^{-t} \cdot K_1^r \\
&= K_4 \cdot K_2^{-t} \cdot K_1^r.
\end{aligned}
\tag{9}
$$

In the following, by Theorems 7 and 8, we will show that our SA verification signature scheme above is secure under our security model; namely, the SA verification protocol described in Algorithm 2 is sound against collusion and adaptive chosen message attacks and also satisfies message-signature privacy against collusion and adaptive chosen message attacks.

Theorem 7. *The SA verification protocol described in Algorithm 2 satisfies soundness against collusion and adaptive chosen message attacks.*

Proof. In order to prove that the SA verification protocol in Algorithm 2 is $(t, q_v, 1/(p - 1))$-sound against collusion and adaptive chosen message attacks, we will show that the adversary can only prove an invalid signature as valid with at most probability $1/(p - 1)$. The challenger and the adversary play the following game.

(i) *Setup.* The challenger generates the system parameter $(\kappa, \mathbb{G}_1, \mathbb{G}_T, p, g, e, H)$, chooses a random element $x \in \mathbb{Z}_p$, and sets $sk = x$ and $pk = g^x$. Then it also computes $K_1 = e(g, g)$. Finally the challenger sends $(\kappa, \mathbb{G}_1, \mathbb{G}_T, p, g, e, H, sk, pk)$ to the adversary.

(ii) *Queries.* The adversary $\mathscr{A}$ is allowed to make at most q_v server-aided verification queries. The challenger $\mathscr{C}$ responds by executing SA-Verify with the adversary $\mathscr{A}$. At the end of each execution, the challenger returns the output of SA-Verify to the adversary.

(iii) *Output.* Eventually, the adversary $\mathscr{A}$ outputs a message m^*. The challenger $\mathscr{C}$ chooses a random element σ^* in $\mathbb{G}_1 \setminus \Omega_{m^*}$, where Ω_{m^*} is the set of valid signatures of m^*. Then they interact with each other as described in the SA-Verify protocol. Concretely, the challenger chooses two random elements $r^*, t^* \in \mathbb{Z}_p^*$, computes $m' = H(m) \cdot g^{t^*}$, $\sigma' = \sigma \cdot g^{r^*}$ and $K_2 = e(g, pk)$, and sends (m', σ') to the adversary. Then the adversary returns K_3^* and K_4^* to the challenger.

In the following, we will show that $K_3^* = K_1^{r^*} \cdot K_2^{-t^*} \cdot K_4^*$ happens with probability $1/(p-1)$.

From (10), we can directly deduce (11):

$$m' = H(m) \cdot g^{t^*},$$

$$\sigma' = \sigma \cdot g^{r^*} \tag{10}$$

$$\Downarrow$$

$$DL_g m' = DL_g H(m) + t^*,$$

$$DL_g \sigma' = DL_g \sigma + r^*. \tag{11}$$

Since (t^*, r^*) is chosen randomly from $\mathbb{Z}_p^*$, from the equation set (7), we can see that the adversary can only obtain m and σ with probability $1/(p-1)$. Furthermore, from the following, (13) can be deduced:

$$K_3^* = K_1^{r^*} \cdot K_2^{-t^*} \cdot K_4^* \tag{12}$$

$$\Downarrow$$

$$DL_{e(g,g)} \left(\frac{K_3^*}{K_4^*} \right) = DL_{e(g,g)} K_1^* \cdot r^* - DL_{e(g,g)} K_2^* \cdot t^*. \tag{13}$$

We can see that K_3^*/K_4^* is determined by r^* and t^*. Since $DL_{e(g,g)} K_1^* \cdot r^* - DL_{e(g,g)} K_2^* \cdot t^*$ is a random element in $\mathbb{Z}_p^*$, the adversary can only give out a pair (K_3^*, K_4^*) satisfying (11) with probability $1/(p-1)$. This completes the proof of Theorem 7. □

Theorem 8. *The SA verification protocol described in Algorithm 2 satisfies message-signature privacy against collusion and adaptive chosen message attacks.*

Proof. In order to prove that the SA verification protocol in Algorithm 2 satisfies $(t, q_v, 1/2)$-message-signature privacy against collusion and adaptive chosen message attacks, we will show that the adversary can only succeed with at most probability $1/2$ in the game with the challenger described as follows.

(i) *Setup.* The challenger generates the system parameter $(\kappa, \mathbb{G}_1, \mathbb{G}_T, p, g, e, H)$, and the secret key $sk = x$ and the public key $pk = g^x$. Then it also computes $K_1 = e(g, g)$. Finally the challenger sends $(\kappa, \mathbb{G}_1, \mathbb{G}_T, p, g, e, H, sk, pk)$ to the adversary.

TABLE 2: Efficiency of the SA verification signature scheme.

Verification	Pairing	Exponentiation	Multiplication	Hash
BLS [18]	2	0	0	1
Our scheme	1	$2(\mathbb{G}_1) + 2(\mathbb{G}_T)$	$2(\mathbb{G}_1) + 2(\mathbb{G}_T)$	1

(ii) *Queries.* The adversary $\mathscr{A}$ is allowed to make at most q_v server-aided verification queries. The challenger $\mathscr{C}$ responds by executing SA-Verify with the adversary $\mathscr{A}$, and, at the end of each execution, returns the output of SA-Verify to the adversary $\mathscr{A}$.

(iii) *Challenge.* $\mathscr{A}$ outputs two pairs (m_0, σ_0) and (m_1, σ_1), where σ_i is a valid signature on m_i for $i = 1, 2$. Then it sends them to the challenger $\mathscr{C}$. $\mathscr{C}$ chooses a bit $b \in \{0, 1\}$ at random and interacts with $\mathscr{A}$ by running SA-Verify$(\mathscr{A}, \mathscr{C}^{m_b, \sigma_b, pk, \text{VString}})$. Concretely, the challenger chooses two random elements $r^*, t^* \in \mathbb{Z}_p^*$, computes $m_b' = H(m_b) \cdot g^{t^*}$, $\sigma_b' = \sigma_b \cdot g^{r^*}$ and $K_2 = e(g, pk)$, and sends (m_b', σ_b') to the adversary. Then the adversary returns K_3^* and K_4^* to the challenger. After the interaction, $\mathscr{C}$ sends the output of SA-Verify$(\mathscr{A}, \mathscr{C}^{(m_b, \sigma_b, pk, \text{VString})})$ to $\mathscr{A}$.

(iv) *Output.* Finally, $\mathscr{A}$ outputs a bit $b' \in \{0, 1\}$. We will show that $\mathscr{A}$ can only succeed with probability $1/2$.

From the equation set (10) in the proof of Theorem 4, we can see that there exist two pairs (m_0, t_0^*) and (m_1, t_1^*) satisfying $m' = H(m_b) \cdot g^{t_b^*}$ for $c = 0, 1$, and there exist two pairs (σ_0, r_0^*) and (σ_1, r_1^*) satisfying $\sigma' = \sigma_b \cdot g^{r^*}$ for $c = 0, 1$. So the adversary cannot obtain anything about (m_b, σ_b), and it only guesses b correctly with probability $1/2$. This completes the proof of Theorem 8. □

Efficiency Analysis. In the following, we will show that our SA verification signature scheme based on BLS signature above is computation saving and efficient. We will analyze the efficiency of SA-Verify algorithm by comparing that of BLS Signature scheme [18]. In BLS signature scheme [18], to verify a message-signature pair (m, σ), the verifier needs to compute $H(m)$ which takes 1 hash function and two bilinear pairings $e(H(m), pk)$ and $e(\sigma, g)$. However, in our server-aided verification signature scheme, the verifier can first precompute a pairing $e(g, g)$ which can be used by multiple SA-Verify protocols. Then in a SA-Verify protocol, the verifier needs to compute totally 2 exponentiations in $\mathbb{G}_1$ and 2 exponentiation in $\mathbb{G}_T$ as well as 2 multiplications in $\mathbb{G}_1$ and 2 multiplications in $\mathbb{G}_T$ and 1 hash function. From the comparison, we can see that our SA verification signature scheme based on BLS signature is computation saving. The concrete computation cost comparison of the verifier in the verification of BLS signature and SA-Verify in Algorithm 2 is shown in Table 2.

5. Conclusion

In this paper, we studied the SA verification signature schemes with message-signature privacy for mobile computing. A power-constrained mobile device can outsource the verification of a signature to a server with powerful resources in order to reduce its computational load. We first present two definitions for privacy of server-aided verification protocol, respectively, named message privacy and message-signature privacy under collusion and adaptive chosen message attacks. Then under our security models, two concrete constructions based on existing signature schemes were presented and proved secure. By efficiency analysis, we showed that the two concrete schemes are both computation saving and efficient.

Conflict of Interests

The authors declare that there is no conflict of interests regarding the publication of this paper.

Acknowledgments

This work is partially supported by the National Natural Science Foundation of China (nos. 61202466, U1135004, 61170080, and 61100224), 973 Program under Grant no. 2014CB360501, Foundation for Distinguished Young Talents in Higher Education of Guangdong, China (no. 2012LYM_0017), and Fundamental Research Funds for the Central Universities (South China University of Technology) (no. 2014ZM0032). An abstract of this paper has been presented in the 4th International Conference on Emerging Intelligent Data and Web Technologies (2013), 414–421 [31].

References

[1] A. Durresi and M. Denko, "Preface: advances in mobile communications and computing," *Mobile Information Systems*, vol. 5, no. 2, pp. 101–103, 2009.

[2] A. Durresi and M. Denko, "Advances in wireless networks," *Mobile Information Systems*, vol. 5, no. 1, pp. 1–3, 2009.

[3] S. Benabbas, R. Gennaro, and Y. Vahlis, "Verifiable delegation of computation over large datasets," in *Advances in Cryptology—CRYPTO 2011*, vol. 6841 of *Lecture Notes in Computer Science*, pp. 111–131, Springer, Berlin, Germany, 2011.

[4] R. Gennaro, C. Gentry, and B. Parno, "Non-interactive verifiable computing: outsourcing computation to untrusted workers," in *Advances in Cryptology—CRYPTO 2010*, vol. 6223 of *Lecture Notes in Computer Science*, pp. 465–482, Springer, Berlin, Germany, 2010.

[5] K. M. Chung, Y. Kalai, and S. Vadhan, "Improved delegation of computation using fully homomorphic encryption," in *Advances in Cryptology—CRYPTO 2010*, vol. 6223 of *Lecture Notes in Computer Science*, pp. 483–501, Springer, Berlin, Germany, 2010.

[6] D. Fiore and R. Gennaro, "Publicly verifiable delegation of large polynomials and matrix computations, with applications," in *Proceedings of the ACM Conference on Computer and Communications Security (CCS '12)*, pp. 501–512, ACM, October 2012.

[7] B. Parno, M. Raykova, and V. Vaikuntanathan, "How to delegate and verify in public: verifiable computation from attribute-based encryption," in *Theory of Cryptography*, vol. 7194 of *Lecture Notes in Computer Science*, pp. 422–439, Springer, Berlin, Germany, 2012.

[8] C. Papamanthou, R. Tamassia, and N. Triandopoulos, "Optimal verification of operations on dynamic sets," in *Advances in Cryptology—CRYPTO 2011*, vol. 6841 of *Lecture Notes in Computer Science*, pp. 91–110, 2011.

[9] P. Béguin and J.-J. Quisquater, "Fast server-aided RSA signatures secure against active attacks," in *Advances in Cryptology—CRYPTO 1995*, vol. 963 of *Lecture Notes in Computer Science*, pp. 57–69, Springer, Berlin, Germany, 1995.

[10] J.-J. Quisquater and M. de Soete, "Speeding up smart card RSA computation with insecure coprocessors," in *Proceedings of the Smart Cards 2000*, pp. 191–197, 1989.

[11] C. H. Lim and P. J. Lee, "Security and performance of server-aided RSA computation protocols," in *Advances in Cryptology—CRYPTO'1995*, vol. 963 of *Lecture Notes in Computer Science*, pp. 70–83, Springer, Berlin, Germany, 1995.

[12] P. Nguyen and J. Stern, "The Béguin-Quisquater server-aided RSA protocol from Crypto '95 is not secure," in *Advances in Cryptology—ASIACRYPT 1998*, vol. 1514 of *Lecture Notes in Computer Science*, pp. 372–379, Springer, Berlin, Germany, 1998.

[13] M. Girault and J. J. Quisquater, "GQ + GPS = new ideas + new protocols," in *Proceedings of the Eurocrypt 2002-Rump Session*, April-May 2002.

[14] M. Girault and D. Lefranc, "Server-aided verification: theory and practice," in *Advances in Cryptology—ASIACRYPT 2005*, vol. 3788 of *Lecture Notes in Computer Science*, pp. 605–623, Springer, Berlin, Germany, 2005.

[15] W. Wu, Y. Mu, W. Susilo, and X. Y. Huang, "Provably secure server-aided verification signatures," *Computers and Mathematics with Applications*, vol. 61, no. 7, pp. 1705–1723, 2011.

[16] G. C. C. F. Pereira, M. A. Simplício Jr., M. Naehrig, and P. S. L. M. Barreto, "A family of implementation-friendly BN elliptic curves," *Journal of Systems and Software*, vol. 84, no. 8, pp. 1319–1326, 2011.

[17] S. Hohenberger and A. Lysyanskaya, "How to securely outsource cryptographic computations," in *Theory of Cryptography*, vol. 3378 of *Lecture Notes in Computer Science*, pp. 264–282, Springer, Berlin, Germany, 2005.

[18] D. Boneh, G. Lynn, and H. Shacham, "Short signature from the Weil pairing," in *Advances in Cryptology—Asiacrypt 2001*, vol. 2248 of *Lecture Notes in Computer Science*, pp. 514–532, Springer, Berlin, Germany, 2001.

[19] D. Boneh and X. Boyen, "Short signatures without random oracles," in *Advances in Cryptology—EUROCRYPT 2004*, vol. 3027 of *Lecture Notes in Computer Science*, pp. 382–400, Springer, Berlin, Germany, 2004.

[20] F. Zhang, R. Safavi-Naini, and W. Susilo, "An efficient signature scheme from bilinear pairing and its applications," in *Public Key Cryptography—PKC 2004*, vol. 2947 of *Lecture Notes in Computer Science*, pp. 277–290, 2004.

[21] B. Waters, "Efficient identity-based encryption without random oracles," in *Advances in Cryptology—EUROCRYPT 2005*, vol. 3494 of *Lecture Notes in Computer Science*, pp. 114–127, 2005.

[22] X. Chen, F. Zhang, Y. Mu, and W. Susilo, "Efficient provably secure restrictive partially blind signatures from bilinear pairings," in *Financial Cryptography and Data Security*, vol. 4107 of *Lecture Notes in Computer Science*, pp. 251–265, Springer, Berlin, Germany, 2006.

[23] X. Chen, F. Zhang, W. Susilo, H. Tian, J. Li, and K. Kim, "Identity-based chameleon hash scheme without key exposure," in *Information Security and Privacy*, vol. 6168 of *Lecture Notes in Computer Science*, pp. 200–215, Springer, 2010.

[24] S. Goldwasser, S. Micali, and R. L. Rivest, "A digital signature scheme secure against adaptive chosen-message attacks," *SIAM Journal on Computing*, vol. 17, no. 2, pp. 281–308, 1988.

[25] D. Chaum, "Blind signatures for untraceable payments," in *Advances in Cryptology*, pp. 199–203, Plenum, 1983.

[26] D. F. Aranha, K. Karabina, P. Longa, C. H. Gebotys, and J. López, "Faster explicit formulas for computing pairings over ordinary curves," in *Advances in Cryptology—EUROCRYPT 2011*, vol. 6632 of *Lecture Notes in Computer Science*, pp. 48–68, Springer, 2011.

[27] J.-L. Beuchat, J. E. Gonzalez-Díaz, S. Mitsunari, E. Okamoto, F. Rodríguez-Henríquez, and T. Teruya, "High-speed software implementation of the optimal ate pairing over Barreto-Naehrig curves," in *Pairing-Based Cryptography—Pairing 2010*, vol. 6487 of *Lecture Notes in Computer Science*, pp. 21–39, Springer, 2010.

[28] D. Hankerson, A. J. Menezes, and M. Scott, "Software implementation of pairings," in *Identity-Based Cryptography*, M. Joye and G. Neven, Eds., pp. 188–206, IOS Press, 2008.

[29] M. Naehrig, R. Niederhagen, and P. Schwabe, "New software speed records for cryptographic pairings," in *Progress in Cryptology—LATINCRYPT 2010*, vol. 6212 of *Lecture Notes in Computer Science*, pp. 109–123, Springer, Berlin, Germany, 2010.

[30] C. Costello, K. Lauter, and M. Naehrig, "Attractive subfamilies of BLS curves for implementing high-security pairings," in *Progress in Cryptology—INDOCRYPT 2011*, vol. 7107 of *Lecture Notes in Computer Science*, pp. 320–342, Springer, 2011.

[31] L. Xu and S. Tang, "Server-aided verification signatures with privacy," in *Proceedings of the 4th International Conference on Emerging Intelligent Data and Web Technologies (EIDWT '13)*, pp. 414–421, September 2013.

8

Enhanced Key Management Protocols for Wireless Sensor Networks

Baojiang Cui,[1] Ziyue Wang,[1] Bing Zhao,[2] Xiaobing Liang,[2] and Yuemin Ding[3]

[1]*School of Computer, Beijing University of Posts and Telecommunications, Beijing 100876, China*
[2]*State Grid Metering Center, Beijing 100192, China*
[3]*Department of Electronic Systems Engineering, Hanyang University, Ansan 426791, Republic of Korea*

Correspondence should be addressed to Baojiang Cui; cuibj@bupt.edu.cn

Academic Editor: David Taniar

With rapid development and extensive use of wireless sensor networks (WSNs), it is urgent to enhance the security for WSNs, in which key management is an effective way to protect WSNs from various attacks. However, different types of messages exchanged in WSNs typically have different security requirements which cannot be satisfied by a single keying mechanism. In this study, a basic key management protocol is described for WSNs based on four kinds of keys, which can be derived from an initial master key, and an enhanced protocol is proposed based on Diffie-Hellman algorithm. The proposed scheme restricts the adverse security impact of a captured node to the rest of WSNs and meets the requirement of energy efficiency by supporting in-network processing. The master key protection, key revocation mechanism, and the authentication mechanism based on one-way hash function are, respectively, discussed. Finally, the performance of the proposed scheme is analyzed from the aspects of computational efficiency, storage requirement and communication cost, and its antiattack capability in protecting WSNs is discussed under various attack models. In this paper, promising research directions are also discussed.

1. Introduction

Wireless sensor networks (WSNs) have been extensively used in various applications, such as homeland security, battlefield surveillance, environmental monitoring, and health care. Through collection and processing of the sensing data from the coverage area, WSNs enable users to access detailed and reliable information at any time and any place, which is a ubiquitous sensing technology.

WSNs have two salient characteristics: (i) it uses wireless communication and anyone within the range of the network can attack it; (ii) it may be deployed in unattended environments or even hostile regions, such as battlefield, where it can be physically attacked or captured [1]. Thus, how to ensure the security of WSNs becomes a significant issue.

Security researches of WSNs mainly focus on key distribution, secure routing protocols, secure transmission, and security defense. In these scopes, using key management mechanisms to settle security issues under the wireless sensor network environment is the most crucial and challenging problem [2].

Although key management mechanisms in the cable network have been deeply studied, the research is still immature in WSNs [3] because of limited communication bandwidth, computing and storage capacity of sensor nodes, and unfixed infrastructures. There is also a contradiction between the maximum security performance and minimum resource consumption.

It is worth noting that, due to the resource limitations, asymmetric encryption algorithms are seldom applied to the sensor network and most of the related works are based on symmetric key systems.

Although a number of classic protocols and schemes have been proposed for WSNs, many protocols concentrated on communication and processing technologies without paying enough attention to security issues, such as TEEN [4] and LEACH [5].

In recent years, scholars have proposed more sophisticated protocols which are mainly divided into two categories: predistribution scheme based on symmetric key and key management scheme based on public key.

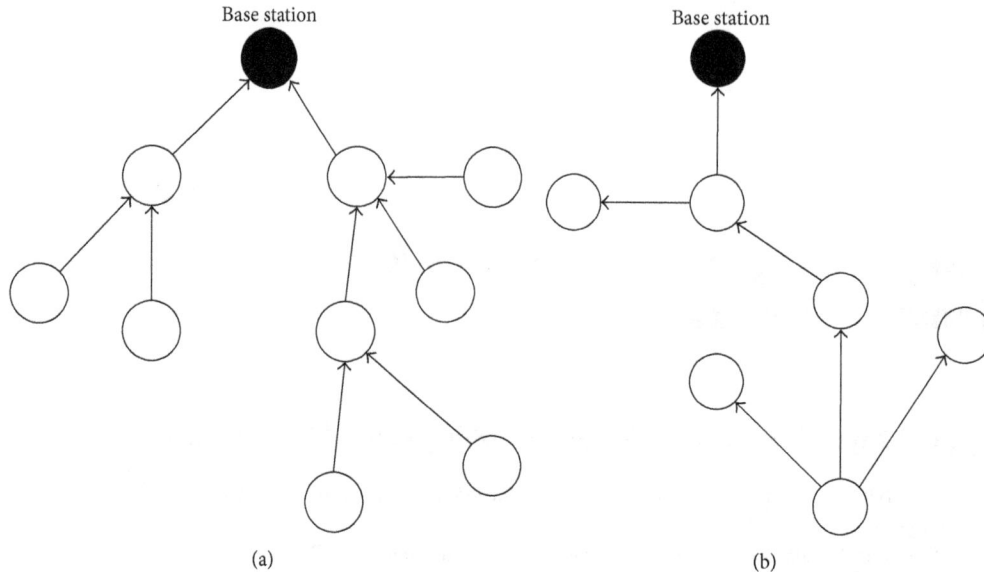

FIGURE 1: Examples of in-network processing.

Among the predistribution schemes, SPINS [6] is recognized as a classical secure protocol for WSNs. It consists of two modules: SNEP for data confidentiality, two-party data authentication and data freshness, and μTESLA for authenticated broadcast. It provides security for the entire network based on a single key and is easy to implement but the expansibility is limited.

To balance the security performance and resource consumption, random key predistribution schemes, polynomial key predistribution schemes, and key predistribution scheme based on deployment knowledge are subsequently proposed.

E&G [7] scheme is one of the earliest random key predistribution schemes. It achieves the establishment of pairwise key in WSNs for the first time based on the idea of preallocated key generation, solves the problem of unpredictable network topology, and provides a probability-based security. After that, the proposed Q-composite scheme [8] improves E&G schemes based on multicommon keys to generate pairwise keys.

Though quite a lot of superior security protocols have been proposed recently, most of them have their own deficiencies. Park proposed a lightweight security protocol (LISP); it can tolerate packet loss but the protocol cannot handle node revocation problem. After that, SRDA [9] proposed a secure data aggregation protocol, which takes the integrity into consideration but ignores the confidentiality of the information. LDP [10] proposes a local key management protocol based on dynamic cluster. It effectively supports the WSN security data fusion but does not give an effective solution of revoking captured nodes and updating keys.

To avoid above deficiencies, LEAP [11] establishes four kinds of keys and provides a strong application and scalability but requires huge amount of communication for key establishment and update. Furthermore, its security is heavily dependent on the initial secure time. ChengY's predistribution scheme [12] is based on clusters with advantages of

the good connectivity, network survivability, and low communications costs. However, the cost for rekeying is significant.

Based on previous studies, this paper proposes improved strategies to overcome some defects. In addition, how to apply the established keys to form security mechanisms to confront kinds of attacks is described in detail.

2. Requirements of Sensor Networks

Many security requirements of WSNs are similar to those of traditional networks, such as data confidentiality, authentication, and integrity. What is more, it should guarantee low energy consumption and high efficiency [13].

It is proved in recent researches that in-network data processing (shown in Figure 1), which mainly includes passive participation and data aggregation, is quite energy-efficient and should be widely employed.

The typical application of in-network processing is to divide the network into multiple clusters where the cluster head node collects and aggregates information from its neighbors and delivers the summary directly to the base station to avoid redundant transmissions and save communication bandwidth.

Generally, the pairwise key performs better over achieving data confidentiality, authentication, and integrity of WSNs, whereas, the cluster key or network-wide key is needed to achieve in-network data processing (shown in Figure 1) [14].

The particularity of the WSNs requires the ability of resistance to physical attacks and trapping. For example, once a node is compromised, the loss of secret information does not threaten remaining security links. Moreover, well-designed security mechanism should have capabilities of key revocation and update.

Therefore, it is fundamental to design a security mechanism, which satisfies above requirements, in order to achieve the security of WSNs.

3. Prerequisite Knowledge

3.1. Notations. The notations used in this paper are given in Notations section.

Note that, in order to simplify the representation in the following discussion, notations A and B are used to represent their node identifiers instead of ID_A and ID_B.

In addition, since keys for various security uses can be derived from the same key k, such as $K_0 = f(K, 0)$ for authentication and $K_1 = f(K, 1)$ for encryption, we just say a message M is authenticated or encrypted with K instead of saying in detail.

3.2. Function and Algorithm Description

One-Way Hash Function. One-way hash function H meets the following properties [15].

(i) Given x, it is easy to compute y using function $y = H(x)$.

(ii) Given y, it is difficult to compute x from function $y = H(x)$.

(iii) Given x, it is difficult to find a y meeting the condition that $y \neq x$ and $H(y) = H(x)$.

One-way hash chain is a sequence of the following hash value $\{x_m, x(m-1), \ldots, x_j, \ldots, x_1\}$, fulfilling the restriction $\{x_j \mid 0 < j \leq m, x_{j-1} = H(x_j)\}$, where x_m is a random selection of key seed. Due to the unidirectional feature, one-way hash key chain is widely used in secure authentication. For example, when x_1 is given, it can be verified that whether x_i is an element of the one-way hash key chain sequence using the equation $x_1 = H^{i-1}(x_i)$.

Key Generation Function. Pseudorandom function f is employed as the key generation function here for its high computational efficiency. When it is used in key establishment process, the computational cost is negligible. Note that, this function is stored in all the network nodes as well as the base station.

Diffie-Hellman Algorithm. Diffie-Hellman provides a method to ensure safety of shared key through insecure networks and it is an integral part of OAKLEY algorithm.

The ingenious point is that two sides of communication can use this method to determine the symmetric key, which can be used for encryption and decryption. Note that the key exchange protocol can only be used for key exchange, without being able to encrypt and decrypt the messages [16].

Since the key exchange algorithm itself is usually limited to be used as key exchange technology for many commercial products, it is usually called Diffie-Hellman key exchange (abbreviated as DH algorithm, key exchange based on DH algorithm is also commonly referred to as *DH exchange*).

The purpose of this key exchange technique is to enable two users to achieve secure key exchange in order to ensure the encryption of subsequent packets. The effectiveness of Diffie-Hellman key exchange algorithm relies on the difficulty of computing discrete logarithms [17]. In short, the discrete logarithm can be defined as follows.

First define primitive root of prime number p, which is integer roots generated from each of its powers from 1 to $p - 1$; that is, if a is a primitive root of prime number p, the values of $a \bmod p$, $a^2 \bmod p, \ldots, a^{p-1} \bmod p$ are all different integers from 1 to $p - 1$ in a certain arrangement.

For an integer b and a primitive root a of prime number p, we can find the unique index i, making $b = a^i \bmod p$, where $0 \leq i \leq (p-1)$, index i is called discrete logarithm or exponent of modulus p which is based to cardinal number a of integer b.

Based on the definition and nature of the primitive root, Diffie-Hellman key exchange algorithm is described as follows [18].

(1) There are two global parameters: prime number p and integer a, where a is a primitive root of p.

(2) Suppose users A and B wish to exchange a key, user A selects a random number X_A ($X_A < p$) as private key and calculates the public key $Y_A = a^{X_A} \bmod p$. The confidentiality store of X_A by user A makes Y_A publicly available to user B. Similarly, user B also selects a random number X_B ($X_B < p$) as private key and calculates the public key $Y_B = a^{X_B} \bmod p$. The confidentiality store of X_B by user B makes Y_B publicly available to user A.

(3) User A calculates shared secret key by $K = (Y_B)^{X_A} \bmod p$, and user B similarly calculates shared secret key K by $K = (Y_A)^{X_B} \bmod p$.

Since

$$K = (Y_B)^{X_A} \bmod p = \left(a^{X_B} \bmod p\right)^{X_A} \bmod p$$

$$= a^{X_B X_A} \bmod p = \left(a^{X_A}\right)^{X_B} \bmod p \qquad (1)$$

$$= \left(a^{X_A} \bmod p\right)^{X_B} \bmod p = (Y_A)^{X_B} \bmod p.$$

Thus, it corresponds that two sides have exchanged the same secret key K. Because X_A and X_B are confidential, an adversary can only use parameters q, a, Y_A and Y_B. Thus, adversary is forced to use discrete logarithm to determine the shared key K. The security of Diffie-Hellman key exchange algorithm relies on the fact that although computing exponent, which takes prime number as module, is relatively easy, computing discrete logarithm is very difficult. For large prime numbers, calculating the discrete logarithm is almost impossible.

3.3. Assumptions. Basic assumptions are as follows.

(i) Topology is unknown before the deployment of the nodes.

(ii) The sensor network is static (sensor nodes are not mobile) after deployment.

(iii) Sensor nodes have similar computational and communication capabilities.

(iv) Transmission power of nodes can be adjusted to control the propagation distance.

(v) The base station has enough energy supply and computing power.

(vi) The attacker has the ability to eavesdrop on all the channels as well as to replay former messages and inject malicious packets.

(vii) Once a node is captured, all the stored information will be obtained by the adversary.

(viii) Every node has enough space to store hundreds of bytes for key establishment materials.

(ix) Each node has some degree of ability to resist attack and it will not be captured with in a limited period of time.

4. Protocol Description

This section introduces the basic protocol in detail, including four kinds of secure key establishment mechanisms to satisfy various secure communication requirements and mechanisms for key erasure and update.

4.1. Overview. As discussed above, the single key mechanism cannot provide appropriate protection to all the required communication in the WSNs. Moreover, the security performance and resource consumption have to be balanced when making use of different kinds of keys.

The degree of sharing keys in the security mechanism has to be taken into consideration. For example, if unique pairwise keys are used for each two nodes in the WSNs to guarantee secure communication, the node captured by an attacker will not reveal any security information of other normal nodes, which is ideal to prevent threat to the entire network. However, it requires significant communication bandwidth and energy resources, which is quite inefficient.

On the contrary, if only a network-wide key is used for authentication and encryption, no communication between nodes is required for establishment of additional keys, and the storage costs and energy consumption can also be minimized. However, the security will be extremely poor. Once any node in the system is captured by an attacker, the whole network suffers an enormous risk.

4.2. Key Establishment. In this section, the establishment of four kinds of keys is discussed in detail as well as their characteristics and abilities to resist attacks.

4.2.1. Individual Key Establishment. Individual key is a unique key of each sensor node that shared with the controller (the base station) which is used for individual authentication and secure communication assurance [19].

For example, individual key can be used to encrypt sensitive information, such as special instructions and rekeying commands, exchanged between a sensor node and the base station. It can also be used for message authentication to get verification of the base station or other nodes.

Since every node in the network shares a unique individual key with the base station, it is neither practical nor efficient to store all these keys for the base station especially when the network scalability is very huge. Thus, it is important to adopt a strategy to reduce the storage overhead, which can be achieved by the key generation function f.

First of all, it is argued that each node holds the key establishment function f and an initial key K_I which is derived from the master key K that is only possessed by the controller; all of them are preloaded in the nodes before the key establishment phase. The generation of individual key for node A (here A indicates the unique ID of node A) is as follows:

$$K_A = f(K_I, A). \tag{2}$$

In the above, the function f for key establishment is a pseudorandom function and it is efficient enough to be used on sensor nodes.

Once the individual key is generated, the related node stores it within its life cycle. Since the base station has full knowledge of the initial key K_I and efficient establishment function f, the storage overhead for individual keys of each sensor node can be reduced.

4.2.2. Pairwise Key Establishment. Pairwise keys of a node indicate the keys shared with each of its direct neighbors, so the storage overhead of such keys for each node depends on the number of its neighbors [20, 21].

In this protocol, pairwise keys have a lot of uses. For example, it can be used for a cluster head to encrypt the cluster key, which has to be transmitted to all of its neighbors, to achieve the distribution security. It is also a component to improve system security.

However, it will impede passive participation, which is important in saving communication energy, if such key mechanism is employed individually. The initial pairwise key establishing progress is shown in the Figure 2.

The generation of pairwise keys for nodes A and B (here A is assumed to be the node that call for key establishment) is as follows:

$$A \longrightarrow * : \text{Nonce}_A$$
$$B \longrightarrow A : B, \text{MAC}_{K_B}(\text{Nonce}_A \mid B). \tag{3}$$

Here, node A broadcasts a nonce to all of its direct neighbors to request establishing pairwise without authenticating its identity, because if it cannot provide its own identity (namely, it does not own the individual key), it will fail to generate the pairwise in the following steps:

$$K_{AB} = f(K_B, A). \tag{4}$$

Since node A possesses both the key establishment function f and the initial key K_I, it can compute K_B independently and then obtains the pairwise key K_{AB} as well.

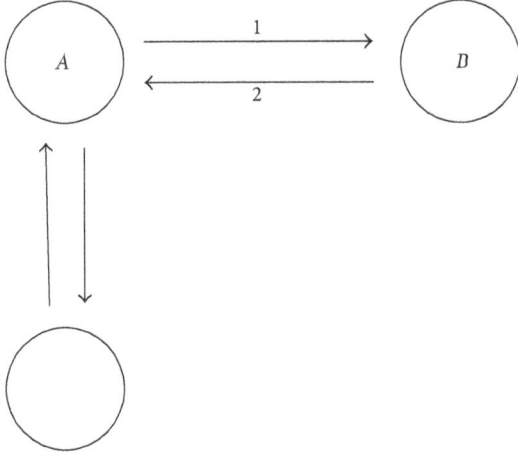

FIGURE 2: Pairwise key establishing phase.

Note that, each node has a timer which conducts it to achieve key erasure when it makes sure that the pairwise keys establishment is finished. This process is significant because all the nodes keep the network-wide initial key K_I to help complete the establishments in the initial period, and once the relatively safe period passes by, it will face great risk that some nodes may be compromised.

So it is suggested that, after a reasonable length of time, the initial key K_I and the neighbors individual master keys stored in the node be all erased (but its own individual master key will always be held).

In this way, when almost the pairwise keys are established successfully, no nodes will possess the necessary generating key materials until there is a new group of nodes to be joined. The key erasure mechanism is so necessary that how to control the key erasing time is worth exploring, but it is not an emphasis in this paper.

In addition, it can also be seen from the above equation that after the establishing time, namely, related key materials are erased, once the node A is compromised by an attacker and a A' broadcasts a nonce for establishing pairwise keys, it cannot success due to such establishment mechanism.

But once the attacker uses A' to take passive joining strategy, the responding node A' will generate the pairwise key with B (here B is one of a new batch of joining nodes that is asking to establish pairwise key with its neighbors including A') as follows: $K_{BA'} = f(K_{(A')}, B)$ and then the attacker will be able to inject erroneous packets into the network at will.

For the new added nodes, an alternative is proposed to establish secure pairwise key:

$$K_{AB} = f(K_B, A) \oplus f(K_A, B). \tag{5}$$

Since the pseudorandom function f is efficient, such improvement could be accepted.

The advantage of above key establishing scheme is that there is no message exchanging between nodes A and B during the computing step which extremely saves communication overhead.

Note that there will be a situation that two nodes want to establish the pairwise key while one of them does not possess

the master key K_I, such as one new added node and an older node which has finished all its pairwise key establishments and erased the master key K_I.

To deal with such situation, a scheme that asks for help from controller is simply presented as follows:

$$A \longrightarrow B : \text{Nonce}_A, A$$

$$B \longrightarrow \text{Base station} : R_{K_{AB}}, A, B, \text{MAC}_{K_B}(R_{AB}, A, B)$$

$$\text{Base station} \longrightarrow A : E_{K_A}(K_{AB}), \text{MAC}_{K_A}(B, E_{K_A}(K_{AB}))$$

$$\text{Base station} \longrightarrow B : E_{K_B}(K_{AB}), \text{MAC}_{K_B}(A, E_{K_B}(K_{AB})). \tag{6}$$

Here A is a new node who calls for establishing pairwise key with its neighbor B. Here B is an older node that has generated all its own pairwise keys and erased the initial key K_I, which makes it unable to generate new pairwise key.

If B wants to verify the identity of node A, the most credible way is asking for help of base station.

However, reducing the use of base station is an important goal here and the improvement is worth further exploring.

4.2.3. Cluster Key Establishment. Cluster key is a key generated by an elected cluster head and shared with its neighbors and it is mainly used for encrypting local broadcast packets. Its most significant advantage is that it enables the in-network processing such as passive participation and data aggregation, which cannot be supported by the pairwise key but could save energy consumption efficiently.

This key establishing process is obvious as follows:

$$A \longrightarrow B_i : E_{K_{AB_i}}(K_A^C). \tag{7}$$

Here node A is the elected cluster head and B_i represents one of its immediate neighbors: $B_1, B_2, \ldots, B_n$ ($1 \leq i \leq n$). Cluster head A first generates a key K_A^C randomly and encrypts it with its pairwise keys and then sends it to each neighbor B_i. Moreover, node B_i decrypts the cluster key and then stores K_A^C in a table.

When any neighbor of A is revoked which means there will be a risk to continue using the old cluster key, cluster head A regenerates and transmits the $K_A^{C'}$ in the same way.

Cluster division and cluster head selection approaches are also worthy of discussion. But it is not an emphasis in this paper. A simple mesh division method is shown in Figure 3 based on virtual cluster idea.

4.2.4. Group Key Establishment. The group key K_g is used for encrypting messages that need to be broadcasted to the whole group. Note that, different from above situations, the key point here is no longer about key establishment or encrypting schemes because there is only one group key shared among the entire network; meanwhile it does not make sense to encrypt a broadcast message using master key of each sensor node separately.

It is also because there is only one group key shared among sensor nodes; once a compromised node is revoked,

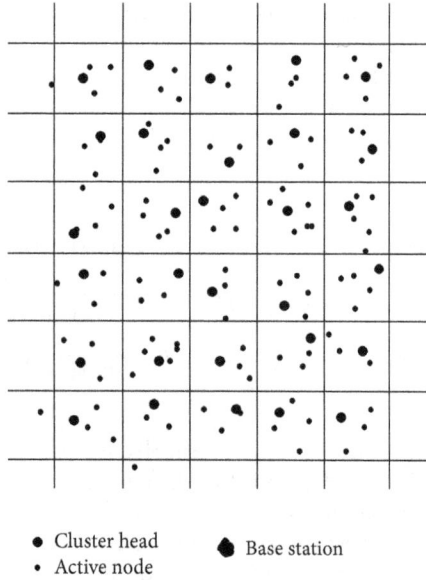

FIGURE 3: Mesh division method.

- ● Cluster head   ◆ Base station
- · Active node

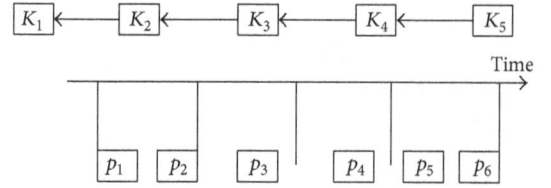

FIGURE 4: Using the one-way hash function for source authentication.

the rekeying and updating mechanism comes to be important.

μTESLA [22] is a widely employed protocol due to the high efficiency and perfect tolerance for packet loss. A one-way hash function H is used here to help achieve the process. Firstly, the controller generates a random seed k_m and uses the function H to get a sequence of the following hash values: $\{k_m, k_{m-1}, \ldots, k_j, \ldots, k_1\}$ that meets the restriction $\{k_j \mid 0 < j \le m, k_{j-1} = H(k_j)\}$.

Then preload this key chain $\{k_m, k_{m-1}, \ldots, k_j, \ldots, k_1\}$ in the base station and use delayed key disclosure to achieve message authentication. Let A be the revoked node and K_g' the new group key; the process is as follows:

$$\text{Base station} \longrightarrow * : A, f\left(K_g', 0\right), \text{MAC}_{k_j}\left(A \mid f\left(K_g', 0\right)\right). \tag{8}$$

When the verification is done, all the nodes will remove related information of node A and restore the group key K_g' in the table.

Note that the initial Group key K_g is preloaded in all the sensor nodes before their deployment like the initial key K_I, but we cannot take K_I also as the group key because it will be erased in a very short time after the pairwise key establishment. The key used for deriving related keys must be protected separately from normal ones.

Figure 4 simply illustrates the authentication mechanism:

$$k_{j-1} = H\left(k_j\right). \tag{9}$$

5. Enhanced Protocol

5.1. Requirements Analysis.
The design of the basic scheme presented in the previous section is motivated by the observation that single keying mechanism is not suitable for meeting all the security requirements of different types of exchanged messages.

The advantage of this scheme is that the captured node does not threat the safety of the other nodes in case the master key K is absolutely safe in time interval $T_{\min}$.

During the time interval $T_{\min}$, all the nodes of the WSN will hold the general master key K and we note that this scheme cannot provide confidentiality when a node is compromised in $T_{\min}$. Because, by using the stolen information like the master key K, an attacker can easily derive the master keys of all the rest normal nodes that are deployed in the same time interval as well as negotiating new pairwise key with normal nodes in any region, which means once a node is compromised in time interval $T_{\min}$, the security of the entire network is extremely dangerous.

5.2. Enhanced Scheme.
Based on the Diffie-Hellman algorithm above, presenting the improved scheme, prior to deployment of the network, each node prestores the large prime number p and its primitive root a instead of the initial key K_I which is derived from the master key K.

Note that the generation of individual key for node A is still same:

$$K_A = f\left(K_I, A\right). \tag{10}$$

Different from the basic scheme, this process is completed once the node is deployed, after that the information of the initial key K_I is deleted. Thus, the attacker cannot get any information about the initial key K_I or the master key K even if it is compromised during the working period.

Since the node no longer keeps initial key K_I, which is required to participate in relevant calculations (function) in the pairwise key generating process, the basic scheme cannot be achieved. For this situation, make the following improvements.

Gain a key evolution function to each node. Takes node A and B for examples:

$$\begin{aligned} X_A &= h\left(A \mid K_A\right) \bmod p \\ X_B &= h\left(B \mid K_B\right) \bmod p. \end{aligned} \tag{11}$$

Then calculate the public message:

$$\begin{aligned} Y_A &= a^{X_A} \bmod p \\ Y_B &= a^{X_B} \bmod p. \end{aligned} \tag{12}$$

The pairwise key generation process is as follows:

$$A \longrightarrow * : \text{Nonce}_A, Y_A$$
$$B \longrightarrow A : \text{MAC}_{K_{AB}}\left(B \mid Y_B\right), B, Y_B. \tag{13}$$

Here, node A broadcasts a nonce to all its direct neighbors and asks to establish pairwise key and broadcasts the public message Y_A at the same time. When its neighbor (take node B, for example) receives the message, it first verifies the legitimacy of Y_A and then calculates the pairwise key using the following function:

$$K_{AB} = \left(Y_A\right)^{X_B} \bmod p. \tag{14}$$

After that, node B sends messages B and Y_B back to the asking node A and sends a message $\text{MAC}_{K_{AB}}(B \mid Y_B)$ to authenticate its identity. If node B cannot respond to node A in this way, it means node B cannot get K_{AB} only taking use of Y_A; then consider node B as untrusted. In addition, node A does not need to send authenticating message back to node B anymore because if it cannot prove its own identity (namely, it cannot get K_{AB} only taking use of Y_B, and it will fail to generate the pairwise key K_{AB}).

Compared with the basic protocol, the most obvious improvement of enhanced protocol is that it takes use of Diffie-Hellman algorithm to generate pairwise keys instead of storing the initial key K_I in a certain period of time. Thus, even if a node is compromised in $T_{\min}$, the attacker can merely get the information of key related to the compromised node, which means only limited security threats can be caused, avoiding the disruption of the entire network caused by losing initial key K_I. Despite the slight increment in the computational overhead, the security of the WSN is greatly improved.

6. Performance Evaluation

The ability of the protocol to fight against kinds of attacks is discussed in detail in above sections. This section analyzes the storage requirement and energy efficiency.

6.1. *Storage Requirement.* In the basic protocol, a node needs to store four types of keys. Considering a node with m neighbors in the WSN, it needs to store one individual key, m cluster keys, m pairwise keys, and one group key. In the enhanced protocol, each node stores the same number of keys as the basic protocol.

When the key establishment is complete in a network having a scale of N, there is an upper limit of the number of keys to be stored in the nodes including N individual keys, $C(N, 2)$ pairwise keys, $N/2$ cluster keys, and N group keys (though there is only one group key in a certain period), which add up to $((5/2)N + (N!/2(N-2)!) = (N^2+3N)/2)$ and average to each node is $(5/2 + (N-1)!/2(N-2)! = N/2 + 2)$.

Note that communication distance of sensor node is limited so that it will not reach a high complexity that each two nodes are connected.

In addition, using an efficient clustering method can reduce the number of required cluster keys and the real storage complexity is much smaller.

Although memory is a quite scarce resource for the current generation of nodes in WSNs, for a reasonable degree, storage is not an issue in our protocol. For example, 100 keys totally take 800 bytes when the key size is 8 bytes.

6.2. *Communication Cost.* In this paper, the average communication cost increases with the connection degree of a sensor network and decreases with the network size N. Efficient preloaded functions are widely used, which greatly reduces the message exchanges in key establishing phase so that to save communication cost. Whats more, the use of located cluster key enables in-network data processing which also helps achieve communication and energy efficiency.

It is worth noting that the communication cost of the enhanced protocol remains at the same level as that of the basic protocol.

6.3. *Computational Cost.* Functions used in the proposed protocols are all of high computational efficiency. For example, pseudorandom function f is employed to be the key generation function, and the computational cost will be negligible when it is used in key establishment process. In the enhanced protocol, although computational cost is slightly increased by using Diffie-Hellman algorithm, for a network of reasonable density, we believe that the computational overhead is applicable for a network of reasonable density in our protocols. For example, for a WSN of size $N = 1000$ and connection degree of 20, the average computational cost is 2.7 symmetric key operations per node per revocation and a larger N will reduce the cost further.

Overall, we conclude that the protocols proposed in this study are scalable and efficient enough in storage, communication, and computation.

7. Security Analysis

This section analyzes the security of the key management protocols. The survivability of the network is discussed when undetected compromised nodes occur and the robustness of proposed schemes is studied in defending against various attacks.

7.1. *Survivability.* Once a sensor node A is compromised, the adversary can launch attacks by utilizing keying materials of node A. If the threat is detected somehow, the protocols can revoke node A efficiently and update the information of nodes quickly throughout the whole network. Basically, each neighbor of compromised node A could delete its pairwise key shared with node A as well as updating the cluster key. The group key could also be updated efficiently by taking use of μTESLA mechanism. When the revocation is completed, the adversary cannot launch further attacks anymore.

However, security detection in WSNs is more difficult than in other systems since sensor systems are often deployed in unattended environments. Thus, the survivability of the network is one of most important security requirements when compromised nodes is not detected.

Firstly, because individual key is only shared between the base station and each sensor node, it usually does not help the attacker launch attacks.

Secondly, obtaining the cluster keys and pairwise keys of a compromised node enables the attacker to establish trust with the neighbor nodes, which can be used by the attacker to inject malicious sensor readings and routing control information into the network. However, in the proposed protocols in this study, the attacker usually has to achieve such attacks by taking use of the identity of the captured node.

Note that a salient feature of the proposed protocols is the ability in localizing possible threats. Because after the deployment of the network and the pairwise key establishing phase, every node will keep a list of trusted neighbor nodes. As compromised node and its copy nodes cannot establish trust relationship with other nodes except its neighbors, the attacker can only damage secure links within limited range.

Finally, obtaining the group key enables the attacker to decrypt messages broadcast by the base station. The broadcast messages, by their nature, are intended to be received by all the nodes in the network. Thus, compromising any single node is enough to possess this message, whatever security mechanism is used. However, obtaining the group key does not allow the attacker to damage the entire network with malicious packets by impersonating the base station because all messages sent from the base station are authenticated by μTESLA mechanism.

7.2. Dealing with the Attacks on Secure Routing. Ciou et al. have described various possible attacks of routing protocols for WSNs [18]. How the proposed schemes can defend against such attacks is shown in this section.

An inside attacker may attempt to alter and replay routing information to make routing loops, attract or repel network traffic, and generate false messages. Moreover, the attacker can launch the selective forwarding attack, in which the captured node suppresses routing packets sent from a few selected nodes while forwarding the other packets reliably.

In this paper, the schemes cannot protect the WSNs from such attacks; however, the schemes can hinder or minimize the consequences caused by such attacks.

First, based on the key establishment and authentication phases of the proposed protocols, it is apparent that such attacks are only possible within a small area of two-hops from the captured node.

Second, since such attacks are localized in a certain zone, the attacker faces a high risk of being detected when launching such attacks. For example, the probabilistic challenge mechanism can help detect the spoofing attack and the detection of altering attack is also possible since the related sending node may overhear the forwarded messages altered by the captured node.

Last but not least, once a compromised node is detected, the group rekeying process of the protocols can efficiently revoke the compromised node from the network.

The proposed protocol can protect WSNs from the following attacks.

Sybil Attacks. In Sybil attacks, the attacker may replicate the captured node and deploy multiple replicas into the original network. With help of the base station, such replica nodes will then try to establish pairwise and cluster keys with normal nodes that are not neighbors of the captured node [23]. If the base station does not know the precise topology of the wireless network, this attack may work in pairwise key establishment. However, it cannot happen for proposed protocols because each normal node keeps a list of its approved neighbors and the base station is not involved for pairwise or cluster key establishments in this study.

HELLO Flood Attack. The attacker may send a HELLO message to all nodes in the network by increasing the transmission power to be high enough to make all the nodes convinced that it is their neighbor. Once this attack succeeds, nodes of the entire network may send their readings and some other packets in vain. However, it cannot succeed in proposed protocols because the attacked does not have a network-wide key for authentication.

It is worth noting that the group key in the protocols is not for authentication purpose but for the distribution of secure messages to the entire network from the base station.

7.3. Defending against Sinkhole and Wormhole Attacks. The combination of the sinkhole and the wormhole attacks is one of the most difficult attacks to be prevented.

In the sinkhole attack, a malicious node tries to attract packets from the neighbor nodes and then drops them. It can launch such attack by advertising information of high reliability or high remaining energy, which is very hard to detect in the WSNs.

In the wormhole attack, two distant malicious nodes conceal their distance information to the network. After placing one such node near the target zone and another one near the base station, the attacker will convince the nodes within the target area, which are usually multiple hops away from the base station, as only one or two hops to create a sinkhole. Moreover, nodes which are multiple hops away may believe that they are neighbors of each other. Since to launch wormhole attack the attacker does not need to compromise any sensor nodes, such attack is very powerful in practice [24].

In the proposed protocols, an outside attacker cannot succeed in launching wormhole attack except in the neighbor discovery process, since a node will know all its neighbor nodes after the pairwise key is established, which means the attacker cannot convince two distant nodes to believe that they are neighbors of each other.

Because the time of neighbor discovery process is very short (usually for seconds), the probability that the attacker achieves such attacks is also quite small. If an inside attacker compromises two or more nodes, it can launch such attacks. However, it cannot convince two distant nodes as neighbors when the neighbor discovery phase is finished. The authenticated neighborhood information is critical to deal with the wormhole attacks.

In the sinkhole attack, if the attacker compromises a node *A* that is close to the base station and another node *B* in the target area, the attacker will succeed in making node A as a sinkhole. Since the number of hops between node B and the base station turns smaller, node B will be especially attractive to surrounding nodes. In practice, the location of base station is usually static. When the network is constructed, topology will be known to the entire network, and then sensor nodes will know the approximate number of hops from the base station. Thus it is difficult for an attacker to make a very attractive sinkhole in the WSN without being detected.

7.4. Conclusion. This paper proposes a basic key management protocol based on initial secure time, which assumes that the attacker cannot compromise a node in a short time. It satisfies various security requirements of WSNs using the combination of four kinds of secure keys. Meanwhile, the erasure and update mechanism of keys is important to support network security.

To further improve the security of the basic scheme, an enhanced protocol based on Diffie-Hellman algorithm is proposed, which avoids storing the master key in sensor nodes so as to restrict the security impact of a captured node to the rest network.

The proposed protocol achieves high communication and energy efficiency by supporting in-network data processing and enhances the network security through strict authentication and encryption mechanisms. Compared to original ideas, the proposed scheme improves not only the network security but also the extensibility of WSNs.

This paper presents a proposal for key establishment and achieves security mainly based on the combining application of four kinds of keys. This is a critical step and how to use such keys to found a protection mechanism is a focus in our future research.

Notations

N:	The number of nodes in the network
A, B:	Two communicating nodes in the network (also represents the node identifier)
$f(K, A)$:	Calculate with parameter A using the key K in pseudorandom function f
$H(K)$:	One-way hash function to generate a chain of keys using the seed K
$MAC_K(m)$:	Message authentication code (MAC) of message m using MAC key K
K:	The master key only possessed by base station
K_A:	Individual key of node A
$E_K(m)$:	Encryption of message m with a symmetric key K
$M_1 \mid M_2$:	Concatenation of the sequences M_1 and M_2
$A \to B : M$:	Node A sends a message M to node B
$A \to * : M$:	Node A sends a local broadcast message M to all its neighbors
$h(m)$:	Calculate hash value of message m.

Conflict of Interests

The authors declare that there is no conflict of interests regarding the publication of this paper.

Acknowledgments

This work was supported by National ratural Science Foundation of China (nos. 61170268, 61100047, and 61272493), International S&T Cooperation Special Projects of China (no. 2013DFG72850), and The National Basic Research Program of China (973 Program) (no. 2012CB724400).

References

[1] I. F. Akyildiz, W. Su, Y. Sankarasubramaniam, and E. Cayirci, "Wireless sensor networks: a survey," *Computer Networks*, vol. 38, no. 4, pp. 393–422, 2002.

[2] X. He, M. Niedermeier, and H. de Meer, "Dynamic key management in wireless sensor networks: a survey," *Journal of Network and Computer Applications*, vol. 36, no. 2, pp. 611–622, 2013.

[3] R. Riaz, A. Naureen, A. Akram, A. H. Akbar, K. H. Kim, and H. Farooq Ahmed, "A unified security framework with three key management schemes for wireless sensor networks," *Computer Communications*, vol. 31, no. 18, pp. 4269–4280, 2008.

[4] C. Intanaonwiwat, R. Govindan, and D. Estrin, "Directed diffusion: a scalable and robust communication paradigm for sensor networks," in *Proceedings of the 6th Annual ACM/IEEE International Conference on Mobile Computing and Networking (MobiCom '00)*, pp. 56–67, ACM/IEEE, Boston, Mass, USA, August 2000.

[5] A. Manjeshwar and D. P. Agrawal, "TEEN: a routing protocol for enhanced efficiency in wireless sensor networks," in *Proceedings of the 15th International Parallel and Distributed Processing Symposium (IPDPS '01)*, pp. 2009–2015, IEEE Computer Society, San Francisco, Calif, USA, April 2001.

[6] A. Perrig, R. Szewczyk, V. Wen, D. Culler, and J. D. Tygar, "SPINS: security protocols for sensor networks," in *Proceedings of the 7th Annual International Conference on Mobile Computing and Networking (Mobicom '01)*, pp. 189–199, Rome, Italy, July 2001.

[7] W. Du, J. Deng, Y. S. Han, and P. K. Varshney, "A pairwise key pre-distribution scheme for wireless sensor networks," in *Proceedings of the 10th ACM Conference on Computer and Communications Security (CCS '03)*, pp. 42–51, ACM Press, Washington, DC, USA, October 2003.

[8] H. Chan, A. Perrig, and D. Song, "Random key predistribution schemes for sensor networks," in *Proceedings of the IEEE Symposium on Security and Privacy*, pp. 197–213, Oakland, Calif, USA, May 2003.

[9] H. O. Sanli, S. Ozdemir, and H. Cam, "SRDA: secure reference-based data aggregation protocol for wireless sensor networks," in *Proceedings of the IEEE 60th Vehicular Technology Conference (VTC '04)*, pp. 406–410, IEEE, Los Angeles, Calif, USA, 2004.

[10] T. Dimitriou and I. Krontiris, "A localized, distributed protocol for secure information exchange in sensor networks," in *Proceedings of the 19th IEEE International Parallel and Distributed Processing Symposium (IPDPS '05)*, pp. 37–45, IEEE, April 2005.

[11] S. Zhu, S. Setia, and S. Jajodia, "LEAP: efficient security mechanisms for large-scale distributed sensor networks," in *Proceedings of the 10th ACM Conference on Computer and Communications Security (CCS '03)*, pp. 62–72, ACM, New York, NY, USA, October 2003.

[12] J. Shen and L. Xu, "Cluster-based key pre-distribution seheme for wireless sensor networks," *Journal of Wuhan University: Natural Science Edition*, vol. 55, no. 1, pp. 117–120, 2009 (Chinese).

[13] X. Huang, M. Yang, and S.-S. Lv, "Secure and efficient key management protocol for wireless sensor network and simulation," *Journal of System Simulation*, vol. 20, no. 7, pp. 1898–1903, 2008.

[14] X. Chen, J. Li, J. Ma, Q. Tang, and W. Lou, "New algorithms for secure outsourcing of modular exponentiations," in *Computer Security—ESORICS 2012: 17th European Symposium on Research in Computer Security (ESORICS '12), Pisa, Italy, September 10–12, 2012*, vol. 7459 of *Lecture Notes in Computer Science*, pp. 541–556, Springer, Berlin, Germany, 2012.

[15] L.-C. Li, J.-H. Li, and J. Pan, "Self-healing group key management scheme with revocation capability for wireless sensor networks," *Journal on Communications*, vol. 30, no. 12, pp. 12–17, 2009.

[16] Z. Ming, W. Suo-ping, and X. He, "Dynamic key management scheme for wireless sensor networks based on cluster," *Journal of Nanjing University of Posts and Telecommunications (Natural Science)*, vol. 32, no. 1, 2012.

[17] G.-J. Wang, T.-T. Lv, and M.-Y. Guo, "Transitory initial key-based key management protocol in wireless sensor networks," *Chinese Journal of Sensors and Actuators*, vol. 20, no. 7, pp. 1581–1586, 2007.

[18] Y.-F. Ciou, F.-Y. Leu, Y.-L. Huang, and K. Yim, "A handover security mechanism employing the Diffie-Hellman key exchange approach for the IEEE802.16e wireless networks," *Mobile Information Systems*, vol. 7, no. 3, pp. 241–269, 2011.

[19] J. Li, X. Chen, J. Li, C. Jia, J. Ma, and W. Lou, "Fine-grained access control system based on outsourced attribute-based encryption," in *Computer Security—ESORICS 2013: 18th European Symposium on Research in Computer Security, Egham, UK, September 9–13, 2013. Proceedings*, vol. 8134 of *Lecture Notes in Computer Science*, pp. 592–609, Springer, Berlin, Germany, 2013.

[20] A. Zhu, S. Xu, S. Setia, and S. Jajodia, "Establishing pairwise keys for secure communication in ad hoc networks: a probabilistic approach," in *Proceedings of the 11th IEEE International Conference on Network Protocols (ICNP '03)*, pp. 326–335, Atlanta, Ga, USA, November 2003.

[21] W. Du, Y. S. Han, J. Deng, and P. K. Varshney, "A pairwise key pre-distribution scheme for wireless sensor networks," in *Proceedings of the 10th ACM Conference on Computer and Communications Security (CCS '03)*, pp. 42–51, Washington, DC, USA, October 2003.

[22] D. Liu and P. Ning, "Multi-level μTESLA: broadcast authentication for distributed sensor networks," *ACM Transactions on Embedded Computing Systems*, vol. 3, no. 4, pp. 800–836, 2004.

[23] J. Li, Q. Wang, C. Wang, and K. Ren, "Enhancing attribute-based encryption with attribute hierarchy," *Mobile Networks and Applications*, vol. 16, no. 5, pp. 553–561, 2011.

[24] Y. S. Lee, J. W. Park, and L. Barolli, "A localization algorithm based on AOA for ad-hoc sensor networks," *Mobile Information Systems*, vol. 8, no. 1, pp. 61–72, 2012.

9

Optimized Group Channel Assignment Using Computational Geometry over Wireless Mesh Networks

Anitha Manikandan and Yogesh Palanichamy

Department of Information Science and Technology, College of Engineering, Anna University, Guindy, Chennai, Tamilnadu 600 025, India

Correspondence should be addressed to Anitha Manikandan; animani.kpm@gmail.com

Academic Editor: Peter Jung

Wireless Mesh Networks (WMNs) are an evolving division in the field of wireless networks due to their ease of deployment and assured last mile connectivity. It sets out a favorable situation to guarantee the Internet connectivity to all the mobile and static nodes. A wireless environment is dynamic, heterogeneous, and unpredictable as the nodes communicate through the unguided links called channels. The number of nonoverlapping channels available is less than the number of mesh nodes; hence, the same channel will be shared among many nodes. This scarcity of the channels causes interference and degrades the performance of the network. In this paper, we have presented a group based channel assignment method to minimize the interference. We have formulated a mathematical model using Nonlinear Programming (NLP). The objective function defines the channel assignment strategy which eventually reduces the interference. We have adapted the cognitive model of Discrete Particle Swarm Optimization (DPSO), for solving the optimization function. The channel assignment problem is an NP hard problem; hence, we have taken the benefits of a stochastic approach to find a solution that is optimal or near optimal. Finally, we have performed simulations to investigate the efficiency of our proposed work.

1. Introduction

Wireless Mesh Network [1–4] is an infrastructure based multihop network that can be deployed even in underprivileged areas like terrains where providing Internet connectivity is a laborious and expensive task. Deployment of WMNs is undergoing a rapid advancement in the area of networking due to the inherent features of WMNs like self-healing and self-organizing. The WMN infrastructure contains different components in three layers. End user devices like laptops, PCs, and so forth are known as mesh clients (MCs) and the MCs form the bottom most layer. Mesh routers (MRs) that forward the data in multiple hops between the source and the destination form the middle layer. MRs with higher configuration are called Mesh Gateways (MGs). These MGs that render the connectivity to the Internet and other networks for the end users form the top layer. Due to the falling-off price of WLAN equipment (e.g., mesh routers), each device is manufactured with multiple radios to support simultaneous transmission and reception of data using multiple orthogonal channels. The IEEE 802.11b/g and IEEE 802.11a standards define 3 and 8 nonoverlapping frequency channels available in the legal frequency spectrum. This composes the entire network to act as multiradio multichannel WMN. This characteristic enhances the network throughput [5, 6] by using the increased parallelism in the network and also decreases the transmission delay. However, these advantages also introduce other problems in the network like network partitioning, timing mismatch, distributed operation, routing issues [7], link scheduling, and channel assignment (i.e., mapping of channels to radios at each node) [8].

The key design issue in multiradio multichannel WMN is the assignment of the orthogonal channels, that is, binding the radios with the channels. Channel assignment schemes should be aware of the trade-off between the network connectivity and the interference. Two nodes in the transmission range can communicate with each other if they share one common channel in any of its radios. However, if the same channel is used in different nodes for communication, it leads to interference. "Interference" is a major setback that occurs in the wireless network where the signals collide and results in packet loss and the end result of this is the performance degradation of the network. On that point there are several solutions suggested in the literature to address these problems [6, 9–12].

The nonoverlapping channels are less in number and hence it becomes necessary to assign the same channel to different radios. Assigning the same channel to different radios results in interference. But this is a fact which cannot be avoided since the number of orthogonal channels cannot be increased beyond a limit that is posed by telecommunication departments of various countries. Hence, it is not possible to avoid interference altogether and it is possible only to minimize the interference. This paper concentrates on how to assign channels in order to minimize the interference. We are invigorated by the job scheduling problem which is an operation research problem that optimally assigns jobs on the shop floor to the machines available to finish the jobs with minimal time consumption [13–15]. We have modelled the channel assignment problem as an optimization problem with the aim of optimally assigning channels to all the links in order to minimize the interference. We have used the nonlinear programming (NLP) technique to formulate our mathematical model. Our channel assignment scheme is a traffic independent algorithm [12, 16, 17]; that is, we do not regard the traffic flow in the network for the channel designation.

The strategy used in this paper has two major steps. The first step clusters the WMN nodes based on their locations using Voronoi diagram [18–20]. We apply the novel idea of membership to select the leader for each cluster or group. Membership of a node refers to the number of groups in which that particular node is present. The node with the maximum value of membership is elected as the leader of that group. The selection of the leaders is responsible for computing the Interference Matrix of that particular group. We have used the protocol model described in [21, 22] to represent the interference in the network. Generally, Conflict Graphs [8, 10, 12, 16] are used to represent the interference present in wireless networks. This research work forms the interference matrices using an enhanced version of the Conflict Graph called Enhanced Multiradio Multichannel Conflict Graph (E-MMCG) [23]. E-MMCG considers the effect of colocated radios assigned with the same channel in the estimation of interference. Besides the novel idea of membership, integrating the protocol model described in [21, 22] and E-MMCG [23] is a novel contribution of this research work.

The second step is a swarm intelligence based approach known as Discrete Particle Swarm Optimization (DPSO) [24] for solving the channel assignment problem in the group. DPSO is a variation of Particle Swarm Optimization (PSO) [25, 26] which is the original version of the approach. The original version was designed to solve continuous problems by Kennedy and Eberhart [27] and was later modified to support discrete problems. We have used the cognitive model of DPSO with the inertia weight to make the particle converge faster. Applying nonlinear programming (NLP) to model the channel assignment problem as an optimization problem with the objective of minimizing the interference is another novel contribution of this research work.

We make the following contributions to achieve our objective:

(i) We have employed the Voronoi diagram for clustering the nodes into small subgraphs in order to define the optimization problem with a smaller number of nodes rather than formulating it for the entire network.

(ii) We have designed a simple algorithm to identify the leaders for each group which has control over the small graph and also it takes the responsibility of assigning the channels to the available links in the small group.

(iii) We have derived a mathematical model to define the objective of our proposed system and also adapted an evolutionary algorithm approach to solve the optimization problem.

(iv) The DPSO algorithm is not applied in its original form but rather we have defined the velocity clamping parameter to suit our channel assignment algorithm and also the DPSO employed in this paper does not follow any social parameter since the graph is already divided into small subgraphs where the required information is exchanged and controlled by the leader of that group.

The remainder of the paper is formed as follows. We have discussed the related work in Section 2 and the network architecture in Section 3 and the assumptions made. In Section 4, we have formulated the problem. In Section 5, we have presented our algorithm and we have discussed the performance and comparison in Section 6. Finally, in Section 7 we have concluded the paper.

2. Related Work

There are many solutions proposed for the channel assignment problem by various researchers. But nevertheless it is a significant issue which requires further refinement to the solutions proposed, as it is an NP hard problem [28]. The authors of papers [22, 29] have done an extensive survey of the solutions provided by eminent researchers. We present a brief study done on the channel assignment problem in multiradio multichannel WMNs.

Subramanian et al. proposed a solution for quasistatic channel assignment problem [10]. Channel assignment is performed using a variation of the graph coloring problem. They have developed centralized and distributed algorithms which are based on heuristic search algorithm and greedy approximation algorithm, respectively. They have used a tabu search and Max k cut problems modelled as semidefinite programming and integer linear programming. They have enforced a constraint known as the interface constraint where the number of channels assigned to each node cannot exceed the number of interfaces it has. This approach considers the traffic load for solving the optimization problem and also preserves the topology. They have obtained a tight lower bound for optimal network interference. But the method proposed in the paper fails to consider the multiple-link case [22].

Avallone and Akyildiz proposed Maxflow-based Channel Assignment and Routing (MCAR) [11], a centralized algorithm that jointly considers the channel assignment and routing. Their proposed heuristic solution aims to maximize the throughput which is modelled as integer linear programming (ILP) as the problem is defined as an NP complete problem. The algorithm is split into link-group binding and group channel assignment. In the first step, links are grouped based on the flows they carry by enforcing a condition that a number of different groups assigned to its links do not exceed the number of radio interfaces. In the second step, the assignment is made to all the links in the group. They have performed extensive simulations and have indicated that their algorithm performs better when compared with Load Aware Channel Assignment (LACA) [5] and Balanced Static Channel Assignment (BSCA) [9] regardless of the traffic demands.

Kyasanur and Vaidya proposed a protocol named Probabilistic Channel Usage Based Channel Assignment (PCU-CA) [30] that defines each node to have both fixed and switchable interfaces. The fixed interfaces are employed to obtain the data transmitted by other nodes through fixed channels while the switchable interfaces change over to the fixed channels to transmit the information. Due to the handling of two interfaces, the channel assignment is done in two stages, namely, assigning channels to fixed interfaces and switchable interfaces. This algorithm has been devised to alleviate the channel oscillation problem and the authors also have proved it experimentally. But the algorithm does not provide an optimal solution for the network performance.

Raniwala et al. [5] proposed a joint routing and channel assignment algorithm named Hyacinth. The routing protocol establishes a tree topology with gateway nodes and mesh nodes, and then the channel assignment is carried out based on the topology. Multiple Network Interface Cards (NICs) are equipped in the 802.11 hardware. A set of NICs operates as UP-NICs and a set of NICs work as DOWN-NICs where the former connects to its parent and the latter connects to its children. Then, these interfaces are bound with their neighbors and then the channels are bound to these interfaces. Each node is responsible for binding only its DOWN-NIC with the channel since its UP-NICs are restricted to following its parent. They have used two metrics, namely, the number of links using this channel within the interference range and the aggregate traffic load on this channel from all links using this channel within the interference range.

Marina et al. [16] proposed a traffic independent approach for channel assignment named Connected Low Interference Channel Assignment (CLICA). The authors have proved that the problem is NP complete. They have formulated an ILP for finding the optimum lower bound. A greedy approach is used to assign channels based on priorities assigned to the nodes based on the Depth First Search (DFS) algorithm. Then, the nodes are traversed in the decreasing order to assign

channels. They have compared their algorithm with simple channel assignment scheme and showed that their algorithm is completed in polynomial time. But there are drawbacks in the algorithm as there is some sacrifice in the fairness and computation of unassigned radios which is added overhead.

In recent years, researchers have introduced game theoretical concept for solving the channel assignment problem. We have expounded some papers where authors have applied game theory in this section.

Bezzina et al. proposed an algorithm named Interference-Aware Game Based Channel Assignment Algorithm (IGCA) [31]. They have formulated the problem based on the game theoretic approach. The proposed algorithm aims to minimize the network interference and improves the number of simultaneous connections in the WMN backhaul network. They have modelled the problem as a common interest game where the mesh routers are termed as players. Duarte et al. proposed a near optimal channel assignment by utilizing partially overlapping channels rather depending on the orthogonal channels [32]. Near Optimal Partially Overlapping Channel Assignment (NPOCA) is modelled as a cooperative channel assignment game. Each player formalizes its own utility function based on the strategy it follows. Then, each player negotiates with other players on their nonindependent strategy. The channels are allocated to the radios as per the strategy. The drawback of their system is that the nodes with low probability during the negotiation phase are not identified. This low probability reduces the average performance of their system.

Nezhad and Cerdà-Alabern proposed an algorithm known as Semidynamic and Distributed Channel Assignment (SICA) [33]. This algorithm does not rely on the central node and assumes that nodes do not possess complete information. This algorithm is designed based on the online learner algorithm that assigns best channels to each radio. The nodes gather information during the channel sensing period and utilize this data for strategy formation. The authors have compared SICA and Urban-X [34] and concluded that their algorithm performs better. Yen et al. [35] proposed a two-stage channel allocation algorithm. Their algorithm assigns channels to radios using the game theoretic approach and then assigns the radio-channel pair to a link using a greedy method. Here, the interfaces are modelled rather than the mesh routers themselves. The proposed algorithm aims to minimize the interference. The authors have considered physical model to represent the interference. They have compared their approach with cooperative channel assignment game (CoCAG) [32] and concluded that their algorithm outperforms the latter.

Game theory has become a conjugal method for finding solutions to many problems. But not all of game theory's success stories are like that [36]. The players need to exchange their strategies and come to terms in order to find the solution for the problem. The dynamic nature of WMN makes the application of game theory a complex process. Each time when there is a change in the communication among the nodes the strategy has to be set and discussed. The next problem with game theory approach is that it is a greedy approach, where each player has his own maximization

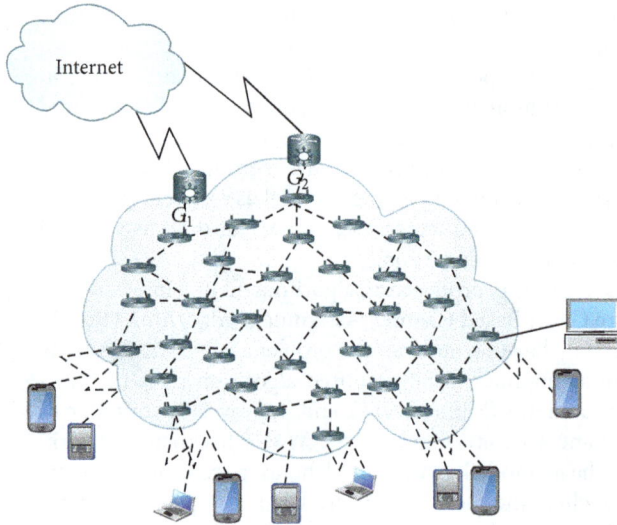

FIGURE 1: Wireless Mesh Network architecture.

goal [37] and tries to achieve his goal irrespective of other participants. The next major issue is providing security. Wireless networks are vulnerable to attacks, and hence game theory may have attackers disguised as players and negotiate their strategy with other participants, where the defender may not recognize that it is conferring with a malicious player. So, game theory requires more security.

Our algorithm considers a group assignment rather than the individual assignment unlike the existing channel assignment algorithms discussed above. We propose a distributed algorithm to avoid burden or bottleneck at one point. We also form groups based on the location, which will not change since the WMN nodes are usually static in nature. We have employed the Hybrid Wireless Mesh Protocol (HWMP) as the routing protocol (default protocol of IEEE 802.11s). Our algorithm is traffic independent; that is, channels are not assigned based on the traffic flow. In the following section, we formally describe our channel assignment method.

3. Network Architecture and Assumptions

This paper considers the WMN architecture shown in Figure 1. In this architecture, it is considered that all the backhaul nodes (middle layer nodes) and gateway nodes are equipped with multiple radios and each radio is capable of supporting multiple orthogonal channels. The number of orthogonal channels is outnumbered by the number of links. This clearly states that interference cannot be avoided in the wireless environment but can be minimized. In the proposed work, the links are grouped (identified by the Link Matrix M_L) so that the number of groups does not exceed the number of available channels of that particular node. The Enhanced Multiradio Multichannel Conflict Graph (E-MMCG) is applied to model the interference, according to the protocol model discussed in [22, 23]. This graph also considers the cochannel interference created by the radios. We have derived a mathematical model based on the job scheduling problem which is an operation research problem

[13]. The proposed optimization model is performed to minimize the interference by proper channel assignment among the nodes in the network. Generally, the channel assignment problem is an NP hard problem. Finding an exact solution to an NP hard problem is a time consuming procedure and sometimes it is not possible. Hence, we propose a heuristic approach for resolving the problem. The DPSO, a variation of the evolutionary algorithm PSO, is employed to solve the optimization problem. The DPSO provides an optimal solution or nearly optimal solution for assigning channels to the communication links.

We have made the following assumptions to define the problem:

(i) All the nodes in the network have the same transmission range T_r, detection range D_r, and interference range I_r. These three follow the inequality relation $T_r < D_r < I_r$. The Euclidean distance between nodes say u and v should be less than or equal to T_r; that is, $d(u, v) \leq T_r$.

(ii) All the nodes are in general position [38, 39] so that no three nodes are collinear and no four nodes fall on the same circle.

(iii) A node u can communicate with node v only if it shares the same channel for communication and if they are in each other's T_r.

(iv) There is more than one gateway node (MG) in the network. Each MG is capable of more storage capacity than the nodes in the backhaul network.

(v) The interference model we use in our work is the two-hop interference model. This model describes that two links interfere if they are sharing a common router [12, 40–42]. When a node is shared by two other nodes, the two links connecting them interfere if the same channel is assigned.

(vi) All the nodes need not have the same number of radios and support the same number of channels.

4. Problem Definition

The network architecture is depicted as an undirected graph $G(V, E)$ where V is the set of vertices of the graph defining the nodes and E is the set of edges defining the link between the nodes of the network architecture. The number of links in each node is greater than the number of available orthogonal channels; hence, interference is present. The interference in the network is represented as a protocol model according to [21]. According to the protocol model, two links say $a \leftrightarrow b$ and $c \leftrightarrow d$ interfere if the Euclidean distances $(d(a, c), d(a, d), d(b, c), d(b, d))$ are less than the transmission range T_r and $a, b, c, d \in V$ and $a \leftrightarrow b$ and $c \leftrightarrow d \in E$. We apply the Conflict Graph to model the interference. The graph is plotted in a Euclidean space $\mathbb{R}$ where each node in the graph knows its 2D coordinate position (x, y). With these coordinates positions each pair of nodes is capable of finding its Euclidean distance d. The Voronoi neighbors are derived based on this Euclidean distance d and the Voronoi diagram $V_D(G)$. The Delaunay Triangulation $D_T(G)$ gives the Connectivity Graph C_G among the neighbours which is a subgraph of $G(V, E)$. Then two matrices are constructed: one is the Link Matrix M_L which describes the interfering links and the second is the Interference Matrix I based on each node's assignment of channels to each link with respect to its radio. Then, the NLP model is formulated as shown in

$$\min \quad \sum_{j=1}^{C} \sum_{i=1}^{L} (M_L * I)_{ij} C_{ij} \tag{1}$$

$$\text{s.t} \quad \sum_{i=1}^{L} C_{ij} \le m_j, \quad j = 1, 2, \ldots, C \tag{2}$$

$$\sum_{i=1}^{L} C_{ij} = 1 \quad j = 1, 2, \ldots, C \tag{3}$$

$$M_L = \begin{cases} 1, & \text{for incident links} \\ 0, & \text{otherwise} \end{cases} \tag{4}$$

$$I_{lc} \ge 0 \quad \forall 0 < l \le L, \ 0 < c \le C, \tag{5}$$

where M_L ⇔ the incidence Boolean square matrix of size $L \times L$, represented using Conflict Graph. L ⇔ set of interfering links present in Conflict Graph C_c. C ⇔ set of orthogonal channels. I ⇔ Interference Matrix of size $L \times C$, computed using triangulation property (Thales theorem). m_j ⇔ the maximum number of k-orthogonal channels available depending on the standard. C_{ij} ⇔ the channel assignment matrix of size $L \times C$; I_{lc} ⇔ the value in Interference Matrix I in row l and column c.

This problem is similar to the optimization problem in standard job assignment problem or a weapon target assignment problem in operation research. The assignment problem is usually an NP hard problem for which we provide a swarm intelligence approach to obtain an optimal or near optimal solution.

The mathematical model formulated consists of an objective function as in (1). The value of this function defines the optimal allocation of channels to links. The goal is to reduce the number of interfering links which subsequently reduces the interference in the network and hence the objective function is defined as a minimization function. The function consists of a binary matrix M_L that describes the interfering links for any $l_i \in L$, where L stands for the set of vertices (which we later term as "link vertices") in the Conflict Graph C_C. The information of interfering links can be obtained from the Conflict Graph C_C as shown in Figure 3. The Interference Matrix I is computed based on the *Thales* theorem. The leader consolidates the number of times each channel $c_j \in C$ is assigned to a link while applying the triangulation property. Here, C represents the set of orthogonal channels available in the network. Multiplication of the Link Matrix and the Interference Matrix gives the matrix of size $L \times C$. This matrix demonstrates the number of channels assigned to a particular link and the number of times the same channel is assigned to that link. The matrix C_{ij} is the channel assignment solution matrix obtained from the particle's position matrix

$(X_k^{t+1}(i, j))$ with binary values, constructed by the DPSO algorithm for each iteration. Initially the values of C_{ij} represented as $(X_k^t(i, j))$ are randomly assigned and then updated for each iteration based on the velocity matrix $V_k^{(t+1)}(i, j)$. For each channel $c_j \in C$ assigned for each link $l_i \in L$, the value of the objective function is computed. If the current value is less than the previous value then the current solution is considered as the optimal solution.

Equations from (2)–(4) are the constraints that must be satisfied for a solution to be a feasible solution. The channels assigned for the links should not exceed the maximum number of orthogonal channels represented as m_j as shown in (2). The next condition in (3) states that each link must be assigned with a single channel. The matrix M_L is assigned with binary values based on the Conflict Graph C_C as shown in (4).

The construction algorithms of matrices and the solution are elaborately explained in the following sections.

5. Computational Geometric Based Channel Assignment

This section formally describes the proposed channel assignment scheme known as Computational Geometric Based Channel Assignment (CGCA). This method has been devised for the channel assignment problem in order to minimize the interference and hence increase the throughput of the network. The goal is to assign channels to each radio of the nodes in the backhaul network that will participate in the data communication. Hybrid Wireless Mesh Protocol (HWMP) has been used in proactive mode as the routing protocol [4]. Minimization problems can be viewed as a two-entry ordered tuple [43], where one entry provides a set of feasible solution S to the problem and the other entry is the objective function f of the problem. Then, for each solution say $s \in S$ the objective function value $f(s)$ is computed. The value $f(s)$ is compared with the previous value say $f(s')$ where s and $s' \in S$. If $f(s) < f(s')$ then the feasible solution $s \in S$ becomes the current optimal solution for the problem. The best solution for the objective function is obtained when the value of f does not change for further iterations. We propose an algorithm that works in two phases:

(i) The first phase identifies the neighboring nodes using computational geometric method. The Voronoi diagram has been applied to identify the neighboring nodes in the transmission range.

(ii) The second phase assigns channels to the radios of the nodes participating in data communication. To solve the channel assignment problem, the problem is represented as a nonlinear programming (NLP) model and the solution is obtained using the stochastic approach called the Discrete Particle Swarm Optimization (DPSO).

In the first phase, each node identifies its neighbors with the help of Voronoi diagram. Then, each node sends its group information like the number of members in the group and the nodes in the group to Mesh Gateway (MG). A node may be present in more than one group which is used as its membership number. A node having maximum number of memberships (i.e., a node present in a number of groups) is considered to be the leader of that group. Each leader is responsible for assigning and updating the channel information to the gateway. These small groups are considered for the channel assignment. A distributed channel assignment strategy is proposed as an NLP model where the objective function is defined as the minimization function. The goal of our programming model is to minimize the number of interfering links which subsequently reduces the interference. The swarm intelligence approach has been employed to solve the nonlinear equations. A detailed explanation is given in the subsequent sections.

5.1. Neighborhood Discovery and Leader Selection. The network is split into small subgraphs to think locally and act locally. To split the network into small subgraphs, Voronoi diagram is used. The Voronoi diagram is applied to the graph to find the neighbor nodes of any node $n \in N \wedge n \in G$. To compute the Voronoi diagram, the mesh routers are viewed and plotted as points in a planar graph. Voronoi diagram partitions the plane and the neighboring nodes are identified by the Voronoi edges which are shared with the neighboring nodes. This clustering is used for making the channel assignment distributed and localized. As a result, the convergence is much faster than assigning channels to the entire network.

We assume that each node knows its location in the space using GPS [44] and broadcasts its position to its neighbors. If the nodes know their position, then the distance between two nodes can be computed. The nodes are assumed to be in general positions so that no three nodes are collinear and no four nodes fall on the same circle [38, 39]. Each mesh router (MR) computes its own Voronoi cell and the entire Voronoi diagram delivers a graph with the neighbors close to each node.

Each node registers its neighborhood group with the gateway node MG. The gateway consolidates these clusters to identify some landmark nodes called leaders based on the occurrence of the node in the groups. That is, a node occurring in maximum number of groups is polled as the leader. These leaders take up the responsibility of assigning channels and transmitting the channel information to gateway MG. MG acts as the central server for storing the information so as to avoid packet loss and to improve the throughput of the network as it is the point which encounters maximum amount of traffic throughout the communication. This safeguards the data from signal degradation due to the interference present in the network.

Figure 2 explains the leader selection process for more clarity. The circles in the diagram represent the routers of the network. The node in red color is the special node called the gateway. The groups created by the node are named G_i where $i = 1, 2, \ldots, N$ and N is the number of nodes. The dotted lines around the nodes denote the group of each node created. For example, we have shown two groups created for node 1 (G_1) and node 2 (G_2). The nodes in groups G_1

and G_2 are computed as $G_1 = \{n_0, n_1, n_2, n_6\}$ and $G_2 = \{n_1, n_2, n_3, n_5\}$ by applying the algorithm in [18]. From this, the membership for nodes n_0, n_1, n_2, and n_6 is computed as 2, 1, 2, and 1, respectively, by the gateway node. All the nodes within the transmission range are considered to be neighbors of a particular node. Here, for simplicity, we have grouped only the one hop neighbors of nodes 1 and 2. The Voronoi diagram of group G_2 is shown in Figure 2(b). The thick lines are the bisectors between the points and the thin lines are the Delaunay Triangulation edges.

Each node $n_j \in N$, where $j = 0, 1, 2, \ldots, N$, is responsible for computing its group G_i, where $0 \leq i \leq |G|$. $G = \{G_0, G_1, \ldots, G_N\}$ is the set of groups in which the subscripts represent the nodes for which the group is formed; for example, G_0 is the group for the gateway node. The groups present in G are not disjoint; that is, a node may be present in more than one group. After the groups are registered, the gateway looks for the prospective leaders by checking the membership of the nodes present in the group. We determine the membership of each node by counting the number of times it is present in G. The maximum occurrence of a node n_j is computed using (6). Each time when a node n_j appears in any group $G_i \in G$ the membership $M(n_j)$ is incremented by 1. Then, MG sends the leader information to all the groups formed in the network:

$$M\left(n_j\right) = \sum_{G_i \in G \wedge n_j \in G_i} 1. \tag{6}$$

For example, we have taken group G_2 and assumed its leader to be node 2 as the node belongs to G_1 and G_2. In this leader selection process, some groups may not have leaders since all the nodes $n_j \in G_i$; they may be the leader of some other group. Thus, this group does not require any leader for itself.

Figure 2(b) shows the connectivity among the nodes that fall in the group formed by node 2. The Delaunay Triangulation (DT) edges show the nodes which share the same Voronoi edge and they are termed to be neighbors. The DT edges provide the connectivity between the nodes and the graph formed is called the Connectivity Graph C_G. Then, the channels are assigned to the radios of the nodes in these small groups $\{G_1, G_2, \ldots, G_N\}$ independently.

Algorithm 1 explains the selection of a leader in each group which is carried out by the MG node after the nodes register their groups with it.

The next step of our channel assignment procedure is to construct the link and interference matrices based on the protocol model to define the interference. Our process of modelling the interference is inspired by the E-MMCG algorithm [23] which models the radios as nodes instead of the routers themselves and also considers the cochannel interference. An elaborate description is given in the next section.

5.2. Channel Assignment and Matrices Construction. The Link Matrix M_L and Interference Matrix I are constructed based on the Connectivity Graph C_G, which is created by

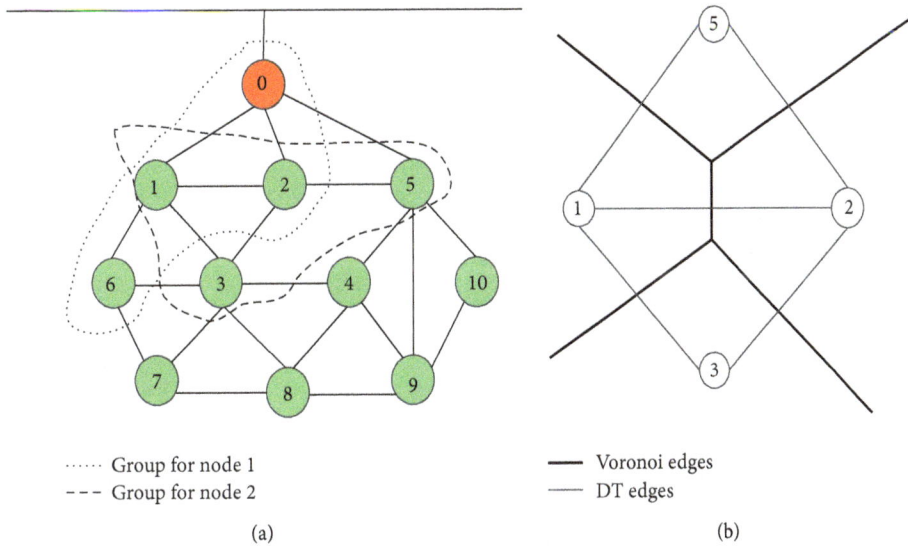

FIGURE 2: (a) Leader selection. (b) Voronoi diagram for node 2 (G_2).

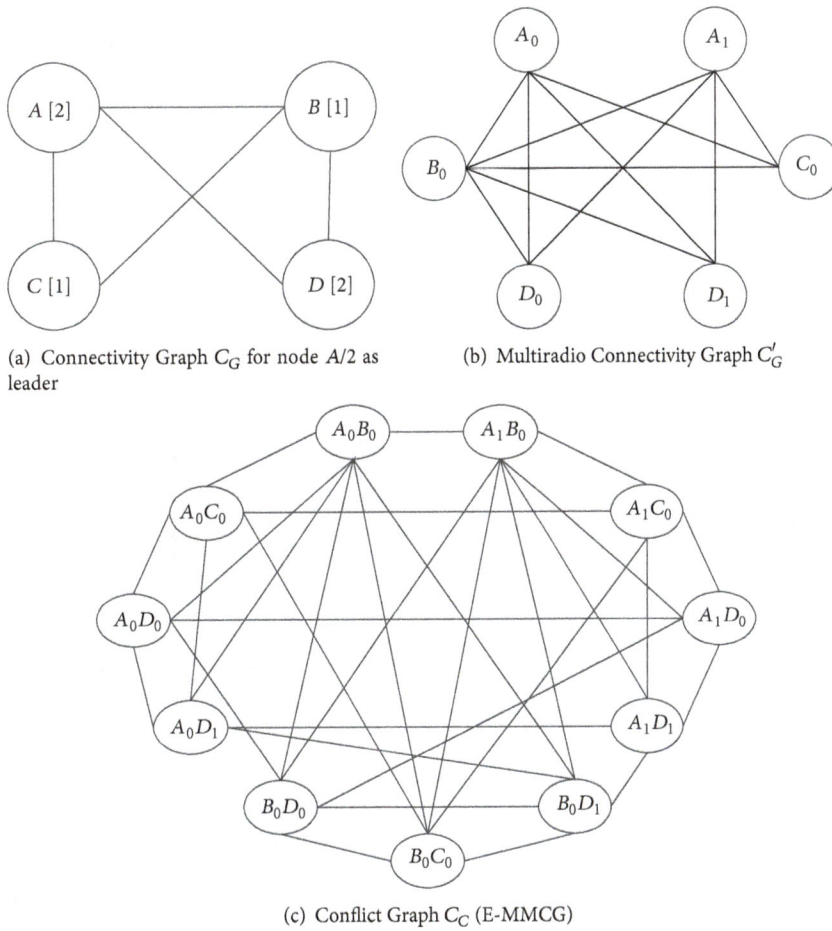

(a) Connectivity Graph C_G for node $A/2$ as leader

(b) Multiradio Connectivity Graph C_G'

(c) Conflict Graph C_C (E-MMCG)

FIGURE 3: Interference model.

Let G be the groups created by N nodes and registered with gateway MG $G = \{G_1, G_2, \ldots, G_N\}$
For each node $n_j \in N$
    For each Group $G_i = \{n_1, n_2, \ldots, n_j\}$
      $k = 0$
      $L = \phi$
      if $(\exists\, n_j \in N : n_j \text{ in } G_i)$
        $k = k + 1$
      $M(n_j) = \max(k, 1)$
    End for
    $L(G_i) = \max(M(n_1), M(n_2), \ldots, M(n_j)), \forall j \leq i$
    $L = L \cup L(G_i)$
End for

ALGORITHM 1: Leader selection.

applying the Delaunay Triangulation (DT) in the Voronoi diagram. The DT connectivity gives the possible neighbors of any node to form a group. Sections 5.2.1, 5.2.2, and 5.2.3 in this section explain the interference model and the steps involved in constructing the link M_L and interference I matrices.

5.2.1. Interference Model.

We have employed E-MMCG [23] to model the interference effect. We have redrawn the graph in Figure 2(b), the Connectivity Graph C_G with the alphabets, to distinguish it from the number of radios that each node contains (2-A, 1-B, 3-C, and 5-D) as in Figure 3(a). Figure 3(a) shows the subtopology generated after identifying the neighbors and the leader of that particular group. The circles represent the node and the number in the square brackets inside each circle represents the number of radios incorporated in that particular node. We represent the topology for group G_2 as graph $C_G(V_G, E_G)$, where V_G and E_G are the vertices and edges of the Connectivity Graph, respectively, for node 2 (A). Each node is incorporated with different number of radios which is represented in the square brackets.

An intermediate graph known as the Multiradio Connectivity Graph (MCG) $C_G'(V_G', E_G')$ is built, where the nodes in C_G are replaced with their radios and the links are established between the radios as the edges as shown in Figure 3(b). Graph C_G' provides the information of interfering radios and is used in constructing the MMCG graph C_C. Algorithm 2 shown below explains the procedure for converting $C_G \rightarrow C_G'$.

Each node in C_G is replaced with its radios; for example, in Figure 3, node A is replaced by its radios A_0 and A_1 and the edges are established between the radios. The Multiradio Connectivity Graph is used to create MMCG graph to model the interference and we have used the E-MMCG algorithm [23] to obtain the Conflict Graph named $C_C(V_C, E_C)$ as shown in Figure 3(c). V_C is the set of vertices named "link vertices" to differentiate it from the set of vertices in V_G'. It represents the link present between two radios/nodes in C_G'. E_C is the set of edges named "link edges" which is established between two link vertices if the interference subsists. The E-MMCG shown in Figure 3(c) includes the cochannel interference; that is, the radios present in a node may use the same channel which

Input: Connectivity Graph
Output: Multi-radio Connectivity Graph
Let $V_G = \{u_1, u_2, \ldots, u_g\}$
Each node $u_g \in V_G$ consists of radios $R_i \mid i = \{1, 2, \ldots, r\}$
$\forall V_G \in C_G$
    $\forall R_i \in u_g$
      Create nodes V_G'
    End for
End for
$\forall (u_i, u_j) \in V_G \bigwedge d(u_i, u_j) \leq T_r \bigwedge i \neq j \; //1 \leq i, j \leq g$
    if $(R_i \in u_i \bigwedge R_j \in u_j)$
      Create links $\emptyset$ R_i and R_j
    End if
End for

ALGORITHM 2: Multiradio Connectivity Graph $(C_G'(V_G', E_G')) \vdash$ Connectivity Graph $(C_G(V_G, E_G))$.

results in interference. Using the two graphs C_G' and C_C, the link and interference matrices are constructed.

5.2.2. Link Matrix.

Link Matrix M_L is a $L \times L$ matrix where $L \subseteq V_C \wedge V_C \in C_C$. The Link Matrix explains the number of links that interfere with a particular link. Construction of M_L is done using Algorithm 3 as described below. The degree of each link vertex belonging to C_C is computed using

$$D(v_C) = \sum_{k=2}^{L} 1 \mid \exists 1 \text{ in } M_L(j, k),$$

$$\text{where } j = 1, 2, \ldots, L \bigwedge \forall v_c \in V_C. \tag{7}$$

The degree of each "link vertex" is incremented by 1 for every link incident on that particular "link vertex" $v_c \in V_C$ present in the Conflict Graph C_C. The parameter L is the number of "link vertices" in the Conflict Graph.

5.2.3. Channel Assignment and Interference Matrix.

Interference matrix I is constructed by consolidating the number of times the same channel is used by the radios. First, we present

```
Input: Conflict Graph
Output: Link Matrix
∀i → 1 to L
    ∀j → 1 to L
        if (i == j)
            M_L[i][j] = 0
        else
            if (∃e_c ∈ E_C ∤ (i, j) ∈ L)
                M_L[i][j] = 1
            End if
        End if
    End for
End for
```

ALGORITHM 3: Link Matrix M_L.

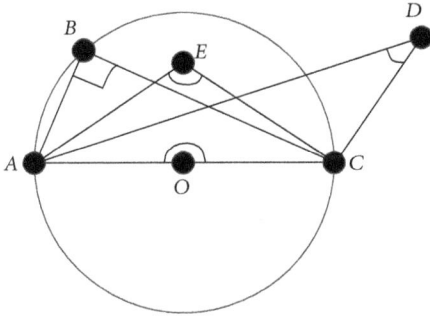

FIGURE 4: Thales theorem.

the mathematical background used to construct the matrix. Then, using this mathematical background, the steps involved in constructing Interference Matrix are described.

Mathematical Background. We have used Thales theorem and associated triangulation property for our derivation of the Interference Matrix I.

Theorem 1. *The triangulation property says that a circumcircle drawn with respect to the longer side of a triangle enforces three cases.*

Case 1. $\angle AOC = 2 \times \angle ABC$ | *Points A, B, and C lie on the perimeter of the circle with O as centre.*

Case 2. $\angle AEC > 90°$ | *Point E lies inside the circle.*

Case 3. $\angle ADC < 90°$ | *Point D lies outside the circle.*

Figure 4 is used to explain the theorem and all the three cases. The points on a circle with one chord $\overline{AC}$ as its diameter provide the relationship with angles and the point location. The theorem states that the angle created by the points on the perimeter of the circle is always a 90° regardless of the point location; that is, $\angle ABC = 90°$.

The other property related to triangulation states that the largest angle is always opposite to the largest side of the triangle. With reference to Figure 4, in $\triangle AEC$ $\angle AEC > 90°$,

TABLE 1: Skewness factor.

Value of skewness	Quality of the $\triangle$
$(S = \pm1)$	Degenerate
$0 < S < \pm1$	Bad
$S = 0$	Excellent

and hence the largest side is $\overline{AC}$. This property implicitly states that the sides holding the largest angle $\overline{AE}$ and $\overline{CE}$ come close to the largest side of the triangle $\overline{AC}$. This property is used to identify the edges that are close to each other. To determine and quantify this closeness factor of the edges we compute skewness S for the triangles using

Skewness (S)

$$= \frac{\text{Optimal triangle size} - \text{Actual triangle size}}{\text{Optimal triangle size}}. \quad (8)$$

The optimal triangle is the equilateral/equiangular triangle [45]. Both the optimal triangle and the actual triangle are computed as in

$$\triangle \text{ size} = \frac{1}{2} \times \det \begin{vmatrix} x_1 & x_2 & x_3 \\ y_1 & y_2 & y_3 \\ 1 & 1 & 1 \end{vmatrix}. \quad (9)$$

Using these properties, we group the links and the channels are assigned to the links by the radios based on the skewness value. Table 1 shows the quality of the triangle based on the skewness [45]. For simplicity, we consider only three cases. First case $(S = \pm1)$ happens when the triangle almost flattens and appears to be in parallel with the diameter of the circle, which can never be used (general position avoiding collinear coordinates). The interpretation of the negative skewness factor S is slightly different. In $+S$, the sides move close to the largest side of the $\triangle$ (Figure 4; for $\triangle AEC$ line segments $\overline{AE}$ and $\overline{CE}$ are coming close to $\overline{AC}$), but in $-S$, line segment $\overline{AD}$ moves closer to line segments $\overline{AC}$ and $\overline{CD}$ for $\triangle ADC$. When $S = 0.9$ to 0.1, we consider it to be bad as this case leads to sides coming closer to the diameter. The last case where $S = 0$ represents the optimal size of the triangle.

The triangulation property is applied to the Multiradio Connectivity Graph C_G'. These chords explained in Figure 4 are analogous to the links of the graph C_G'. Steps involved in grouping the links and assigning channels to the links are listed as follows.

Step 1. Identify the link vertex $v_c \in C_C$ for each radio $v_G' \in V_G'$ of the leader node.

Step 2. Compute the degree of v_c using (7).

Step 3. Identify the corresponding radios in C_G' for the incident links equal to $D(v_c)$.

Step 4. Check whether triangles are formed by the identified radios in C_G'.

Leaders in the graph LDR_m, where $m = \{1, 2, \ldots, m\}$, $m < N$
Number of radios per node R_r, where $r = \{1, 2, \ldots, r\}$ //r depends on the product
Number of k-orthogonal channels C, where $C = \{1, 2, \ldots, k\}$ //depends on the frequency spectrum of the radios
$R_{INL} = \emptyset$ //associated radios of incident links in C'_G
$I[V_C][\text{Channels}] = 0$ //rows = link vertices in C_C and columns = available orthogonal channels
Input: Multi-radio Connectivity Graph (C'_G) and E-MMCG (C_C)
Output: Interference Matrix I
$\forall LDR_m$
   $\forall R_{ldr}$
      Identify its corresponding link vertex $v_c \in V_C$ in C_C //arbitrary selection
      $d = D(v_c)$ //using (7)
      INL = set of link vertices incident to v_c
      $\forall INL \in C'_G$
         R_{INL} = Identify the associated radios in C'_G
         $\bigcup_1^d R_{INL}$
      End for
      Identify the triangles formed with other radios $r \in R_{INL}$ in C'_G with reference to R_{ldr}
      $\forall \triangle$
         Check the necessary condition //using Thales theorem
         Compute skewness S //using (8)
         $\triangle_{sides}$ = sides of the triangle
         $\forall R_{INL}$
            if ($S == 0$) //Assign same channel to all the links
               $\forall i \rightarrow 1$ to $|V_C|$
                  $\forall side \in \triangle_{sides} \bigwedge side \in V_C$
                     $I[i][j] = I[i][j] + 1$ //j may be assigned any arbitrary value between 1 to $|C|$
                  End for
               End for
            elseif ($0 < S < 1$)
               $\forall i \rightarrow 1$ to $|V_C|$
                  $\forall side \in \triangle_{sides} \bigwedge side \in V_C$
                    $j = k$
                    $I[i][j] = I[i][j] + 1$
                    $k = k + 1$
                  End for
               End for
            End if
         End for
      End for
   End for
End for

ALGORITHM 4: Interference Matrix I construction.

Step 5. Compute the skewness of the triangles.

Step 6. Based on the skewness value, assign channels to the links.

Step 7. If the skewness $S = 0$, then the leader assigns same channel to all the links as the triangle formed is an equilateral triangle.

Step 8. If the skewness $0.1 > S > 0.9$, then each radio assigns channel independently to all its links that form the triangle and sends the information to its leader.

Step 9. Leader consolidates the number of times the same channel is used for that particular link as the Interference Matrix.

Algorithm 4 explains the detailed procedure for constructing the Interference Matrix based on the steps described above.

Notations section explains the notations used in our algorithm.

5.3. Discrete Particle Swarm Optimization.

Particle Swarm Optimization (PSO) is a stochastic approach based on search algorithms to solve NP hard problem. Generally heuristic algorithms are employed to solve problems, where finding an exact solution is time consuming and sometimes not possible. Originally, the PSO was developed to solve a continuous problem, and then the discrete version was introduced by Kennedy and Eberhart [46]. PSO is used to solve the multidimensional problem and the convergence time taken is much less and the number of parameters to be tuned is minimal when compared with other evolutionary algorithms. The fundamental notion of PSO is the creation of a particle swarm and each particle provides a feasible solution for the problem. The solutions are evaluated based on some fitness function. There are some standard fitness functions, but in our proposed work, the fitness function is the objective function defined by the NLP model as in (1). During successive iterations, the algorithm calculates the objective function value for the feasible solution and assesses the fitness of the solution. After assessing the fitness and based on the optimization problem, the velocity and the position of a particle that are represented in (10) and (11) are updated. The velocity parameter reflects the experimental knowledge and socially exchanged information among the neighborhood:

$$V_{id}(t+1) = V_{id}(t) + \varphi_1 r_1 \left(P_{id} - X_{id}\right)$$
$$+ \varphi_2 r_2 \left(P_{gd} - X_{id}\right), \tag{10}$$

$$X_{id}(t+1) = X_{id}(t) + V_{id}(t+1). \tag{11}$$

In (10), there are three terms; the first term defines the momentum of the particle to retain its current direction itself. Equation (10) does not have the inertia weight constant in the original velocity parameter definition. The later version consists of an explicit inertia weight parameter (ω) which ranges from 0.8 to 1.2 defined in previous studies [47]. This inertia weight ω reduces the trade-off between the exploitation and exploration of the particle. The second term and the third term are called self-knowledge and social knowledge, respectively. The self-knowledge term is used to know how much the current position is close to its own best position and the social knowledge is used to know how much the current solution is close to its best neighborhood solution l_b or the global solution g_b. φ_1 and φ_2 are called the cognitive coefficient and social coefficient, respectively. The technical study done on the parameter of the PSO provides information on selecting the value for the positive constants $[\varphi_1, \varphi_2]$. Generally, the value of these two constants is 2 and their range is between 0 and 4. Utilizing the inertia weight and the two constants φ_1 and φ_2 together is a straight forward method in selecting the values for the parameters. Inequality (12) yields the relation between the three parameters as in [48]

$$\omega > \frac{1}{2}(\varphi_1 + \varphi_2) - 1. \tag{12}$$

This condition is used to avoid the cyclic behavior of the particles in the search space. There are different velocity models with the aim of providing variation in PSO. The velocity model we have employed in our work is the cognitive-only velocity model [48] with inertia weight. The social component is not considered in this model. The behavior of the particles tends to search locally rather than exploring globally. This behavior is applied to our local graph to assign channels locally. Since this model is slightly slower, to make it converge faster we have introduced the inertia weight to the velocity equation which is statically initialized and checked for the condition given in (12).

The topology of the swarm in PSO has an impact on the performance of the evolution process. There are different ways of communication done among the neighborhood inside a swarm. We have considered a star topology for the communication among the neighbors, since we have employed a leader based clustering to obtain the neighborhood. In our star topology, the central is selected based on the membership. In the proposed work, each radio in the leader suggests some feasible solution for assigning channels to the links in Algorithm 3. The leader then combines all the suggested solutions and the total impact of the assignment is fortified as the Interference Matrix I. Thus, there is no further requirement of the neighborhood particles n_b. Hence, the current particle position is initialized by the leader node which influences all the neighbor nodes. We use this model, since our search area is within the neighbors with a leader identified by the Voronoi diagram. The velocity equation of the cognitive model used in this work is represented in

$$V_k^{(t+1)}(i,j) = \omega V_k^t(i,j) + \varphi_1 r_1 \left(\text{Pb}_k^t(i,j) - X_k^t(i,j)\right). \tag{13}$$

In (13), $V_k^t(i,j)$ is the element in ith row (links) and jth column (channels) of the kth velocity matrix at time $t+1$. The particle is of D dimension and moves around the D dimension search space with a velocity parameter and each particle gives a feasible solution. The velocity clamping is used to restrict the particles movement. The velocity parameter is defined with D dimension in the range of $V_{\max} \rightarrow -V_{\max}$ to $+V_{\max}$ in order to avoid the particle moving too far away from search space and the value of $V_{\max}$ is preset based on the problem under consideration. Normally, $V_{\max}$ is in the range of $[-X_{\max}, +X_{\max}]$. But for our problem the position matrix is a binary matrix and hence we determine $V_{\max}$ by

$$V_{\max} = \left\lceil \frac{\text{NN} \times \text{NR}}{\text{NC}} \right\rceil, \tag{14}$$

where NN is the number of nodes in the cluster, NR is the number of radios in the cluster, and NC is the number of k-orthogonal channels available in the radio spectrum which is supported by the radios of each router. In the example graph shown in Figure 3(a), NN = 4, NR = 6, and NC = 3. The $V_{\max}$ value is $\lceil (4 \times 6)/3 = 8 \rceil$ and this fixes up the upper and lower bound of the velocity value and hence the range of velocity matrix is $[-8, 8]$ for the problem. The velocity can take values up to 16 [13]. Initially, the velocity matrix $V_k(i,j)$ is generated randomly at time t where i is the number of links in C_G' and j is the number of orthogonal channels. This matrix provides the information with respect to the particles movement. If $V_k^{t+1}(i,j) > V_k^t(i,j) \ \forall i, j$ it indicates that the particle's position $X_k^{t+1}(i,j)$ is close to the particle's best position. Otherwise,

Initialize the following parameters:
    Particle index k
    Particles best position matrix $\text{Pb}_k^t(i, j)$
    Particles current position $X_k^t(i, j)$
    Velocity clamping range $V_{\max}$
    Velocity matrix $V_k^t(i, j)$
    Inertia weight ω
    Cognitive constant φ_1
    Stochastic random number r_1
Initialize the D dimensional swarm with P particles
repeat
    $\forall$ particle $k \rightarrow 1$ to P
      Calculate $f(X_k^t(i, j)) \,\&\, f(\text{Pb}_k^t(i, j))$ // $f(\cdot)$ is the fitness function
      if $(f(X_k^t(i, j)) < f(\text{Pb}_k^t(i, j)))$
        $\forall i, j \leftrightarrow \text{Pb}_k^t(i, j) = X_k^t(i, j)$ //current assignment is best
      elseif $(f(X_k^t(i, j)) > f(\text{Pb}_k^t(i, j)))$
        $\forall i \in L \rightarrow 1$ to L
          $\forall j \in C \rightarrow 1$ to C
            if $(\text{Pb}_k^t(i, j) > X_k^t(i, j))$
              $V_k^{(t+1)}(i, j) = \omega \times V_k^t(i, j) + \varphi_1 \times r_1$
            else
              $V_k^{(t+1)}(i, j) = \omega \times V_k^t(i, j) - \varphi_1 \times r_1$
              if $(\max < V_k^{t+1}(i, j))$
                $\max = V_k^{t+1}(i, j)$
                $X_k^{t+1}(i, j) = 1$
              End if
            End if
          End for
        End for
      End if
    End for
Until

ALGORITHM 5: Discrete Particle Swarm Optimization for CGCA.

$X_k^{t+1}(i, j)$ is moving away from the best position and if there is no change it is already in the best position:

$$X_k^t(i, j)$$

$$= \begin{cases} 1, & \text{if } j\text{th channel is assigned to } i\text{th link} \\ 0, & \text{if there is no channel assignment,} \end{cases} \quad (15)$$

$$\text{Pb}_k^t(i, j)$$

$$= \begin{cases} 1, & \text{if } j\text{th channel is assigned to } i\text{th link} \\ 0, & \text{if there is no channel assignment.} \end{cases} \quad (16)$$

Equations (15) and (16) are used to initialize the particle position and its best position, respectively. In consecutive iterations, $V_k^{t+1}(i, j)$ is updated based on (13). Then, particle's position $X_k^{t+1}(i, j)$ is updated based on the updated velocity $V_k^{t+1}(i, j)$ using (17). In each iteration, if $f(X_k^t(i, j)) < f(\text{Pb}_k^t(i, j))$, where f is the fitness function, then the best position matrix is replaced by the current position. Otherwise, the velocity and position matrices are updated and the position is evaluated. This particle tends to move towards the best particle Pb to satisfy the fitness function. At any particular iteration, the particle reaches Pb, which becomes the termination condition of the algorithm. We have employed PSO for optimally allocating the k-orthogonal channels to the links in order to reduce the interference in the network, which will further increase the throughput of the data transmission. The particle position is updated by

$$X_k^{t+1}(i, j)$$

$$= \begin{cases} 1 & \text{if } \left(V_k^{(t+1)}(i, j) = \max\left\{V_k^{(t+1)}(i, j)\right\}\right), \quad \forall j \in \{1, 2, \ldots c\} \\ 0 & \text{otherwise.} \end{cases} \quad (17)$$

To update the position matrix, we identify the maximum value in each row and 1 is placed in that index of $X_k^{t+1}(i, j)$. Algorithm 5 shows the steps involved in optimizing our objective function. The solution to this optimization problem states the possible combination of channels that can be assigned to the links in order to minimize the interference.

6. Experimental Evaluation

Computational Geometry Based Channel Assignment (CGCA) is a traffic independent channel assignment

(a) 25 nodes' topology

(b) 50 nodes' topology

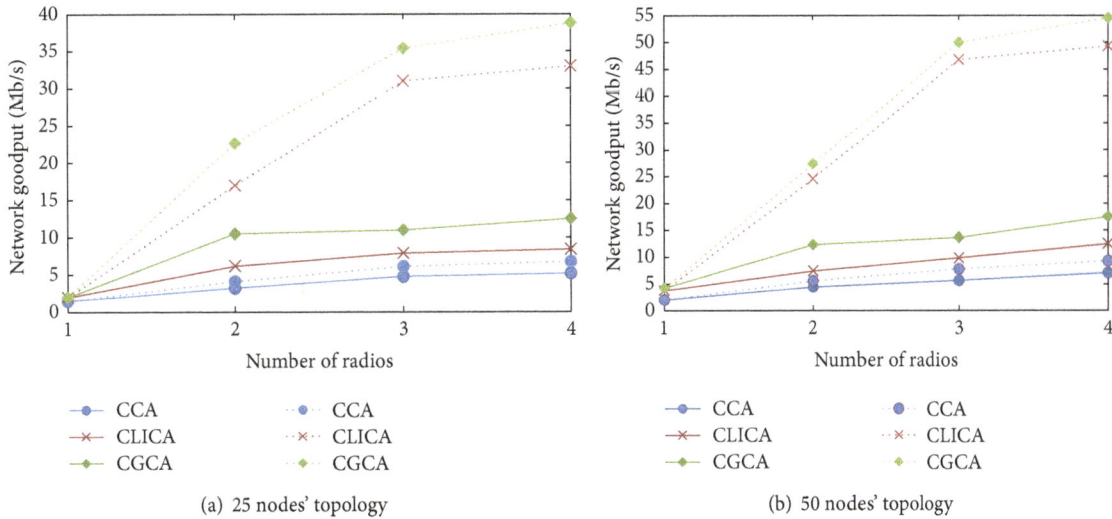

FIGURE 5: Radios versus goodput.

solution that attempts to achieve maximum throughput by minimizing the interference in the network. We have used the default routing protocol (HWMP) for routing and measured the performance of our system with Connected Low Interference Channel Assignment (CLICA) and Common Channel Assignment (CCA) [49]. CLICA uses greedy approach to assign channels. So, the algorithm has to compute the unassigned radios for further assignment. As the algorithm uses greedy approach, fairness is sacrificed [22]. CLICA methodology uses the nodes to construct the Conflict Graph for interference model and hence they do not consider interference caused by colocated radios. The proposed algorithm CGCA addresses the problems and hence performs better than CLICA. We performed simulations to evaluate the performance of CGCA and the results suggest that the distributed channel assignment performs well.

6.1. Simulation Environment. Our scheme is evaluated by simulation using the simulator tool Network Simulator [50] with an extension Hyacinth [51] to support multiple interfaces. We have used the HWMP, the default routing protocol of IEEE 802.11s. Table 2 shows the parameters used in our simulation environment.

6.2. Performance Analysis. CGCA presents an optimal or near optimal solution for each feasible solution. The performance shows that it minimizes the interference by optimally utilizing the available orthogonal channels in the network. In this subsection, we have used various parameters to evaluate our algorithm and have compared it with CLICA and CCA. The graphs show the comparison among all the three traffic independent channel assignment algorithms and show that our algorithm performs well compared to the other two algorithms.

Graphs in Figure 5 show the change of goodput of the network with increasing number of radios in 25 nodes' and 50

TABLE 2: Parameters used in simulation.

Parameter (s)	Value (s)
Simulation time	250 s
Node Distribution Model	Uniform 0, 100
Transmission range (Tr)	250 m
Interference range (Ir)	550 m
Network area size	$1000 \times 1000 \, \mathrm{m}^2$
Network	25 nodes' and 50 nodes' scenarios
MAC standard used	IEEE 802.11 n
Routing protocol	HWMP
Packet size	1500 bytes
Propagation Model	Two-rays
Data stream type	CBR
Mobility	Static nodes

nodes' topologies. Goodput is defined to be the ratio between the number of packets sent and the number of packets received per second. The performance is tested for both the 25 nodes' and the 50 nodes' scenarios eventually increasing the number of orthogonal channels from 3 to 12. From the results shown in graphs, it is observed that even by increasing the number of radios for each node it does not improve the goodput much, but increasing the number of orthogonal channels shows high improvement in the goodput. This is because the number of orthogonal channels C available in the network is limited [5]. This simulation study shows that CGCA and CLICA adjust themselves based on the number of available channels.

The results in Figure 6 show the measurement of average throughput over 25 nodes' and 50 nodes' topologies for different number of available channels. CGCA shows up to 50% and 15% increase in the throughput when compared with CCA and CLICA, respectively. The average throughput of CCA is less because the algorithm requires as many radios as the number of channels available. The throughput generally

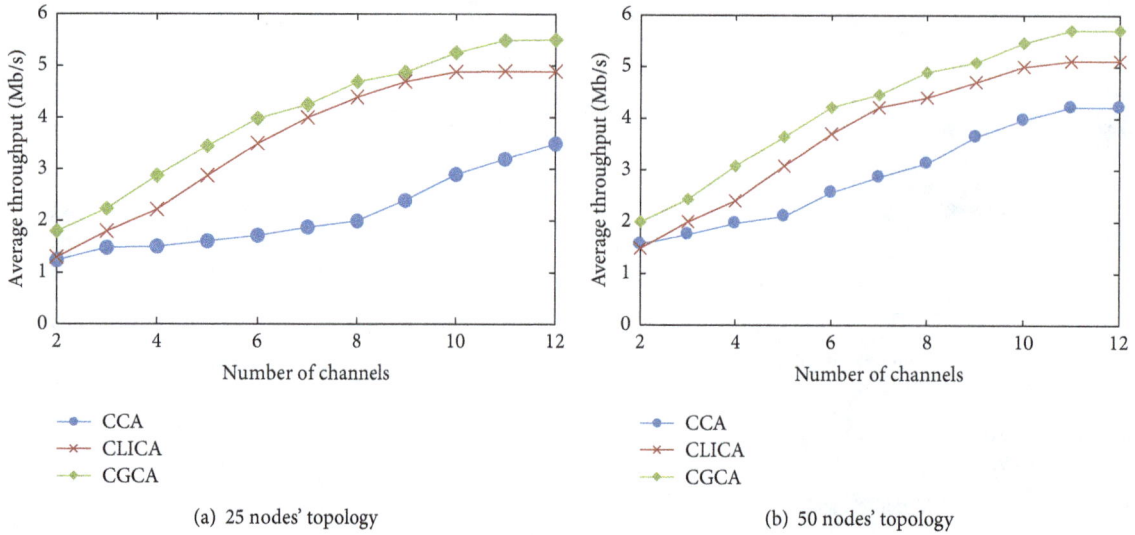

(a) 25 nodes' topology

(b) 50 nodes' topology

FIGURE 6: Orthogonal channels versus average throughput.

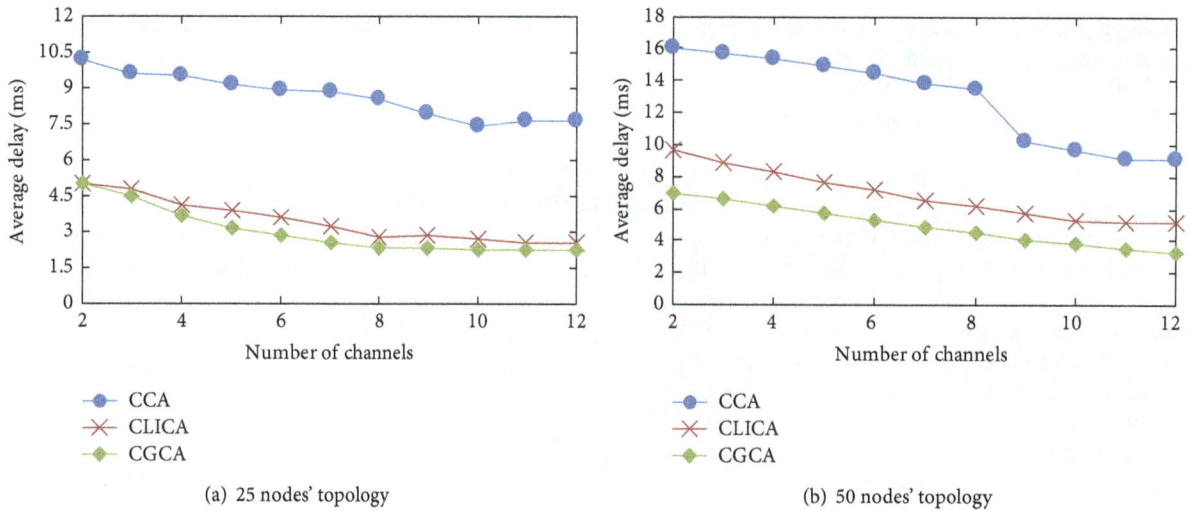

(a) 25 nodes' topology

(b) 50 nodes' topology

FIGURE 7: Channels versus average delay.

increases when the number of radios eventually increases. The throughput of CGCA has increased because the channels are assigned locally and each group has more number of channels. CGCA is not a greedy algorithm and in CGCA the links are assigned with channels by applying the triangulation property. The nodes forming a triangle independently choose any channel $c_j \in C$ if the triangle is not equilateral. The selection of the channel is random from the available set of orthogonal channels.

Figure 7 shows the average delay concerning the number of radios per node in the network which participate in forwarding the data. We have evaluated our algorithm for different traffic flows and the graph plotted against the number of radios and average delay illustrate that CGCA delay is reduced to 2.27 ms for 25 nodes' topology and 3.24 ms for 50 nodes' topology, respectively, when compared to CCA and CLICA. Delay is reduced in CGCA since the channel

switching delay is reduced due to the fact that there are more channels available for assignment and the CGCA is designed based on the static channel assignment strategy. As a result of this strategy, the channels for the links in the path for the data to flow through the network are predetermined which reduces the time taken for selecting and binding the channel with the links in the path.

The throughput of the proposed work is high because of the increase in the packet delivery ratio, which is evident from the results shown in Figure 8. The result of this experiment also indicates that the packet loss during the transmission of data is less when compared to the other two systems. We have also shown the result produced in both the scenarios. As the number of nodes increases, there is more assurance provided by the network in forwarding and receiving packets in WMN. But the problem here is that as the number of nodes increases the network incurs more interference. But

(a) 25 nodes' topology

(b) 50 nodes' topology

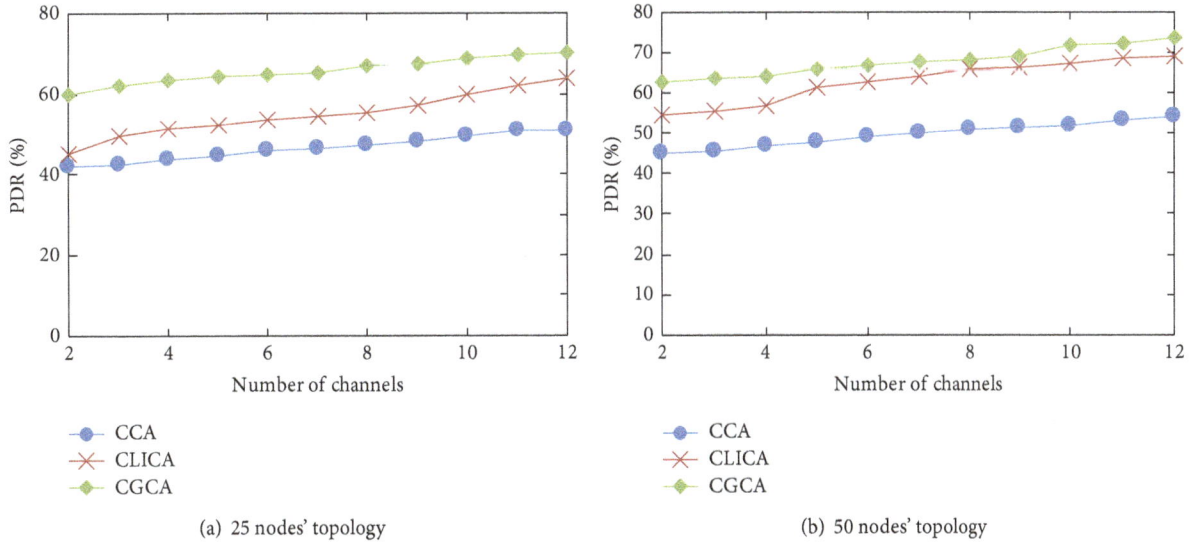

FIGURE 8: Channels versus packet delivery ratio.

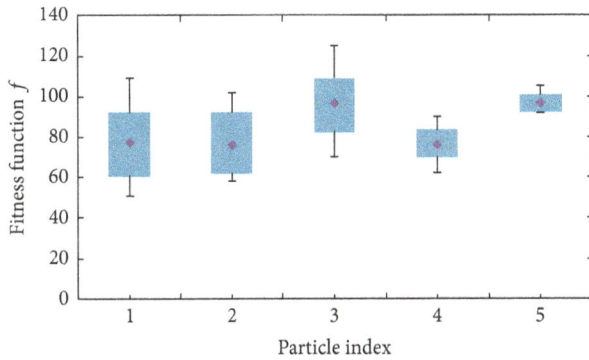

FIGURE 9: Particle indices versus fitness function.

the proposed algorithm minimizes the number of interfering links which leads to less interference in the network. Due to this, reduced interference PDR increases.

The box and whisker chart shown in Figure 9 explain how the fitness function varies based on particle P_k where $k = 1, 2, \ldots, 5$. We have evaluated the fitness function f (objective function) for 5 particle indices in 15 iterations. The stopping criterion for this experiment is the number of iterations. The mean value of the fitness function is plotted inside every box. The mean value lies between 76 and 96 approximately and does not exceed the range. This is because the particle position is fixed by the maximum velocity. The range for the velocity to vary is fixed by velocity clamping parameter. Velocity clamping ensures that the velocity does not fluctuate beyond the range.

Graph in Figure 10 shows the number of times a channel is used in each group. The number of groups considered for our experiment ranges from 4 to 10 formed using the Voronoi diagram as explained in Section 5.1. We tested our algorithm in both spectrums, namely, 2 GHz and 5 GHz, where the former has only 3 nonoverlapping channels and

the latter consists of 8 nonoverlapping channels. Both results in Figure 10 describe that the channels assigned to links are independent and do not depend on the traffic flows; that is, each radio independently chooses the channels irrespective of other radios. There is no control imposed on the radios for choosing the channels. The selection of channels need not follow any constraints unless the channel is supported by the standard and belongs to C. This is shown by the mean value (number of times the channel is used) of each channel that moves to and fro from high to low and low to high. The mean value of the same channel allocated in 5 GHz is less when compared to 2 GHz radio since the number of orthogonal channels is increased. This implicitly states that the increase in the orthogonal channels improves the performance of the network by reducing the interference and increasing the throughput. The best channel allocation is determined by the fitness function shown in (1).

The interference range is always greater than the transmission range for each node; that is, $I_r > T_r$. This helps the wireless nodes to transmit the data and sense the neighboring nodes; that is, say u and v are the nodes and the distance between them should be less than or equal to the transmission range $d(u, v) \le T_r$. But there is always a trade-off between the connectivity and the interference between the nodes in the network. Communication between nodes happens only if it falls in the transmission range of each other. If the connectivity is better, then the interference is more due to the signal overlap. This signal overlapping can be reduced to some extent by employing the nonoverlapping channels but cannot be completely avoided as the number of orthogonal channels is less. Figure 11 shows the probability of packet loss with respect to the interference range. The packet loss probability increases for increased interference range because when the interference range increases the links between the nodes interfere and the signal collides if links use the same orthogonal channel (which cannot be completely avoided). The packet loss probability also increases even if the

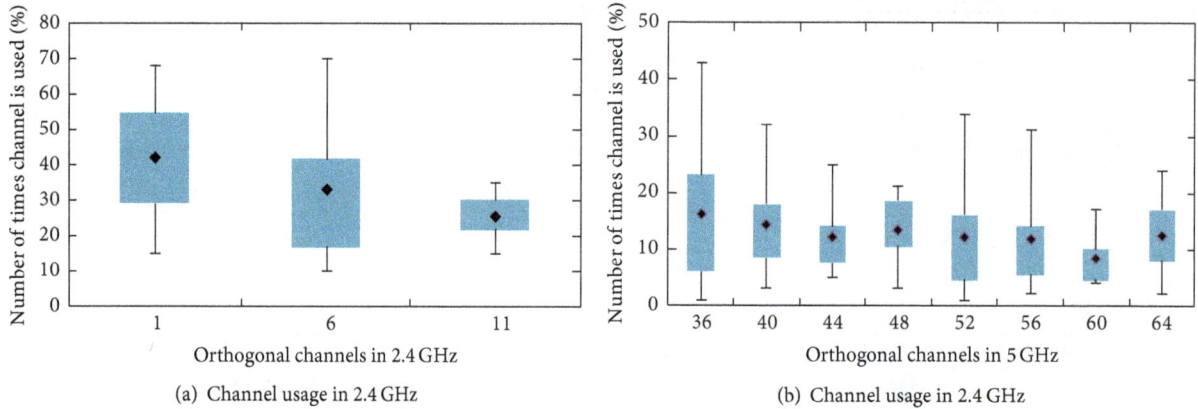

(a) Channel usage in 2.4 GHz

(b) Channel usage in 2.4 GHz

FIGURE 10: Orthogonal channels versus channel usage.

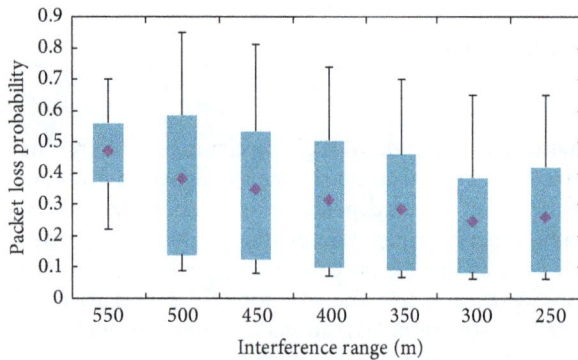

FIGURE 11: Interference versus packet loss.

interference range decreases since the connectivity between the nodes is reduced and hence the nodes are not able to transmit the packets completely.

The interference range is varied with respect to the transmission range. We have fixed the transmission range to be 250 m. The first box shows that the transmission range and interference range are same and the last box shows that the interference range is equal to transmission range. The mean value in the graph states that the packet loss is high when the interference range is too high and too low with respect to the transmission range of the node.

7. Conclusion

In this paper, we have formulated a programming model for the channel assignment problem and we have employed the stochastic approach to solve the problem. We have done the channel assignment for a small group rather than assigning for the entire network. The groups are created using the Voronoi diagram and then we use DPSO, a bioinspired algorithm which produces optimal or near optimal solutions, to solve our channel assignment problem. The channel assignment problem has been formulated as a nonlinear programming model. We have done an exhaustive experimental study using *NS2* simulation and observed that our algorithm achieves good throughput and less packet loss by optimally assigning the channels to the links. The control overhead is more in our algorithm during the leader selection process and also there is a lack of binding among the groups. Our future direction is to reduce the control overhead and achieve binding among the groups.

Notations

N:	Number of nodes in the network
M_L:	Link Matrix derived from the graph $C_C(V_C, E_C)$ of size $V_C \times V_C$
I:	Interference Matrix derived from the graphs $C'_G(V'_G, E'_G)$ and $C_C(V_C, E_C)$
$C_G(V_G, E_G)$:	Connectivity Graph with V_G vertices and E_G edges
$C'_G(V'_G, E'_G)$:	Multiradio Connectivity Graph with V'_G vertices and E'_G edges
$C_C(V_C, E_C)$:	Conflict Graph with V_C link vertices and E_C link edges
R_i:	Number of radios in each node
LDR_m:	Leader of mth group where $m = 1, 2, \ldots, m$ and $m \leq N$
R_{ldr}:	Radios in the leader node
R_{INL}:	Radios associated with the incident link vertices V_C
L:	Number of links $L \subseteq V_C$
INL:	Incident links for the leader
$D(v_c)$:	Degree of the node $v_c \in R_{ldr}$
$\triangle$:	Triangle
$\triangle_{sides}$:	Sides of the triangle
⅃:	Between
⊢:	Derived from
d:	Euclidean distance between nodes u and v
$\angle$:	Angle.

Conflict of Interests

The authors declare that there is no conflict of interests regarding the publication of this paper.

References

[1] I. F. Akyildiz, X. Wang, and W. Wang, "Wireless mesh networks: a survey," *Computer Networks*, vol. 47, no. 4, pp. 445–487, 2005.

[2] M. Seyedzadegan, M. Othman, B. Mohd Ali, and S. Subramaniam, "WirelesMesh networks: WMN overview, WMN architecture," in *Proceedings of the International Conference on Communication Engineering and Networks (IPCSIT '11)*, vol. 19, pp. 12–18, 2011.

[3] D. Benyamina, A. Hafid, and M. Gendreau, "Wireless mesh networks design—a survey," *IEEE Communications Surveys & Tutorials*, vol. 14, no. 2, pp. 299–310, 2012.

[4] IEEE Standard Association, *Part 11: Wireless LAN Medium Access Control (MAC) and Physical Layer (PHY) Specifications Amendment 10: Mesh Networking*, IEEE Standard Association, New York, NY, USA, 2011.

[5] A. Raniwala, K. Gopalan, and T. Chiueh, "Centralized channel assignment and routing algorithms for multi-channel wireless mesh networks," *ACM SIGMOBILE Mobile Computing and Communications Review*, vol. 8, no. 2, pp. 50–65, 2004.

[6] A. Raniwala and T.-C. Chiueh, "Architecture and algorithms for an IEEE 802.11-based multi-channel wireless mesh network," in *Proceedings of the 24th Annual Joint Conference of the IEEE Computer and Communications Societies (INFOCOM '05)*, vol. 3, pp. 2223–2234, Miami, Fla, USA, March 2005.

[7] Z. Zaichen and Y. Xutao, "A simple single radio multi-channel protocol for wireless mesh networks," in *Proceedings of the 2nd International Conference on Future Computer and Communication (ICFCC '10)*, vol. 3, pp. V3441–V3445, Wuhan, China, May 2010.

[8] S. Chieochan and E. Hossain, "Channel assignment for throughput optimization in multichannel multiradio wireless mesh networks using network coding," *IEEE Transactions on Mobile Computing*, vol. 12, no. 1, pp. 118–135, 2013.

[9] M. Kodialam and T. Nandagopal, "Characterizing the capacity region in multi-radio multi-channel wireless mesh networks," in *Proceedings of the 11th Annual International Conference on Mobile Computing and Networking (MobiCom '05)*, pp. 73–87, Cologne, Germany, September 2005.

[10] A. P. Subramanian, H. Gupta, S. R. Das, and J. Cao, "Minimum interference channel assignment in multiradio wireless mesh networks," *IEEE Transactions on Mobile Computing*, vol. 7, no. 12, pp. 1459–1473, 2008.

[11] S. Avallone and I. F. Akyildiz, "A channel assignment algorithm for multi-radio wireless mesh networks," *Computer Communications*, vol. 31, no. 7, pp. 1343–1353, 2008.

[12] K. N. Ramachandran, E. M. Belding, K. C. Almeroth, and M. M. Buddhikot, "Interference-aware channel assignment in multi-radio wireless mesh networks," in *Proceedings of the 25th IEEE International Conference on Computer Communications (INFOCOM '06)*, vol. 6, Barcelona, Spain, April 2006.

[13] H. Izakian, B. T. Ladani, A. Abraham, and V. Snášel, "A discrete particle swarm optimization approach for grid job scheduling," *International Journal of Innovative Computing, Information and Control*, vol. 6, no. 9, pp. 4219–4234, 2010.

[14] E. G. Coffman, *Computer and Job-Shop Scheduling Theory*, John Wiley & Sons, New York, NY, USA, 1976.

[15] R. Zhang and C. Wu, "Bottleneck machine identification based on optimization for the job shopscheduling problem," *ICIC Express Letters*, vol. 2, no. 2, pp. 175–180, 2008.

[16] M. K. Marina, S. R. Das, and A. P. Subramanian, "A topology control approach for utilizing multiple channels in multi-radio wireless mesh networks," *Computer Networks*, vol. 54, no. 2, pp. 241–256, 2010.

[17] K. Bong-Jun, V. Misra, J. Padhye, and D. Rubenstein, "Distributed channel assignment in multi-radio 802.11 mesh networks," in *Proceedings of the IEEE Wireless Communications and Networking Conference (WCNC '07)*, pp. 3978–3983, Kowloon, Hong Kong, March 2007.

[18] W. Alsalih, K. Islam, Y. Núñez-Rodríguez, and H. Xiao, "Distributed Voronoi diagram computation in wireless sensor networks," in *Proceedings of the 20th ACM Annual Symposium on Parallelism in Algorithms and Architectures (SPAA '08)*, p. 364, ACM, Munich, Germany, June 2008.

[19] B. A. Bash and P. J. Desnoyers, "Exact distributed Voronoi cell computation in sensor networks," in *Proceedings of the 6th International Symposium on Information Processing in Sensor Networks (IPSN '07)*, pp. 236–243, Cambridge, Mass, USA, April 2007.

[20] H. Chin Jang, "Applications of Geometric Algorithms to Reduce Interference in Wireless Mesh Network," *International Journal on Applications of Graph Theory in wireless Ad Hoc Networks and Sensor Networks*, vol. 2, no. 1, pp. 62–85, 2010.

[21] P. Gupta and P. R. Kumar, "The capacity of wireless networks," *IEEE Transactions on Information Theory*, vol. 46, no. 2, pp. 388–404, 2000.

[22] W. Si, S. Selvakennedy, and A. Y. Zomaya, "An overview of Channel Assignment methods for multi-radio multi-channel wireless mesh networks," *Journal of Parallel and Distributed Computing*, vol. 70, no. 5, pp. 505–524, 2010.

[23] S. M. Kala, R. Musham, M. P. Kumar Reddy, and B. Reddy Tamma, "Interference mitigation in wireless mesh networks through radio co-location aware conflict graphs," *Wireless Networks*, 2015.

[24] X. Zhuang, H. Cheng, N. Xiong, and L. T. Yang, "Channel assignment in multi-radio wireless networks based on PSO algorithm," in *Proceedings of the 5th International Conference on Future Information Technology (FutureTech '10)*, pp. 1–6, Busan, Republic of Korea, May 2010.

[25] R. C. Eberhart and J. Kennedy, "A new optimizer using particle swarm theory," in *Proceedings of the 6th International Symposium on Micro Machine and Human Science (MHS '05)*, vol. 43, pp. 39–43, Nagoya, Japan, October 1995.

[26] R. C. Eberhart and Y. Shi, "Particle swarm optimization: developments, applications and resources," in *Proceedings of the Congress on Evolutionary Computation*, vol. 1, pp. 81–86, IEEE, Seoul, Republic of Korea, May 2001.

[27] J. Kennedy and R. Eberhart, "Particle swarm optimization," in *Proceedings of the IEEE International Conference on Neural Networks*, vol. 4, pp. 1942–1948, Perth, Wash, USA, December 1995.

[28] A. Raniwala, K. Gopalan, and T. Chiueh, "Centralized channel assignment and routing algorithms for multi-channel wireless mesh networks," *ACM SIGMOBILE Mobile Computing and Communications Review*, vol. 8, no. 2, 2004.

[29] J. Crichigno, M.-Y. Wu, and W. Shu, "Protocols and architectures for channel assignment in wireless mesh networks," *Ad Hoc Networks*, vol. 6, no. 7, pp. 1051–1077, 2008.

[30] P. Kyasanur and N. H. Vaidya, "Routing and interface assignment in multi-channel multi-interface wireless networks," in *Proceedings of the IEEE Wireless Communications and Networking Conference (WCNC '05)*, vol. 4, pp. 2051–2056, IEEE, March 2005.

[31] A. Bezzina, M. Ayari, R. Langar, G. Pujolle, and L. Saidane, "Interference-aware game-based channel assignment algorithm for MR-MC WMNs," in *Proceedings of the IFIP Wireless Days (WD '14)*, pp. 1–6, IEEE, Rio de Janeiro, Brazil, November 2014.

[32] P. B. F. Duarte, Z. M. Fadlullah, A. V. Vasilakos, and N. Kato, "On the partially overlapped channel assignment on wireless mesh network backbone: a game theoretic approach," *IEEE Journal on Selected Areas in Communications*, vol. 30, no. 1, pp. 119–127, 2012.

[33] M. A. Nezhad and L. Cerdà-Alabern, "Adaptive channel assignment for wireless mesh networks using game theory," in *Proceedings of the 8th IEEE International Conference on Mobile Adhoc and Sensor Systems (MASS '11)*, pp. 746–751, Valencia, Spain, October 2011.

[34] W. Kim, A. J. Kassler, M. Di Felice, and M. Gerla, "Urban-x: towards distributed channel assignment in cognitive multi-radio mesh networks," in *Proceedings of the IFIP Wireless Days (WD '10)*, Venice, Italy, October 2010.

[35] L.-H. Yen, Y.-K. Dai, and K.-H. Chi, "Resource allocation for multi-channel multi-radio wireless backhaul networks: a game-theoretic approach," in *Proceedings of the IEEE Wireless Communications and Networking Conference (WCNC' 13)*, pp. 481–486, Shanghai, China, April 2013.

[36] http://www.symposium-magazine.com/game-theory-is-useful-except-when-it-is-not-ariel-d-procaccia.

[37] N. M. Dhande and V. S. Kapse, "Interference aware channel assignment in WMN using a game theoretic approach," *International Journal of Engineering Research & Technology*, vol. 2, no. 9, 2013.

[38] S. C. Nandy, *Advanced Computing and Microelectronics*, Lecture notes on Voronoi Diagrams, Unit Indian Statistical Institute, Kolkata, India, http://www.tcs.tifr.res.in/~ghosh/subhas-lecture.pdf.

[39] M. De Berg, O. Cheong, M. van Kreveld, and M. Overmars, "Voronoi diagrams," in *Computational Geometry Algorithms and Applications*, Springer, 3rd edition, 2008.

[40] Y. Yang, J. Wang, and R. Kravets, "Interference-aware load balancing for multihop wireless networks," Tech. Rep. UIUCDCS-R-2005-2526, Department of Computer Science, University of Illinois at Urbana-Champaign, 2005.

[41] J. Padhye, S. Agarwal, V. N. Padmanabhan, L. Qiu, A. Rao, and B. Zill, "Estimation of link interference in static multi-hop wireless networks," in *Proceedings of the 5th ACM SIGCOMM Conference on Internet Measurement (IMC '05)*, p. 28, Berkeley, Calif, USA, October 2005.

[42] R. Draves, J. Padhye, and B. Zill, "Routing in multi-radio, multi-hop wireless mesh networks," in *Proceedings of the 10th Annual International Conference on Mobile Computing and Networking (MobiCom '04)*, pp. 114–128, ACM, Philadelphia, Pa, USA, October 2004.

[43] R. Mendes, *Population topologies and their influence in particle swarm performance [Ph.D. thesis]*, University of Minho, Braga, Portugal, 2004.

[44] B. Hofmann-Wellenhof, H. Lichtenegger, and J. Collins, *Global Positioning System: Theory and Practice*, Springer, 4th edition, 1997.

[45] https://www.sharcnet.ca/Software/TGrid/html/ug/node380.htm.

[46] J. Kennedy and R. C. Eberhart, "A discrete binary version of the particle swarm algorithm," in *Proceedings of the IEEE International Conference on Systems, Man, and Cybernetics*, vol. 5, pp. 4104–4108, IEEE, Orlando, Fla, USA, October 1997.

[47] A.-Q. Mu, D.-X. Cao, and X.-H. Wang, "A modified particle swarm optimization algorithm," *Natural Science*, vol. 1, no. 2, pp. 151–155, 2009.

[48] A. P. Engelbrecht, *Computational Intelligence—An Introduction*, Wiley, 2nd edition, 2007.

[49] A. Adya, P. Bahl, J. Padhye, A. Wolman, and L. Zhou, "A multi-radio unification protocol for IEEE 802.11 wireless networks," in *Proceedings of the IEEE 1st International Conference on Broadband Networks (BroadNets '04)*, pp. 344–354, October 2004.

[50] http://www.isi.edu/nsnam/ns/.

[51] http://personales.unican.es/aguerocr/.

10

An Enhanced Distributed Scheme for WSNs

Tarek R. Sheltami

Computer Engineering Department, King Fahd University of Petroleum and Minerals, Dhahran 31216, Saudi Arabia

Correspondence should be addressed to Tarek R. Sheltami; tarek@kfupm.edu.sa

Academic Editor: David Taniar

This paper investigates data processing schemes that define the distribution of decision making that affects system accuracy and energy consumption. There exist two typical schemes, namely: centralized and distributed schemes. In a centralized scheme, nodes collect samples and send them to a "fusion center." This scheme provides optimal decision accuracy; however, it consumes considerable energy. In contrast, distributed schemes allow nodes to make local 1-bit decisions, which are sent to the fusion center to make the final decision. In a hybrid scheme, the network specifies the level of accuracy required for the whole system. This can be achieved by manipulating the scheme to work interchangeably as centralized or distributed. Most of the energy consumed is in the transmission process; therefore, this paper proposes an energy-saving hybrid scheme that focuses on optimizing transmission energy. In this proposed scheme, each node is able to alternate between centralized and decentralized scheme according to its location and path length. To validate the proposed approach, it is simulated and the results are compared with the hybrid scheme.

1. Introduction

The main goal of WSNs is to detect certain events in the environment. It is important to try to achieve maximum detection accuracy and to minimize false alarms. At the same time, availability of WSN resources and accessibility limitations should be considered. The solution is a trade-off between two factors: accuracy and energy efficiency. With sensor capabilities a tremendous societal benefit is achieved when sensors are integrated into available devices, machines, and environments. They can help to avoid infrastructure failure disasters, protect precious natural resources, enhance security, and enable new "smart" applications such as context-aware systems and home technologies. Wireless sensor networks are based on numerous advanced technologies such as very large scale integration (VLSI), microelectromechanical systems (MEMS), and wireless communications. The development of these technologies is contributing to a wider application of WSNs. For example, with the enhancement of MEMS technology, sensors are becoming smaller, and developments in semiconductor technologies are producing smaller microprocessors with higher processing capacities. The improvement of computing and sensing technologies is

enabling the development of flexible WSNs, which can be widely applied [1, 2].

Monitoring environmental changes and detecting specified events is the main function of sensor networks. This function is achieved through four basic components of a sensor network [3]: distributed or localized sensors, an interconnecting network (most often wireless based), a central point of information clustering a set of computing resources at the central point or network core to handle data collecting, event trending, status querying, and data mining. WSNs use centralized fusion centers (sinks), which work as cluster gateways, and many distributed sensors (motes) [4]. These sensors sense and send observations to the centralized unit. The centralized unit decides if an event is initiated or not.

Most of the power consumed in a network is used in processing, transmitting, and sensing. Until now limited power resources for sensors has been the main constraint in WSNs. It is very important to reduce sensor power consumption while maintaining acceptable detection accuracy according to application requirements. Many researchers have focused on the above three processes [5], attempting to enhance the power consumption efficiency of the sensors for each of them. Some schemes enhance the operating system and reduce

TABLE 1: Wireless communication protocols.

Standards	IEEE 802.15.4 Low power wireless personal area network (LP-WPAN)				IEEE 802.51.1 (WPAN)	IEEE 802.11 (WLAN)	IEEE 802.16 (WWAN)
	ZigBee	6LoWPAN	WirelessHART	ISA 100.11a	Bluetooth	WiFi	WiMAX
Range in meters	100	50			100	100	15 Km
Data rate	250–500 Kpbs	250 Kbps	250 Kbps	250 Kbps	1–3 Kbps	250 Kbps	250 Kbps
Frequencies	2.4 GHz	2.4 GHz	2.4 GHz	2.4 GHz	2.4 GHz	2.4, 3.7 and 5 GHz	2.3, 2.5 and 3.5 GHz
Network topology	Star, mesh, cluster tree	Star, mesh, cluster tree IPv6	Star, mesh, cluster tree	Star, mesh, cluster tree	Star	Star, tree, P2P	Star, tree, P2P
Applications	WSNs (monitoring and control)	WSNs, Internet, automation and entertainment	Wirelessly process monitoring and control application	Wireless systems of automation	WSNs (monitoring and control)	PC-based data acquisition mobile Internet	Mobile Internet

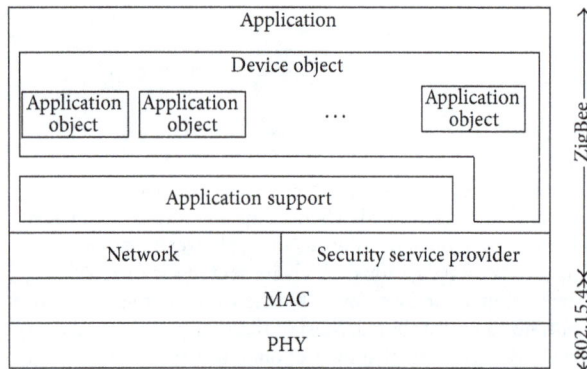

FIGURE 1: ZigBee protocol stack.

FIGURE 2: Elements of a typical WirelessHART installation.

the required processing cycles; other schemes optimize the RF part including collision space and noise filtering. This thesis focuses on schemes which study decision processing and transmitting where those schemes define how to collect observations (sampling rate), where to process them (locally or centralized), and the data to be sent from nodes to the fusion center, which will affect the degree of loss of data and accuracy [6]. The main goal of this paper is to produce an optimum controlling scheme that extends network/sensor lifetime by reducing power consumption and maximizing the network efficiency and accuracy. Our proposed scheme balances between the reduction of data transmission and processing by distributing these two activities among the nodes and the central unit (sink).

Existing wireless communication protocols are shown in Table 1, including IEEE 802.11 (WiFi), IEEE 802.15.1 (Bluetooth), IEEE 802.15.4 (ZigBee, 6LoWPAN, WirelessHART, and ISA-100.11a), and IEEE 802.16 (WiMAX).

The main WSN solutions such as ZigBee, WirelessHART, 6LoWPAN, and ISA-100 are based on this standard, which offers a complete networking solution by developing the remaining upper communication layers.

ZigBee. ZigBee [7–9] is a simple, low cost, low power wireless technology used in LR-PANs embedded applications. ZigBee

provides the network layer and the framework for the application layer. The MAC sublayer and lower layers are based on the IEEE 802.15.4 standards (see Figure 1).

The ZigBee network layer supports star, tree, and mesh topologies. It is utilized in three types of devices: ZigBee coordinator, ZigBee routers, and end devices.

WirelessHART. Based on IEEE 802.15.4 standard highway addressable remote transducer (HART) foundation, WirelessHART [10–12] was developed for low-power 2.4 GHz operation. Similar to ZigBee, WirelessHART specifies four principal devices (Figure 2): network manager, gateways, field devices, and handhelds, as well as adapters, which allow existing HART field devices to be integrated into the network.

6LoWPAN. 6LoWPAN [13, 14] is the abbreviation of IPv6 over low power wireless personal area networks, in which standard IPv6 packets are enabled to communicate over an IEEE 802.15.4-based network. Using 6LoWPAN, low power devices have all the benefits of IP communication and management. It is targeted at wireless IP networking applications in home, office, and factory environments.

ISA100.11a. The core wireless technology employed by ISA100.11a [15] is IEEE 802.15.4, which sorts the 2.4 GHz unlicensed band into 16 channels.

As shown in Table 6, the network layer is based on IETF RFC4944 (6LoWPAN):

(i) IP connectivity through compressed IPv6 and UDP packets,

(ii) addressing scheme:

    (a) EUI-64 (64 bits),

    (b) IPv6 (128 bits),

    (c) short address (16 bits IEEE 802.15.4).

The entire ISA100.11a stack is constructed employing widely accepted and proven industry standards.

2. Related Work

As mentioned earlier, each node has the responsibility of collecting, processing, transmitting, and receiving data. The common functions of every detection scheme are that (1) nodes collect observations, and (2) the fusion center (sink) takes the final decision.

There are two traditional detection schemes: the centralized detection scheme and the distributed detection scheme, the methodologies of which will be covered in details. In our approach, we use a tree topology where nodes are "independently and identically distributed" (i.i.d.). They are connected to the Fusion Center (FC) through a multihop route, where nodes also act as hops to receive data from child nodes and forward it to the FC with any processing, encryption or encoding. The focus here is only on accuracy and energy consumption, and it is assumed that lower layers are working perfectly and that there are no efficiency problems caused by RF or by packet collisions; that is, there is no data retransmission.

2.1. Bayesian Decision Theory

2.1.1. Binomial Distribution.
A Bernoulli trial can result in a success with probability p and a failure with probability $q = 1 - p$. Equation (1) gives the probability distribution of the binomial random variable x. The number of successes in n independent trials is

$$f(x; n, p) = \Pr(x \mid p) = \binom{n}{x} p^x (1-p)^{n-x}. \tag{1}$$

The probability that two events, A and B, will both occur will be $P(A \cap B) = P(B \cap A) = P(B)P(A \mid B) = P(A)P(B \mid A)$. From this formula the main equations (2) of Bayes' rule can be derived:

$$P(A \mid B) = \frac{P(A) P(B \mid A)}{P(B)},$$
$$P(A) = \sum_{i=1}^{k} P(B_i \cap A) = \sum_{i=1}^{k} P(B_i) P(A \mid B_i). \tag{2}$$

From Figure 3 we find $P(E \cap A) = P(E)P(A \mid E) = (2/3)(3/50) = 0.04$.

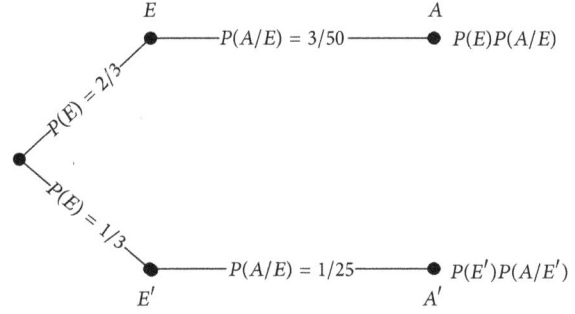

FIGURE 3: Conditional probability example.

The distribution of H, given D, which is called the posterior distribution, where $P(D)$ is the marginal distribution of D, is given by

$$P(H \mid D) = \frac{P(D \mid H) P(H)}{P(D)}. \tag{3}$$

Binary assumption is defined by H, which represents event occurrence: $(H = H_1)$ if an event happens and $(H = H_0)$ if not; $\widehat{H}$ is the actual event status: H' is the final decision; K is node counts; D is the collected samples, T is sample count at each node, and L is the node path length. For each sample S_{ij} where i is varied from 1 to K and J from 1 to T collected in a node i, $P(S_{ij} = 1 \mid \widehat{H} = H_1) = p_1$, $P(S_{ij} = 1 \mid \widehat{H} = H_0) = p_0$.

Bayes Decision. Choose event happened if $P(H_1 \mid D) \geq P(H_0 \mid D)$; otherwise choose not happened.

2.2. Centralized Detection Schemes.
In centralized detection schemes a network will have K number of nodes. These nodes will collect T samples of observations from the environment every specific period, and they will send T samples together at the end of the period. At the fusion center $D = [T * K]$ samples will be received.

According to Bayes, a final decision can be calculated as shown in the following:

$$\widehat{H} = \begin{cases} 1, & P(H_1 \mid D) \geq P(H_0 \mid D) \\ 0, & \text{otherwise.} \end{cases} \tag{4}$$

The probability of error can be calculated using the following equation:

$$P_e = \left(p * P_{\text{false positive}}\right) + \left((1-p) * P_{\text{false negative}}\right)$$
$$\longrightarrow P_e = p * P\left[H' = H_1 \mid H_0\right]$$
$$+ (1-p) * P\left[H' = H_0 \mid H_1\right] \tag{5}$$
$$\longrightarrow P_e = p * \left[1 - P(n \geq \gamma_{\text{FC}} \mid H_1)\right]$$
$$+ (1-p) * P(n \geq \gamma_{\text{FC}} \mid H_0).$$

To calculate power consumption for the whole network in (6), the following equation can be used, where E = total energy, E_T = transmission energy, E_R = receiving energy, and E_P = processing energy:

$$E = E_T + E_R + E_P$$

$$\longrightarrow E = \sum_{i=1}^{K} \left(L_i * T * e_t\right) + \sum_{i=1}^{K} \left(L_i * T * e_r\right) \tag{6}$$

$$\longrightarrow E = \sum_{i=1}^{K} \left(L_i * T * \left(e_t + e_r\right)\right).$$

2.3. Distributed Detection Scheme. In this scheme nodes collect data and make local decisions according to these observations and conclude the event appearance as a 1-bit result. This result is sent to the fusion center to make a final decision according to the collected 1-bit results from all nodes. In this scheme data accuracy between nodes and the fusion center has been lost.

We propose K number of nodes; these nodes will collect T samples of environmental observations with (n_i = number of 1s for node i) and will send a 1-bit local decision every specific period.

At the fusion center, $D = [1 * K]$ samples will be received.

According to Bayes, a local decision can be calculated as shown in the following:

$$\widehat{H_i} = \begin{cases} 1, & P\left(H_1 \mid n_i\right) \geq P\left(H_0 \mid n_i\right) \\ 0, & \text{otherwise.} \end{cases} \tag{7}$$

From (1) and (4), we can calculate local decision as shown in the following:

$$P\left(H_1 \mid n_i\right) \geq P\left(H_0 \mid n_i\right)$$

$$\longrightarrow \frac{P\left(n_i \mid H_1\right) P\left(H_1\right)}{P\left(n_i\right)} \geq \frac{P\left(n_i \mid H_0\right) P\left(H_0\right)}{P\left(n_i\right)}$$

$$\longrightarrow \frac{P\left(n_i \mid H_1\right)}{P\left(n_i \mid H_0\right)} \geq \frac{P\left(H_0\right)}{P\left(H_1\right)}$$

$$\frac{\binom{T}{n_i} p_1^{T} \left(1 - p_1\right)^{T-n_i}}{\binom{T}{n_i} p_0^{T} \left(1 - p_0\right)^{T-n_i}} \geq \frac{\left(1 - p\right)}{p} \tag{8}$$

$$\longrightarrow n_i \geq \frac{\ln\left(\left(1 - p\right)/p\right) + T \ln\left(\left(1 - p_0\right)/\left(1 - p_1\right)\right)}{\ln\left(p_1\left(1 - p_0\right)/p_0\left(1 - p_1\right)\right)}$$

$$= \gamma_{\text{local}}$$

$$\longrightarrow \widehat{H} = \begin{cases} 1, & n_i \geq \gamma_{\text{local}} \\ 0, & \text{otherwise.} \end{cases}$$

For final decision we collect b = total 1s if local decision

$$\widehat{H} = \begin{cases} 1, & P\left(H_1 \mid b\right) \geq P\left(H_0 \mid b\right) \\ 0, & \text{otherwise.} \end{cases} \tag{9}$$

From (1) and (4), we can calculate final decision as shown in the following:

$$P\left(H_1 \mid b\right) \geq P\left(H_0 \mid b\right)$$

$$\longrightarrow \frac{P\left(b \mid H_1\right) P\left(H_1\right)}{P\left(b\right)} \geq \frac{P\left(b \mid H_0\right) P\left(H_0\right)}{P\left(b\right)}$$

$$\longrightarrow \frac{P\left(b \mid H_1\right)}{P\left(b \mid H_0\right)} \geq \frac{P\left(H_0\right)}{P\left(H_1\right)},$$

$$P_D = \sum_{i=\gamma_{\text{local}}}^{T} \binom{T}{i} p_1^{T} \left(1 - p_1\right)^{T-i},$$

$$P_F = \sum_{i=\gamma_{\text{local}}}^{T} \binom{T}{i} p_0^{T} \left(1 - p_0\right)^{T-i},$$

$$\frac{\binom{K}{b} p_D^{K} \left(1 - p_D\right)^{K-b}}{\binom{K}{b} p_F^{K} \left(1 - p_F\right)^{K-b}} \geq \frac{\left(1 - p\right)}{p}$$

$$\longrightarrow b \geq \frac{\ln\left(\left(1 - p\right)/p\right) + K \ln\left(\left(1 - p_F\right)/\left(1 - p_D\right)\right)}{\ln\left(p_D\left(1 - p_F\right)/p_F\left(1 - p_D\right)\right)}$$

$$= \gamma_{\text{FC}}$$

$$\longrightarrow \widehat{H} = \begin{cases} 1, & b \geq \gamma_{\text{FC}} \\ 0, & \text{otherwise.} \end{cases} \tag{10}$$

The probability of error can be calculated using

$$P_e = \left(p * P_{\text{false positive}}\right) + \left(\left(1 - p\right) * P_{\text{false negative}}\right)$$

$$\longrightarrow P_e = p * \left[1 - P\left(b \geq \gamma_{\text{FC}} \mid H_1\right)\right] \tag{11}$$

$$+ \left(1 - p\right) * P\left(b \geq \gamma_{\text{FC}} \mid H_0\right).$$

To calculate power consumption for the whole network (12) can be used, where E = total energy, E_T = transmission energy, E_R = receiving energy, and E_T = processing energy:

$$E = E_T + E_R + E_P$$

$$\longrightarrow E = \sum_{i=1}^{K} \left(L_i * 1 * e_t\right) + \sum_{i=1}^{K} \left(L_i * 1 * e_r\right)$$

$$+ \sum_{i=1}^{K} \left(T * e_p\right) \tag{12}$$

$$\longrightarrow E = \sum_{i=1}^{K} \left(L_i * \left(e_t + e_r\right)\right) + \sum_{i=1}^{K} \left(T * e_p\right).$$

2.4. Hybrid Detection Scheme. Neither the centralized nor the distributed detection scheme is flexible enough for designers to choose between detection accuracy and energy consumption. Yu et al. [16] proposed a hybrid scheme that balances detection accuracy and total energy consumption. According to a defined level of accuracy, the nodes will vary between sending all collected data and sending a 1-bit result. Thus, such schemes attempt to balance accuracy and energy consumption.

In this scheme, assume there are K number of nodes. These nodes will collect T samples of environmental observations with (n_i = number of 1s for node i). There will be upper and lower bounds N_0 and N_1, where $0 \leq N_0 < N_1 \leq T$. The node result will be 0 if the number of 1s is less than N_0. In other words, if the number of 0s collected is greater than or equal to $T - N_0$, 1 will be sent if the number of 1s is greater than or equal to N_1. Otherwise all the collected data will be sent, as shown in (17):

$$\text{Result}_i = \begin{cases} 1, & n_i \geq N_1 \\ n_i, & N_0 < n_i < N_1 \\ 0, & n_i \leq N_0. \end{cases} \tag{13}$$

From (12), assume that out of K sensor nodes, t nodes send 1s, s nodes send 0s, and $K-s-t$ nodes send all their observations. The total data sent, Ω, will be

$$\Omega = \{1, \ldots, 1; n_1, \ldots, n_{K-s-t}; 0, \ldots, 0;\}. \tag{14}$$

From (12) we can derive the following probability:

$$P[b = 0 \mid H_\theta] = \sum_{i=0}^{N_0} \binom{T}{i} p_\theta^T (1 - p_\theta)^{T-i},$$

$$P[b = 1 \mid H_\theta] = \sum_{i=N_1}^{T} \binom{T}{i} p_\theta^T (1 - p_\theta)^{T-i}, \tag{15}$$

$$\text{PD} = \binom{K}{s}\binom{K-s}{t} (P[b = 0 \mid H_1])^s (P[b = 1 \mid H_1])^t$$

$$\cdot (P[n_i = 1 \mid H_1])^{k-s-t}, \tag{16}$$

$$\text{PF} = \binom{K}{s}\binom{K-s}{t} (P[b = 0 \mid H_0])^s (P[b = 1 \mid H_0])^t$$

$$\cdot (P[n_i = 1 \mid H_0])^{k-s-t}. \tag{17}$$

Following from (16) and (17), the final probability of error can be determined using

$$P_e = p * [1 - \text{PD}] + (1 - p) \times \text{PD}. \tag{18}$$

To calculate power consumption for the whole network (16) can be used, where E = total energy,

TABLE 2: Telos typical operation conditions [24].

	Min	Nom	Max	Unit
Supply voltage	2.1		3.6	V
Supply voltage during flash memory programming	2.7		3.6	V
Operation free air temperature	−40		85	°C
Current consumption: MCU on, radio RX		21.8	23	mA
Current consumption: MCU on, radio TX		19.5	21	mA
Current consumption: MCU on, radio off		1800	2400	μA
Current consumption: MCU idle, radio off		54.5	1200	μA
Current consumption: MCU standby		5.1	21.0	μA

E_T = transmission energy, E_R = receiving energy, and E_T = processing energy:

$$E = E_T + E_R + E_P$$

$$\longrightarrow E = \left[\sum_{i=1}^{s+t} (L_i * 1 * e_t) + \sum_{i=1}^{K-s-t} (L_i * T * e_t) \right]$$

$$+ \left[\sum_{i=1}^{s+t} (L_i * 1 * e_r) + \sum_{i=1}^{K-s-t} (L_i * T * e_r) \right]$$

$$+ \left[\sum_{i=1}^{K} (T * e_p) \right]$$

$$\longrightarrow E = \sum_{i=1}^{s+t} (L_i * (e_t + e_r)) + \sum_{i=1}^{K-s-t} (L_i * T * (e_t + e_r))$$

$$+ \sum_{i=1}^{K} (T * e_p). \tag{19}$$

3. System Model (Telos)

In order to evaluate and develop our scheme, the behavior of WSN sensors should be understood, and a power consumption model should be defined [17]. For this purpose the typical operation conditions of TelosB (Figure 4) have been selected as a basis for our power model [18, 19].

I use the datasheet of Telos working at 250 kbps and 2.4 GHz (ultralow power IEEE 802.15.4 compliant wireless sensor module), which is one of the most widely known WSN nodes, and I use the power details from Table 2.

From the power consumption model in Figure 5, it can be concluded that transmission is responsible for a large amount of node power consumption.

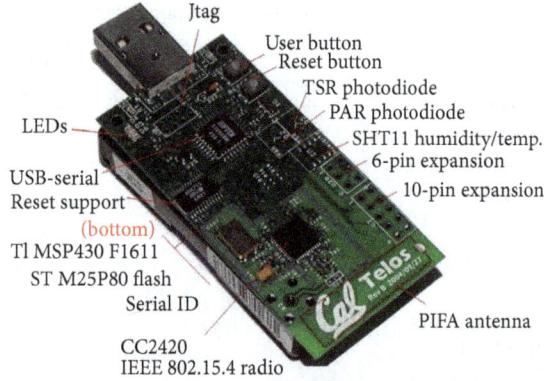

FIGURE 4: TelosB by the University of California.

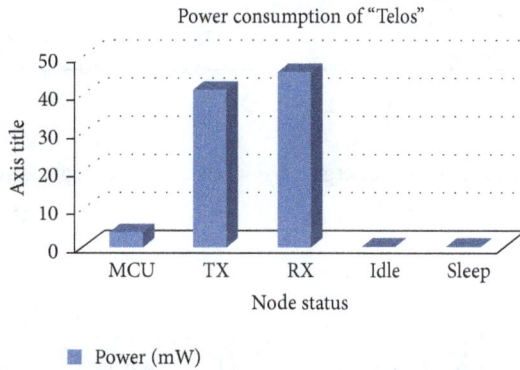

FIGURE 5: Power consumption of "TelosB."

So far, only the behavior of a single node has been discussed. Now the power consumption of all the nodes in the network should be addressed.

The formula for calculating the overall power consumed while sending a single bit from source (sensor node) to destination (fusion center) through N hop nodes equals the energy required to send and receive this bit between all of the N hops, where E = total energy, E_T = sending energy, and E_R = receiving energy:

$$\text{Energy consumed per node: } E = E_T + E_R$$
$$\longrightarrow E = (N * e_t) + ([N - 1] * e_r). \tag{20}$$

So, for the special case where a node is directly connected to the FC, energy used = $e_t + e_r$.

For instance, if hops count = 10, energy used will = $10e_t + 9e_r$.

Thus, the length of a route affects the overall power consumed in the network and the network life time as well.

3.1. Power Calculation. We use TelosB model to calculate power consumption [20, 21], with the following:

  (i) MCU current consumption: 2.4 mA,

  (ii) Rx current consumption: 23 mA,

  (iii) Tx current consumption: 21 mA.

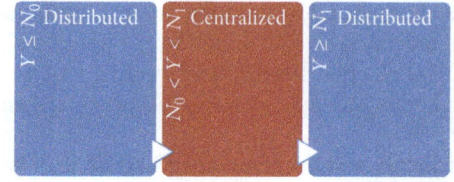

FIGURE 6: Node behaviour depends on N_0 and N_1.

3.2. Processing Energy. The MSP430 [22] is running at a clock rate of 8 MHz, Instruction Cycle Time = 200 ns, so in every cycle we can finish 40 Instructions/Cycle, $[(8 * 10^{-6})/(200 * 10^{-9})]$.

Let us assume that we need 2,000 instructions for the measurement, for data processing, and for preparing a packet for transmission over the network. This would result in $2000/(40 * 8 * 10^6) = 6.25 * 10^{-6}$ s; the amount of energy consumed per sample at a current of 2.4 mA would be

$$E_P = 2.4 * 6.25 * 10^{-6} = 15 * 10^{-6} \text{ mA/s}. \tag{21}$$

3.3. Transmission Energy. Using CC2420, sending data takes place at a speed of 250 kbps. If one node collects from a local sensor or from all the children 1 bit of data to be forwarded to the parent node, power consumption can be calculated as in the following equations:

$$\frac{1}{250 * 10^3} = 4 * 10^{-6} \text{ s}, \tag{22}$$

$$E_R = 23 * 4 * 10^{-6} = 92 * 10^{-6} \text{ mA/bit}, \tag{23}$$

$$E_T = 21 * 4 * 10^{-6} = 84 * 10^{-6} \text{ mA/bit}. \tag{24}$$

3.4. Simulation Network Design. In order to derive more realistic data from our simulation model, we will apply TelosB properties in our network, with maximum coverage = 100 m. A total of N sensors are randomly deployed in the region of interest (ROI) which is a square area of a^2. I have selected a to be equal to 300 m; hence the ROI = 0.09 km^2. The locations of sensors are unknown before deployment time. However, it is known that all sensors are i.i.d., and every sensor locations (x, y) will follow a uniform distribution in the ROI.

4. Enhanced Hybrid Detection Scheme

The hybrid scheme adjusts the behavior of the network to vary between centralized and distributed schemes and also establishes N_0 and N_1 parameters to define that behavior, and Y is the number of observations equal to 1. We propose to enhance the hybrid scheme by dynamically choosing the N_0, N_1 parameter instead of it being static.

In the hybrid scheme if Y is between N_0 and N_1 the node will act as centralized; otherwise it will act as distributed, as shown in Figure 6. For the special case $N_1 - N_0 \leq 1$ the node will always act as distributed, and the node will act more centralized if $N_1 - N_0$ becomes larger (until $N_0 = 0$ and $N_1 = \text{MAX}$).

However, N_0 and N_1 can be made dynamic, with a preference for a distributed orientation for nodes with longer route paths; for the remaining nodes it can remain more centralized, as in the original hybrid scheme, according to the requirements of the application.

For every sensor S varied from 1 to K we will assign specific $\{N_{0i}, N_{1i}\}$. These sensors will be classified according to the route path:

$$S_i = \begin{cases} 1, & n_i \geq N_{1i} \\ n_i, & N_{0i} < n_i < N_{1i} \\ 0, & n_i \leq N_{0i}. \end{cases} \tag{25}$$

Since $0 \leq N_0 < N_1 \leq T$, we will have a finite number of combinations for N_0 and N_1. These combinations—or groups of them should be mapped to all sensors, depending on sensor path weight.

The above mapping is a normal n-to-one mapping problem, which in our case can be solved experimentally by testing it in different deployed wireless sensor applications.

4.1. Network Deployment. Similar to most wireless sensor networks, the FC is deployed in an accessible location, while the sensor nodes are deployed randomly in the targeted area. This can be considered as a mesh network. After deploying the nodes, discovering the network is the first action to be taken. The FC broadcasts discover packets to all nodes in the covered area. These nodes then broadcast to further nodes until all the nodes are covered and routing paths; next hop and node configuration are defined.

To simplify the problem, nodes discover paths to the FC based on SPF (shortest path first) and select subsequent hops to forward received packets to. This information is crucial.

Since the nodes have already collected the broadcast packets, which include source, destination, and path, the nodes are able to select their routes based on the path length.

4.2. Nodes Configuration. For each node there are fixed configurations such as Tx/Rx power, frequencies, and calculation process. Some parameters need to be configured by the manufacturer or the user in order to optimize network efficiency. On the other hand, some parameters may be configured by the network itself.

In our scheme, every node has certain parameters that are defined during the manufacturing phase or by the developer during network deployment, that is, sampling rate (T), CP_{min}, and CP_{max}.

CP is the probability that the system works as centralized, which can be calculated from N_0 and N_1. In our scheme each node will select N_0 and N_1 depending on its path length. N_0 and N_1 should generate CP where $CP_{Min} \leq CP_{N_0,N_1} \leq CP_{Max}$.

From those inputs $\{T, p_0, p_1, \max(\text{path\_length}), CP_{min}, CP_{max}\}$ each node is able to calculate the N_0 and N_1 that satisfy application requirements.

From T we can find (N_0, N_1) combinations $= 2^T$, which is our ROT. Every node should be able to map the proper

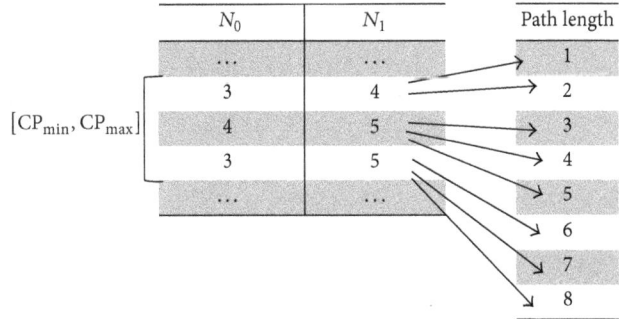

FIGURE 7: Mapping (N_0, N_1) combinations to available path length.

(N_0, N_1) from both (CP_{min}, CP_{max}) and $\max(\text{path\_length})$, as can be seen in Figure 7.

We can map these ranges by means of a simple 1-to-many mapping technique, where we index the initial range and assign the remaining ones equally to the next range, as applied in Algorithm 1.

4.3. N_0, N_1 Selection. To be able to select N_0 and N_1 at each node the problem is divided into simpler problems as follows:

(1) Define all combination of N_0 and N_1 and calculate the equivalent CP.

(2) Define valid N_0 and N_1 based on CP_{min} and CP_{max}.

(3) Define max path length, and map valid (N_0, N_1) to every path zone.

4.3.1. All N_0, N_1 Combinations. A selection of N_0 and N_1 will define when the node will act as more centralized or more distributed. For T collected samples we will have $0 \leq N_0 < N_1 \leq T$. We can have C different combinations for those N_0 and N_1, where $C = T!/((T-2)!2!)$.

For each combination we can calculate its CP_{CENT}, and we will have the C list of CP, which should be sorted according to its values.

CP_{CENT} can be calculated as in the following:

$$CP_{CENT} = \text{binocdf}(N_1 - 1) - \text{binocdf}(N_0), \tag{26}$$

where cdf is the cumulative distribution function which can be calculated as in the following:

$$y = F(x \mid n, p) = \sum_{i=0}^{x} \binom{n}{i} p^i (1-p)^{(n-i)} I_{(0,1,\dots,n)}(i). \tag{27}$$

For example, for $T = 5$, $P[S = 1] = 0.45$, we generate all possible combinations of (N_0, N_1) as shown in Table 3.

4.3.2. Valid N_0 and N_1 Combinations. For example, we can select initial $CP_{min} = 0.50$, $CP_{max} = 0.90$.

First of all we find the valid (N_0, N_1) combinations. The selection of $CP_{min} = 0.50$, $CP_{max} = 0.90$ gives us 4 valid combinations (Table 4).

```
Define {CP_min, CP_max}
for each {N_0, N_1}
        CP_i = binCDF(N_1i - 1, T, P[S = 1]) - binCDF(N_0i, T, P[S = 1]);
        if CP_min < CP_i < CP_max
                add CP_i to CP;
        end if
end for
seed = ⌈count(Path)/count(CP)⌉
for i = 1  →  count(Path)
        cp_seed = min(ceil(i/seed), size(CP, 1));
        CP_i = CP(cp_seed);
end for
```

ALGORITHM 1: Local N_0 and N_1 selection.

TABLE 3: CP values for all (N_0, N_1) combinations.

N_0	N_1	CP
0	5	0.93
0	4	0.82
1	5	0.73
1	4	0.61
0	3	0.54
2	5	0.39
1	3	0.34
2	4	0.28
0	2	0.21
3	5	0.11
0	1	0.00
1	2	0.00
2	3	0.00
3	4	0.00
4	5	0.00

TABLE 4: (N_0, N_1) valid combination list.

CP	N_0	N_1
0.82	0	4
0.73	1	5
0.61	1	4
0.54	0	3

4.3.3. Map (N_0, N_1) Combinations to Available Zones. In our scheme, every node has certain parameters that are defined during the manufacturing phase or by the developer during network deployment, that is, sampling rate (T), CP_{min}, and CP_{max}. All nodes are classified according to path length to Z zones. We have here a random WSN where maximum path = 6, which means 6 different zones. Figure 8 shows an example of WSN nodes classified into zones.

We define minimum CP_{min} and maximum CP_{max}, where CP_{max} is the maximum CP combination that can be assigned

for zone 1, and CP_{min} is the minimum CP that can be assigned to the next zones (where all $CP \le CP_{min}$).

For every zone z, we assign $CP(z) = CP_i$ from (23), which can be used to map between zones and (N_0, N_1) combinations (Table 5):

$$i = \left\lceil z \div \left\lceil \frac{\text{Count of all Zones}}{\text{Count of CPs between } CP_{min} \text{ and } CP_{max}} \right\rceil \right\rceil,$$

$$CP_1 = \left\lceil \frac{1}{\lceil 6/4 \rceil} \right\rceil = \left\lceil \frac{1}{2} \right\rceil = 1,$$

$$CP_2 = \left\lceil \frac{2}{\lceil 6/4 \rceil} \right\rceil = \left\lceil \frac{2}{2} \right\rceil = 1,$$

$$CP_3 = \left\lceil \frac{3}{\lceil 6/4 \rceil} \right\rceil = \left\lceil \frac{3}{2} \right\rceil = 2, \qquad (28)$$

$$CP_4 = \left\lceil \frac{4}{\lceil 6/4 \rceil} \right\rceil = \left\lceil \frac{4}{2} \right\rceil = 2,$$

$$CP_5 = \left\lceil \frac{5}{\lceil 6/4 \rceil} \right\rceil = \left\lceil \frac{5}{2} \right\rceil = 3,$$

$$CP_6 = \left\lceil \frac{6}{\lceil 6/4 \rceil} \right\rceil = \left\lceil \frac{6}{2} \right\rceil = 3.$$

In this scheme we find that (N_0, N_1) are dynamically selected by the nodes, depending on the path length, and the entire procedure described above requires minimal extra processing at the node, since the whole process is part of the network discovery phase.

Out of K sensor nodes, t nodes send 1s, s nodes send 0s, and $k - s - t$ nodes send all their observation so total send data Ω will be $\Omega = \{1, \ldots, 1; n_1, \ldots, n_{k-s-t}; 0, \ldots, 0;\}$. Our final decision, based on Bayes' rule, is $H' = H_1$ if $P[H_1 \mid \Omega] \ge P[H_0 \mid \Omega]$. From the above rule we can derive the following relations:

$$\frac{P[\Omega \mid H_1]}{P[\Omega \mid H_0]} \ge \frac{1 - p}{p}, \qquad (29)$$

TABLE 5: All (N_0, N_1) combinations.

N_0	N_1	CP
0	1	0.00
1	2	0.00
2	3	0.00
3	4	0.00
4	5	0.00
3	5	0.11
0	2	0.21
2	4	0.28
1	3	0.34
2	5	0.39
0	3	0.54
1	4	0.61
1	5	0.73
0	4	0.82
0	5	0.93

TABLE 6: ISA100.11a stack (Internet source).

Session	ISA native and legacy protocols (tunnelling)
Transport	UDP (IETF RFC 768)
Network	6LoWPAN (IETF RFC 4944)
Data link	Upper data link ISA100.11.a
	IEEE 802.15.4
Physical	IEEE 802.15.4

where we can calculate $P[\Omega \mid H_1]$ and $P[b_i = 0 \mid H_\alpha]$, $P[b_i = 1 \mid H_\alpha]$, and $P[b_i = 1 \mid H_\alpha]$ as follows:

$$P[\Omega \mid H_\alpha] = \binom{k}{s}\binom{k-s}{t} P[b = 0 \mid H_\alpha]^s$$

$$\cdot P[b = 1 \mid H_\alpha]^t \times P[n_i \mid H_\alpha]^{k-s-t},$$

$$P[b_i = 0 \mid H_\alpha]$$

$$= \sum_{n=T-N_{0i}}^{T} \binom{n}{T-N_{0i}-1}\left(1-p_\alpha\right)^{T-N_{0i}} p_\alpha^{i-(T-N_{0i})}, \quad (30)$$

$$P[b_i = 1 \mid H_\alpha] = \sum_{n=N_{1i}}^{T} \binom{n-1}{N_{1i}-1}\left(1-p_\alpha\right)^{n-N_{1i}} p_\alpha^{N_{1i}},$$

$$P[n_i \mid H_\alpha] = \binom{T}{n_i}\left(1-p_\alpha\right)^{1-n_i} p_\alpha^{n_i}.$$

5. Simulation Results

5.1. Enhanced Hybrid Detection Scheme

5.1.1. Power Consumption. By applying "Telos" power model and using (18), (19), and (21), power calculation is

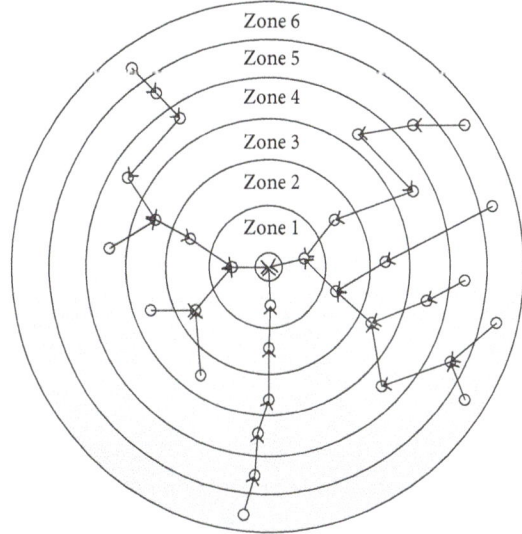

FIGURE 8: Network zones classification.

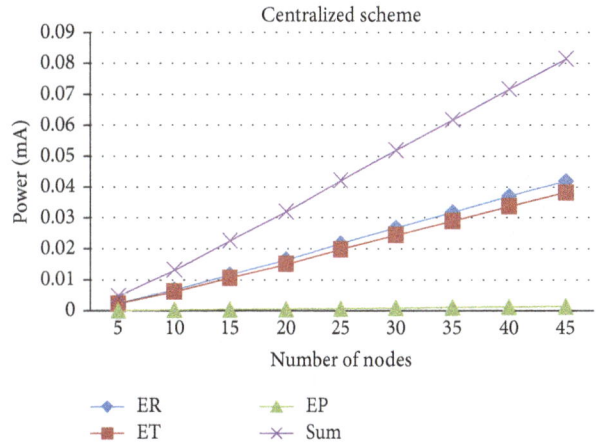

FIGURE 9: Centralized scheme power consumption.

processing power $= 15 * 10^{-6}$ mA/sample, Rx power $= 92 * 10^{-6}$ mA/bit, Tx power $= 84 * 10^{-6}$ mA/bit, and $T = 5$.

Centralized. In centralized scheme, the RF transmission is the most power consuming process (Figure 9).

Distributed. Similar to the centralized scheme, in the RF scheme, the most power consuming process in is the transmission part, (Figure 10).

Hybrid ($N_0 = 1, N_1 = 4$). We assume the following: $p = 0.5$, $p_0 = 0.2$, $p_1 = 0.7$, and $T = 5$; K is varied from 10 to 30 (Figure 11).

Enhanced Hybrid. We define the probability of node behavior as centralized CP; we get CP values out of possible combination of N_0 and N_1.

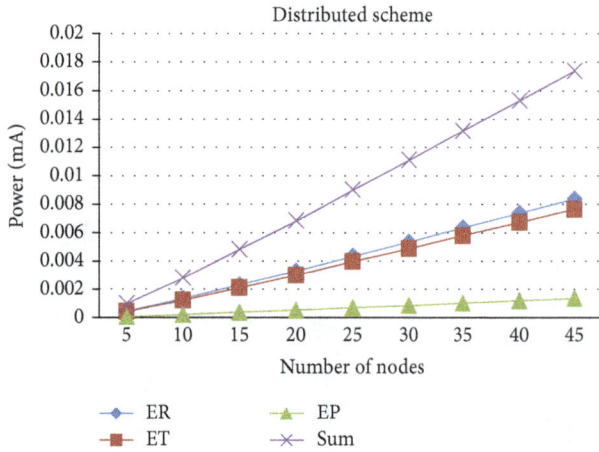

FIGURE 10: Distributed scheme power consumption.

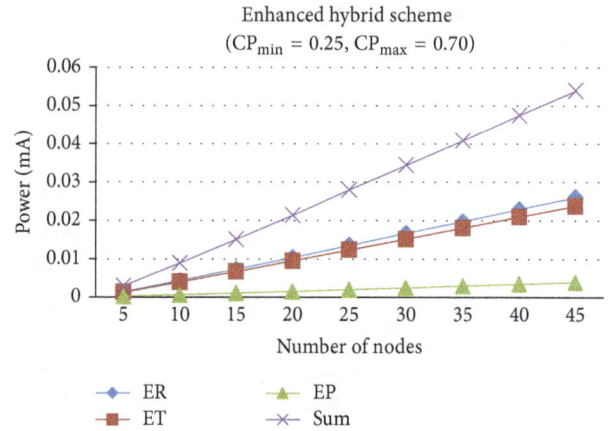

FIGURE 12: Enhanced hybrid scheme power consumption.

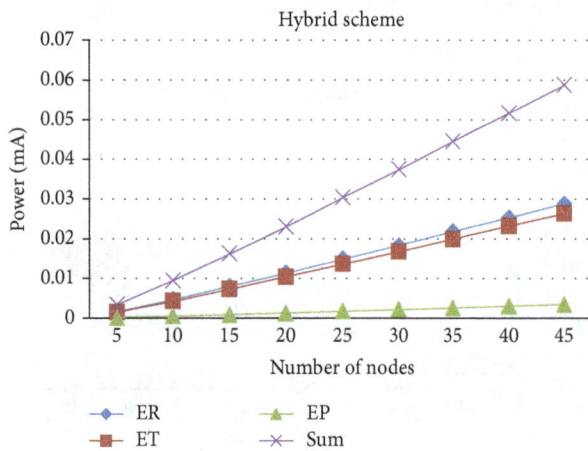

FIGURE 11: Hybrid scheme power consumption.

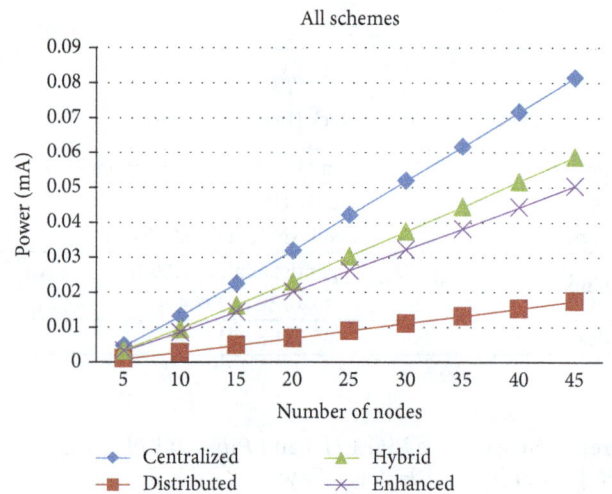

FIGURE 13: Power consumption of all schemes.

Figure 12 shows the power consumption of enhanced hybrid scheme, where we assign $CP_{min} = 0.25$ and $CP_{max} = 0.70$.

All Energy Schemes. The performance of the four schemes is shown in Figure 13, as we can see that enhanced scheme outperforms other schemes.

Accuracy Behavior. In our simulation we are able to get the same result which is given in [16], where we used hybrid $(N_0, N_1) = (1, 3), (1, 4)$, and $(0, 4)$ as shown in Figure 14.

The performance of the four schemes is shown in Figure 15, as we can see that enhanced scheme outperforms other schemes. In comparison to hybrid scheme our scheme gives a very close level of accuracy, as illustrated in Figure 15.

6. Conclusions

The main purpose of this paper is to enhance the event collection and detection capability of current WSN schemes. Two of the available schemes—centralized and distributed

schemes—are basic and have no flexibility. The third scheme—the hybrid scheme—uses the two previous schemes to balance accuracy and energy; nevertheless all the nodes remain with fixed configuration and limited flexibility. On the contrary, the proposed scheme is designed to be more flexible in order to balance the power consumption and the detection accuracy at the node level. Every node is flexible in deciding how to behave, that is, whether to be more centralized or distributed.

To be able to compare the tradeoff between accuracy and power consumption in these schemes the model used should be realistic. This is one of the most significant weaknesses of the previous research, in which all power consumption calculations have been based solely on an assumed model. In the proposed scheme this weakness has been overcome by using a real WSN model, which produces realistic results.

As can be seen from the simulation results, our scheme saves a substantial amount of energy compared to the hybrid scheme, while retaining accuracy to almost the same degree. In addition, our scheme deals more efficiently with larger network area and denser node-number.

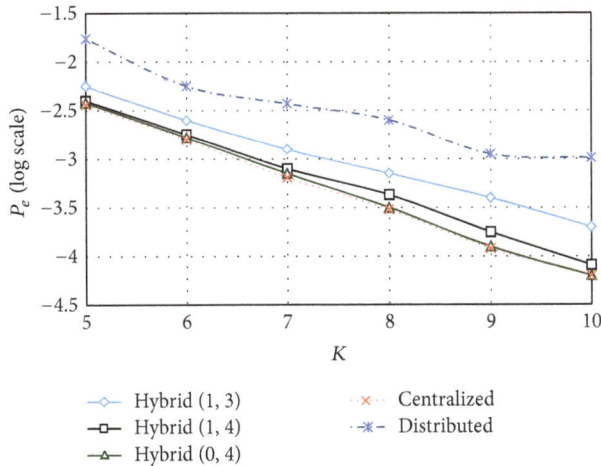

FIGURE 14: Comparison of three schemes in detection accuracy.

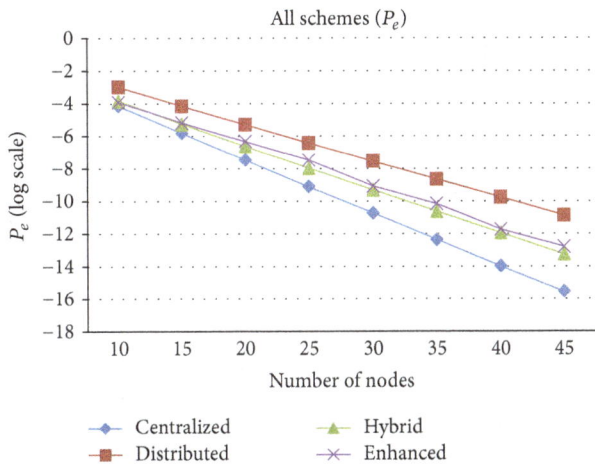

FIGURE 15: Probability of error for all schemes.

Although the focus of this paper has been on reducing overall power consumption, other future enhancements are also possible, for example, enhancing overall network lifetime by improving (N_0, N_1) selection techniques where the following factors can be considered:

(i) individual node power level,

(ii) event occurrence probability at each node,

(iii) next-hop distance and real Tx, Rx power.

Conflict of Interests

The author declares that there is no conflict of interests regarding the publication of this paper.

Acknowledgment

This research is supported by Computer Engineering Department at King Fahd University of Petroleum & Minerals (KFUPM).

References

[1] I. Khemapech, A. Miller, and I. Duncan, *Simulating Wireless Sensor Networks*, University of St Andrews, St Andrews, Scotland, 2005.

[2] W. Dargie and C. Poellabauer, *Fundamentals of Wireless Sensor Networks: Theory and Practice*, John Wiley & Sons, London, UK, 2010.

[3] M. I. Shukur, L. S. Chyan, and V. V. Ya, "Wireless sensor networks: delay guarentee and energy efficient MAC protocols," *World Academy of Science, Engineering and Technology*, vol. 50, no. 2009, pp. 1062–1066, 2009.

[4] T. Haenselmann, Sensornetworks, 2006.

[5] J. S. Wilson, Ed., *Sensor Technology Handbook*, Newnes, New York, NY, USA, 2005.

[6] C. Townsend and S. Arms, *Wireless Sensor Networks: Principles and Applications*, Micro-Strain Inc, 2004.

[7] https://docs.zigbee.org/zigbee-docs/dcn/09/docs-09-5262-01-0rsc-zigbee-rf4ce-specification-public.pdf.

[8] ZigBee Alliance, Zigbee Specification, 2011, http://www.zigbee.org/.

[9] F.-C. Jiang, H.-W. Wu, F.-Y. Leu, and C.-T. Yang, "Toward green sensor field by optimizing power efficiency using D-policy MG1 queuing systems," *Mobile Information Systems*, vol. 9, no. 4, 2013.

[10] E. H. Callaway, *Wireless Sensor Networks—Architectures and Protocols*, Auerbach Publications, 2004.

[11] WirelessHART Data Sheet, May 2007.

[12] S. Tanabe, S. Hirano, and E. Saitoh, "Wearable Power-Assist Locomotor (WPAL) for supporting upright walking in persons with paraplegia," *NeuroRehabilitation*, vol. 33, no. 1, pp. 99–106, 2013.

[13] N. Kushalnagar, G. Montenegro, and C. Schumacher, *IPv6 over Low-Power Wireless Personal Area Networks (6LoWPANs): Overview, Assumptions, Problem Statement, and Goals*, RFC, 2007, http://datatracker.ietf.org/doc/rfc4919/?include_text=1.

[14] A. J. Jara, M. A. Zamora, and A. Skarmeta, "Glowbal IP: an adaptive and transparent IPv6 integration in the Internet of Things," *Mobile Information Systems*, vol. 8, no. 3, pp. 177–197, 2012.

[15] ISA, ISA100 Wireless Compliance Institute, 2011, http://www.isa100wci.org.

[16] L. Yu, L. Yuan, G. Qu, and A. Ephremides, "Energy-driven detection scheme with guaranteed accuracy," in *Proceedings of the 5th International Conference on Information Processing in Sensor Networks (IPSN '06)*, pp. 284–291, April 2006.

[17] http://www.memsic.com/products/wireless-sensor-networks/wireless-modules.html.

[18] J. Polastre, R. Szewczyk, and D. Culler, "Telos: enabling ultra-low power wireless research," in *Proceedings of the 4th International Symposium on Information Processing in Sensor Networks (IPSN '05)*, pp. 364–369, April 2005.

[19] TelosB Mote TPR2420: Datasheet, 2003.

[20] Telos: PRELIMINARY Datasheet, 2004.

[21] T. Hanselmann, Sensornetworks, 2006.

[22] http://www.ti.com/lit/ds/symlink/msp430c1111.pdf.

[23] Industrial ethernet book, http://www.iebmedia.com/.

[24] J. Yick, B. Mukherjee, and D. Ghosal, "Wireless sensor network survey," *Computer Networks*, vol. 52, no. 12, pp. 2292–2330, 2008.

11

Interface Assignment-Based AODV Routing Protocol to Improve Reliability in Multi-Interface Multichannel Wireless Mesh Networks

Won-Suk Kim and Sang-Hwa Chung

Department of Computer Engineering, Pusan National University, Busan 609-735, Republic of Korea

Correspondence should be addressed to Sang-Hwa Chung; shchung@pusan.ac.kr

Academic Editor: David Taniar

The utilization of wireless mesh networks (WMNs) has greatly increased, and the multi-interface multichannel (MIMC) technic has been widely used for the backbone network. Unfortunately, the ad hoc on-demand distance vector (AODV) routing protocol defined in the IEEE 802.11s standard was designed for WMNs using the single-interface single-channel technic. So, we define a problem that happens when the legacy AODV is used in MIMC WMNs and propose an interface assignment-based AODV (IA-AODV) in order to resolve that problem. IA-AODV, which is based on multitarget path request, consists of the PREQ prediction scheme, the PREQ loss recovery scheme, and the PREQ sender assignment scheme. A detailed operation according to various network conditions and services is introduced, and the routing efficiency and network reliability of a network using IA-AODV are analyzed over the presented system model. Finally, after a real-world test-bed for MIMC WMNs using the IA-AODV routing protocol is implemented, the various indicators of the network are evaluated through experiments. When the proposed routing protocol is compared with the existing AODV routing protocol, it performs the path update using only 14.33% of the management frames, completely removes the routing malfunction, and reduces the UDP packet loss ratio by 0.0012%.

1. Introduction

Wireless Mesh Networks and Related Technics. Wireless mesh networks (WMNs) are regarded as a next-generation technology because they can provide high network extensibility and are economical. WMNs research based on IEEE 802.11s, which is an IEEE 802.11 amendment for mesh networking, is actively progressing [1]. WMNs have been selected as backbone networks in many places due to ability to provide network stability and reliability of data transfer in wireless sensor networks, smart city applications, and so forth.

One of the ways to improve network efficiency is the multi-interface multichannel (MIMC) technic, which uses multiple wireless channels via multiple interfaces. The MIMC technic minimizes interference on the same channel and prevents degradation of throughput, even though the number of hops increases [2–5]. In addition, the quality of each link in the network has improved through the use of directional antennas [6, 7].

The ad hoc on-demand distance vector (AODV) routing protocol defined in the IEEE 802.11s standard is widely employed in ad hoc networks and WMNs. The basic operations of AODV using the on-demand routing scheme are as follows. First, a source node that wants to create a new path broadcasts a path request (PREQ) frame including information on the target node. All nodes receiving that management (MGMT) frame generate a backward path toward the source node. If the receiving node is the target of the PREQ, it will send a path reply (PREP) frame to the source node via unicast; otherwise, the PREQ is simply rebroadcast. The PREP sent by the target node is delivered to the source node, and the mesh path is created when the source node receives the PREP. In addition, the source node performs a path update by sending a PREQ periodically in order to find a path with a better metric [8].

Research Motivation and Contribution. As mentioned earlier, WMNs have been utilized as backbone networks in various fields. To enhance the valuation of WMNs, not only must capacity improve, but also efficiency and reliability must be increased. One of the best ways to enhance WMNs is to take advantage of the MIMC technic and the directional antenna. In WMNs that adopt the MIMC technic and directional antenna (MIMC WMNs), additional technical requirements for the routing protocol improve efficiency and reliability.

Actual traffic flows should also be considered, but most routing protocols take into consideration only traffic between the mesh portal and the mesh routers. If network scale expands or a specific application that generates a lot of internal network traffic is in service, traffic flow may not be directed to the mesh portal [9, 10]. In this case, the overhead of the path update process will greatly increase, so it must be handled by the routing protocol.

In this paper, we define the problems that occur when an existing AODV is used in MIMC WMNs, and we present a new routing protocol that improves routing efficiency and network reliability. The existing issues and the research contributions for MIMC WMNs are as follows.

(i) In the original AODV using the single-target PREQ (ST-PREQ), a structural problem exists where the PREQ frame cannot be delivered through the normal path. Because of this, the mesh path will change unintentionally. However, the proposed interface assignment-based AODV (IA-AODV) can resolve this problem by using a multitarget PREQ (MT-PREQ), which is the MGMT frame performing a path update with multiple targets at the same time.

(ii) When a PREQ is received through multiple interfaces, the random receiving order of the PREQ shows a low correlation between the routing metric and the PREQ receiving order. Because of this problem, frequent replacement of the interface responsible for next-hop communication occurs, and network reliability will be degraded from the decrease in packet delivery ratio and an increase in the routing malfunction ratio. The PREQ random receiving order problem is resolved by a PREQ prediction scheme in IA-AODV.

(iii) If the network scale is enlarged and internal network paths increase, the efficiency of routing will decline due to increasing of the MGMT frames for the routing update process. On the other hand, IA-AODV is able to maximize routing efficiency by using MT-PREQ and a PREQ prediction scheme, handling exceptional situations such as PREQ loss.

(iv) The simple rule of the PREQ sender decision reduces efficiency of the routing protocol, but IA-AODV improves routing efficiency by using a PREQ sender assignment scheme.

Organization. The remainder of this paper is organized as follows. Section 2 presents background and related works, and Section 3 presents the system model for MIMC WMNs and defines the problem. In Section 4, IA-AODV is explained in detail, and in Section 5 the experimental results are analyzed. Finally, we conclude this paper in Section 6.

2. Background and Related Works

2.1. Background of the AODV Routing Protocol. The MIMC WMNs presented in this paper use a directional antenna, so the link quality of most mesh links within MIMC WMNs is guaranteed to be high; therefore, the result of the update process to find a better path will be diminished. So, when the legacy AODV routing protocol is employed in MIMC WMNs, there are many points to consider, unlike single-interface single-channel (SISC) WMNs. First, the routing update process is an essential factor to find the best path during communications, but if the wireless channel information is not frequently changed then the number of MGMT frame transmissions in the update process will be an overhead. Therefore, a scheme that minimizes the MGMT frames in every update period is required. Second, routing malfunctions occur frequently due to the random receiving order of the PREQs and the PREQ loss, so additional technics are required to solve this problem.

The MT-PREQ, which efficiently reduces the number of PREQ frames, was presented in IEEE 802.11s. If WMNs take advantage of MT-PREQ, then the objective of path discovery or path update with multiple nodes will be achieved with a single PREQ. In IEEE 802.11s, the fields for the MT-PREQ are *Target Count*, which indicates the number of target nodes, and *Per Target*, which includes information on each target, such as 1 byte for flag, 6 bytes for medium access control (MAC) address, and 4 bytes for sequence number. (The *Per Target* field is referred to as the target list in this paper.) The main technics of the IA-AODV proposed in this paper are based on MT-PREQ.

2.2. Related Works. Currently, a lot of research into routing schemes or broadcast schemes for ad hoc networks and WMNs has been in progress. The main goals of these works are interference avoidance within the network, improvement in packet delivery ratio, and increasing end-to-end bandwidth. Other researches focus on MIMC WMNs with a directional antenna for high-performance backbone networks.

From our own previous work on MIMC WMNs, the channel load aware routing protocol, the multipath AODV protocol for fast recovery, and implementation of IEEE 802.11n multihop have been studied to increase of link bandwidth [11–13]. The channel load aware routing protocol selects the most efficient multihop path via channel usage in multiple paths between the source node and the target node [11], and the multipath AODV protocol searches for an alternative path when the original path is compromised [12]. In this paper which extends earlier research [14], an efficient and novel routing protocol focuses on maximizing routing efficiency and network reliability while minimizing overhead in the routing process.

Several studies of routing protocols were carried out to improve the performance of mobile ad hoc networks (MANETs) and WMNs [15–20]. There is implementation of AODV in the earliest forms of WMNs [18, 20], and a routing protocol using a cache to increase the packet delivery ratio in an ad hoc network with high mobility was proposed [17]. In addition, the hybrid wireless mesh protocol (HWMP) was studied for communication with the mesh portal as well as internal communication in WMNs [19]. The node type aware AODV routing protocol in hybrid WMNs, which includes not only a mesh router but also clients, was proposed [15]. On the other hand, the proposed routing protocol achieves improvement of network throughput by maximizing routing efficiency and network reliability.

A vast amount of research concentrates on routing protocols for MIMC WMNs to improve performance through avoidance of interference [4, 5, 21–25]. T. C. Tsai and S. T. Tsai [25] proposed a cross-layer routing protocol that controls the transmit power of each network interface in MIMC WMNs to increase performance. Paschoalino and Madeira [23] demonstrated a scalability and link quality aware shortest path routing protocol in order to improve throughput. In addition, Subramaniam et al. [21] and Liu and Liao [24] presented a routing protocol using signal-to-interference-plus-noise ratio (SINR) as interference value and expected transmission time (ETT) in MIMC WMNs. In contrast, IA-AODV modifies the existing AODV to improve network reliability and supports scalability without management overhead.

Anker et al. [26] proposed an AODV routing protocol to improve network reliability by preventing a selfish node from concentrating only on power savings without data delivery in a MANET. As a result, it achieved quality of service (QoS) guarantees and an improved packet delivery ratio and was then verified through simulation of the amount of remaining packets in the buffer and the routing malfunction rate. Boice et al. [27] demonstrated a routing protocol that efficiently ensures buffer operation for intermittently connected networks. This routing protocol operates according to the current status of the connection and the buffer. In this way, it increases packet delivery ratio and reliability, but it also reduces the number of routing messages. Unlike other researches [26, 27], a specialized routing protocol for MIMC WMNs with directional antennas as a high-performance wireless backbone network is proposed and was evaluated in terms of reliability of the network and performance increase through actual implementation.

As described above, research into routing protocols to improve performance of MIMC WMNs has been plentiful. However, most of these studies assume ideal conditions and do not consider practical issues. IA-AODV considers features of MIMC WMNs employed in backbone networks in various fields and achieves network reliability through a propagation method for routing MGMT frames based on interface assignment. In addition, it also achieves routing efficiency by decreasing routing overhead based on designation by the management frames' sender.

3. System Model and Problem Statement

3.1. System Model of MIMC WMNs.
A mesh node in MIMC WMNs can have multiple interfaces that conduct communication through different channels. Also, because it uses a directional antenna, each interface is responsible for connection with only one neighbor node. A mesh link is created

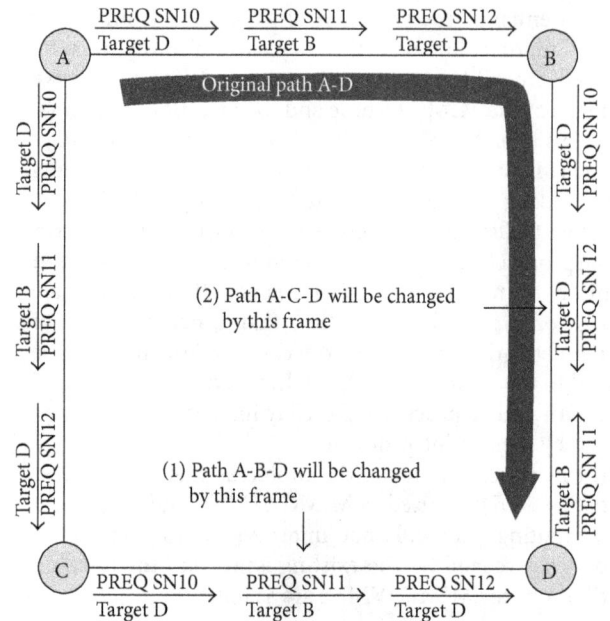

FIGURE 1: The structural problem of AODV using ST-PREQ.

between two interfaces, so the multihop link consists of the number of interfaces, which is the same as double the number of hops. In terms of path changes through information update in the routing mechanism, considerable differences exist in MIMC and SISC. With SISC, change of the next-hop through a path change generates overhead to modify only the receive address (RA) of the MAC header of each frame in the transmit queue. On the other hand, modification of the next-hop in MIMC adds overhead that not only changes the RA of each frame but also resets the appropriate interface for the RA. A single mesh engine controls and manages the multiple interfaces. So, the overhead of above operation that requires a locking mechanism is greatly increased. In addition, the broadcast frame in MIMC WMNs is copied as the number of interfaces in the mesh node and is then sent over each interface separately.

MIMC WMNs employed as a backbone are based on a wireless link, so link quality will absolutely affect the value of the network. Therefore, most industrial MIMC WMNs use a directional antenna. When a directional antenna is used, collision and interference on the same channel are reduced and the reach of the signal is increased, so it can significantly improve link quality. In addition, the logical topology becomes equal to the physical topology.

3.2. Problem Statement

3.2.1. Structural Problem of AODV Using ST-PREQ.
If an existing AODV using ST-PREQ is adopted in MIMC WMNs, the structural problem of routing will occur, as shown in Figure 1. In Figure 1, the active paths are indicated as A-B and A-D; in particular, the path between nodes A and D is formed by A-B-D, which has a better metric. In order to update the path information, node A sends two PREQs every

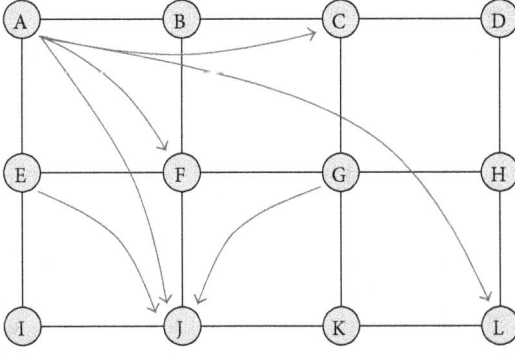

FIGURE 2: An example of MIMC WMN topology and traffic flow.

update period, and each PREQ includes target nodes B and D, respectively. Every PREQ will be sent through all interfaces of node in order to update mesh paths. The PREQ, which has sequence number (SN) 10, is transferred to node D as the target. It will pass through nodes B and C and will arrive at node D. The PREQ with SN 11 is broadcast to node B as the target; one of the copies of the PREQ is directly delivered to node B and another reaches node B via nodes C and D. At this time, node D receives the PREQ, which has the latest SN of node A from only node C. Therefore, node D sets the next-hop towards node A to node C, because it misunderstands that the path A-B-D is not valid anymore. After that, node A sends the PREQ with SN 12 to node D as the target through nodes B and C; the path A-D will be replaced with A-B-D with a better metric. This structural problem from ST-PREQ may occur more frequently in a topology including a multipath. Thus, when the network scale is larger, this problem will occur multiple times during one update period. In addition, this problem may become a serious situation when one node takes charge of sending PREQ/PREP at the same time.

3.2.2. Increasing Routing Complexity Problem of AODV Using ST-PREQ.

Figure 2 shows an example of PREQ transmissions to update and maintain paths in small-scale MIMC WMNs. When the existing AODV routing protocol is applied, the number of PREQ frames required within one path update period is expressed in

$$N_{\text{PREQ}} = \sum_{i}^{N_{\text{src}}} N_{\text{target}(i)}. \tag{1}$$

In (1), N_{src} is the number of mesh nodes responsible for transmitting the PREQ in MIMC WMNs. $N_{\text{target}(i)}$ means the number of target nodes that communicate with the ith source node. For example, in the topology of Figure 2, there are 6 paths (A-C, A-F, A-J, A-L, E-J, and G-J). In that case, the mesh nodes responsible for transmission of the PREQ are nodes A, E, and G. At this time, N_{src} is 3; $N_{\text{target}(A,E,G)}$ is 4, 1, and 1. Therefore, the number of total PREQ frames within the routing update period is $4+1+1 = 6$. It is equal to the number of paths.

In a general broadcast mechanism (flooding—if the broadcast packet is received, it will just be rebroadcast),

the total number of transmissions in the propagation process of the ith PREQ to the whole network is defined in

$$N_{\text{broadcast}(i)} = \sum_{j=1}^{n} N_{\text{IF}(X_j)} - N_{\text{IF}(\text{target}(i))}. \tag{2}$$

In (2), $N_{\text{IF}(X_j)}$ indicates the interface number of the jth mesh node, and $N_{\text{IF}(\text{target}(i))}$ reflects the interface number of the target node that receives the ith PREQ. Each mesh node in MIMC WMNs has the interface as the number of links; therefore, the sum of interfaces of all mesh nodes is twice the number of all mesh links. In Figure 2, the total number of broadcasts in the path update period of the path A-C is $34 - 3 = 31$. Consider

$$N_{\text{total} \, tx} = \sum_{i}^{N_{\text{PREQ}}} N_{\text{broadcast}(i)}. \tag{3}$$

Equation (3) presents the total number of transfers for PREQ propagation within the path update period. Using the example in Figure 2, $N_{\text{total} \, tx}$ is $31 + 30 + 31 + 32 + 31 + 31 = 186$. Along with the structural problems from ST-PREQ, the total number of transfers of MGMT frames will greatly increase when the network scale becomes larger or the internal network paths increase. In particular, each mesh node in the MIMC WMN has the interface as the number of links; therefore, copies of the broadcast frame as the number of interfaces are transmitted. As a result, delay in operation and transmission, waste of network resources, and complexity of the network management process increase significantly.

3.2.3. The Random Receiving Order of PREQ and the PREQ Loss Problem.

In addition to the above problems, the random receiving order of PREQ problem exists. When a PREQ is received through multiple interfaces, the random receiving order of PREQ problem has a low correlation between the routing metric and PREQ receiving order. In other words, when the path update is performed, the next-hop towards the PREQ source node changes according to the first receipt of the latest PREQ via the poor path; then the next-hop towards the PREQ source changes again with the next receipt of the same PREQ via the good path. This problem will occur more frequently if there are more PREQ transmissions and multipaths.

Routing failure caused by PREQ loss also exists. The PREQ is propagated as broadcast by default; it is transmitted after physical carrier sensing is conducted in the MAC layer. Also, the PREQ frame does not require acknowledgement (ACK). Because of these factors, the loss ratio of the PREQ appears quite high. Commonly, PREQ loss is recovered in the next path update, but loss in a link near the source node makes the routing failure affect the entire network. Therefore, PREQ loss may degrade network reliability. The relation of routing error rate and loss rate per link is highly dependent on the location where the link loss occurs.

4. Interface Assignment-Based AODV (IA-AODV) Routing Protocol

IA-AODV is based on MT-PREQ in IEEE 802.11s. The basic operation for MT-PREQ is as follows. When the node receives the PREQ, it checks the target list. If the target list contains the address of the node receiving the PREQ, the node will send a PREP to the source node and remove its own address from the target list. After that, if any addresses remain in the target list, then the PREQ is rebroadcast. Also, the node that has the responsibility for transmission of the PREQ can conduct a path update for all targets with transmission of a single MT-PREQ. With this mechanism, $N_{\text{target}(i)}$ in (1) is replaced with 1, so (1) can be simplified to

$$N_{\text{PREQ}} = N_{\text{src}}. \qquad (4)$$

It means that the node that has responsibility for transmission of the PREQ can transmit only one MT-PREQ rather than the number of targets. Therefore, the PREQ frames up to the number of the PREQ source nodes are propagated during one path update period.

4.1. PREQ Prediction Scheme

4.1.1. Basic Operation of PREQ Prediction. In an existing AODV using MT-PREQ, the use of a simple method of operation as mentioned above is inefficient in MIMC WMNs. In MIMC WMNs, only one neighbor is connected by one interface, so there is no need to transmit the PREQ via an interface that already received another PREQ that has a better metric from the same source node. In addition, it is possible to transmit a PREQ that contains integrated information of received PREQs from other interfaces, and it is also possible to send frames that include predicted information of the PREQ and which are received periodically. The proposed PREQ prediction scheme for MT-PREQ, along with the above concepts, has the following features.

(i) When the interface is initially assigned, the role of the interface is assigned as receiving or transmitting the PREQ, and then the number of PREQ frames can be reduced efficiently through this role. The interface assigned the role of PREQ receiving is called an IN interface; otherwise, it is an OUT interface.

(ii) All OUT interfaces are configured to send the required target list which is combined with information of the target list from each IN interface.

(iii) In the path update process, the predictive PREQ will be transmitted via all OUT interfaces, even if the PREQ is not received through all IN interfaces. In this way, the PREQ is rapidly propagated, and the PREQ random receiving order problem can be resolved.

(iv) With the IN interface, due to the routing, protocol can immediately handle the change of routing information or the PREQ loss, and improvement of routing efficiency and network reliability can be achieved.

TABLE 1: Example of a PREQ information table.

Interface	Direction	Targets	SN	Metric	Exp_time
0	IN	F, G, L	11	468	300 ms
1	OUT	F, L	11	612	301 ms
2	OUT	F, L	11	612	301 ms
3	IN	E, F, L	11	407	320 ms

Figure 3 represents the flow of PREQ propagation in MIMC WMNs. Figures 3(a) and 3(c) show PREQ propagation when the PREQ prediction scheme is not used, and Figures 3(b) and 3(d) indicate the PREQ propagation with the PREQ prediction scheme. In Figure 3(a), node A broadcasts the PREQ, which has node L as a target; all forwarding nodes receiving that frame rebroadcast it using all their own interfaces. As shown in Figure 3(b), the PREQ is propagated in an efficient way by the role assignment of interfaces. This is possible because the role of interfaces is preallocated by classifying the PREQ that has a good metric and the other PREQ.

The process of path update with multiple targets is represented in Figure 3(c). The A-(E)/F/(G)/L means that the inclusion of nodes E and G was not determined in the corresponding PREQ, if the PREQ is received via a path that has a good metric relatively late; then the other PREQ, which has the same SN and a different target list, will be transmitted earlier. This issue will occur more frequently at the node far from the PREQ source node, and it brings on degradation of network reliability by increasing routing complexity. However, Figure 3(d) presents the PREQ prediction scheme that affects the path update with multiple targets; thus, all PREQs are propagated through the whole network through the most suitable path.

The PREQ prediction scheme adopted to propagate MT-PREQ more efficiently on MIMC WMNs can reduce the number of transmissions during the path update period by removal of unnecessary and duplicated broadcasts in each link. As a result, this scheme contributes to (2), which means the total number of transmissions to propagate the ith PREQ can be replaced with (5). Consider

$$N_{\text{broadcast}} = \frac{\sum_{i=1}^{n} N_{\text{IF}(X_i)}}{2}, \qquad (5)$$

$$N_{\text{total } tx} = N_{\text{PREQ}} \times N_{\text{broadcast}}, \qquad (6)$$

where $N_{\text{broadcast}}$ is the same as the number of all links in the network; it can be an independent value from the source or target node of the PREQ. Thus (3) can be simplified to (6), and the $N_{\text{total } tx}$ in Figure 2 is $3 \times 17 = 51$.

4.1.2. PREQ Information Table. For the PREQ prediction scheme, classification by the source node of the information of the received PREQ is required. The PREQ information table (PIT) is utilized for the classification, and it will exist in all mesh nodes. The number of PITs in each node is the same as the number of the PREQ source nodes. In a specific mesh node, the PIT for node A, which is the source node of the PREQ, is shown in Table 1. The PIT has data entries as the

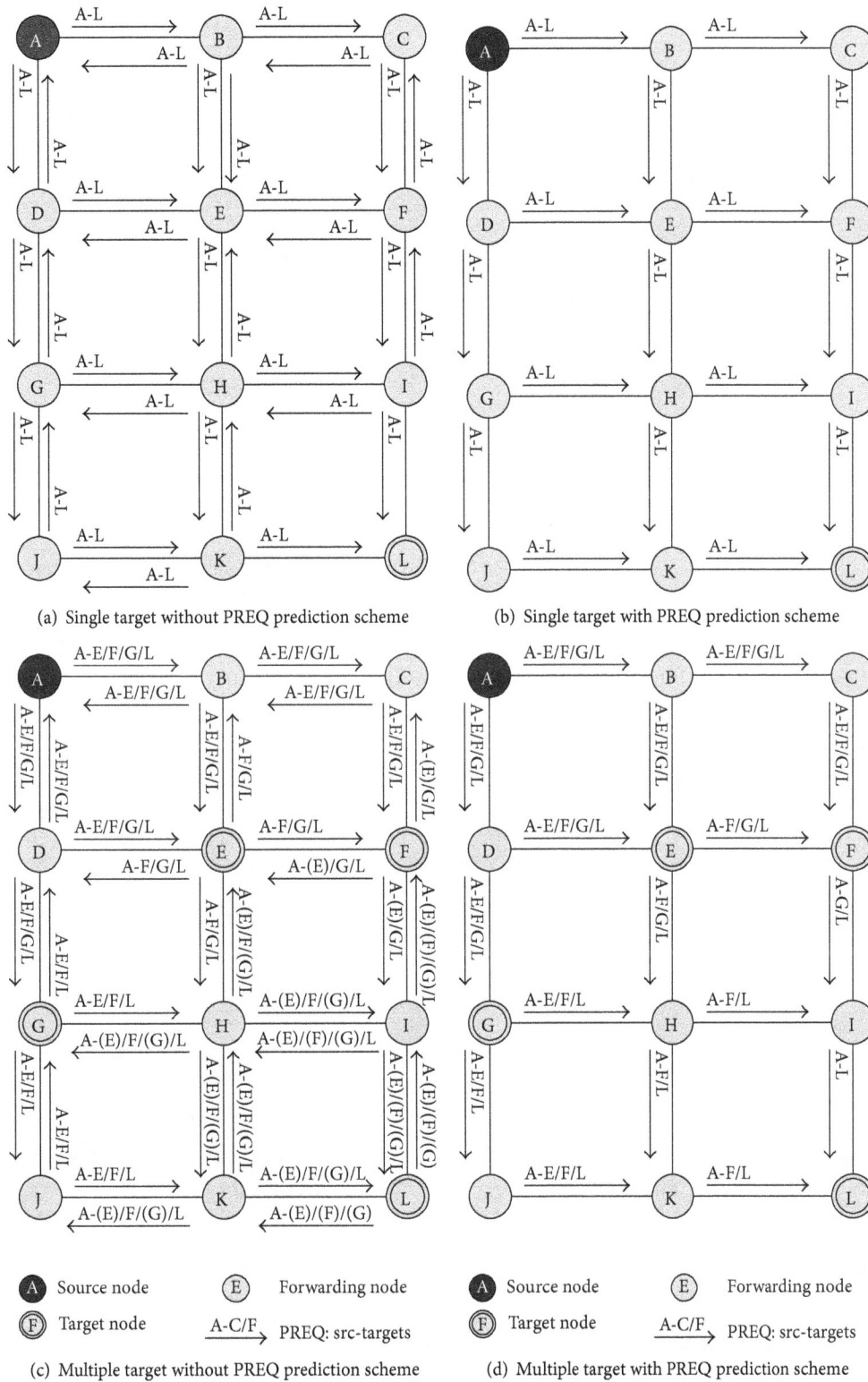

(a) Single target without PREQ prediction scheme

(b) Single target with PREQ prediction scheme

(c) Multiple target without PREQ prediction scheme

(d) Multiple target with PREQ prediction scheme

FIGURE 3: Applying the PREQ prediction scheme.

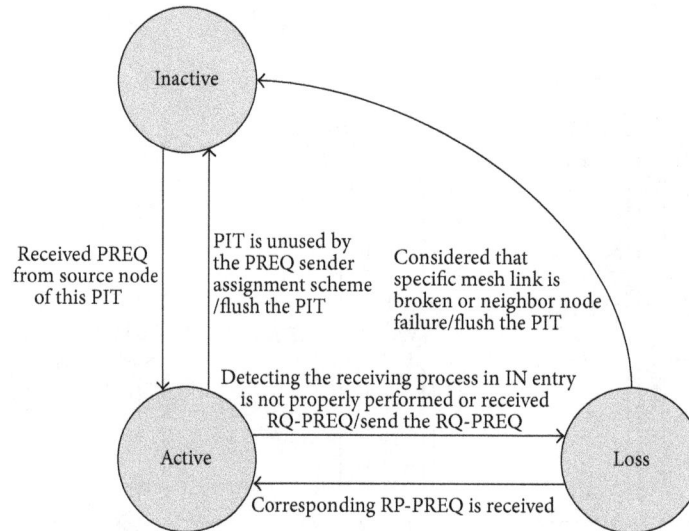

FIGURE 4: The state transition diagram of the PIT state.

number of the interfaces of the node, and the meaning of each field is as follows.

(i) *Interface*. The index of the interface within the mesh node.

(ii) *Direction*. The role assigned to the interface; it can have one of three values: NULL, IN (receiving), and OUT (transmitting).

(iii) *Targets*. The calculated target list of the recent PREQ.

(iv) *SN*. The sequence number of the received or transmitted PREQ.

(v) *Metric*. The routing metric of the PREQ that passed through this interface.

(vi) *Exp_time*. The elapse time from the last PREQ.

In particular, each entry of the PIT is mapped to a specific interface by the interface field. The direction field of entry sets the role of the corresponding interface.

4.1.3. States for PIT. All PITs within mesh nodes in MIMC WMNs using the PREQ prediction scheme have one of three states, as seen in Figure 4. When the PREQ is received, if the PIT that is relevant to the source node of the PREQ does not exist, the PIT and entry will be set, and the state of the PIT becomes active. In the active state, the initial setup of the PIT and the assigning of the interface are performed, and the PIT is continually updated based on the information of the received PREQ in every path update. If the PREQ is not received through the IN interface within the time limit, the state of the PIT will be changed to a loss state, and then a recovery request PREQ (RQ-PREQ) frame will be sent through the appropriate interface. In the loss state, if the recovery reply PREQ (RP-PREQ) is not received within the time limit, it is possible that the specific mesh link is broken or a neighbor node has failed. So, all entries of the PIT will be flushed, and the state will change to inactive. The behaviors

relating to the loss state will be covered in more detail in Section 4.2. In addition, the state of the PIT may change to inactive through the PREQ sender assignment scheme. This will be discussed in detail in Section 4.3.

The PREQ prediction scheme can react immediately against the change in routing information from the PIT state transition. In addition, it can distinguish a temporary PREQ loss and a topology change from a broken link or a node failure and so can greatly improve network reliability.

4.1.4. Processing Details for PREQ Prediction Scheme. The algorithms in the remainder of the paper use the expressions defined in Expression by PREQ Prediction Algorithms. The behaviors of the PIT depend on the state of the PIT. In the active state, which is the initial state, the setting of the PIT related to the source node of the PREQ is completed in all mesh nodes within only one period of PREQ propagation. The algorithm that as PIT in the active state follows is shown in Algorithm 1.

The algorithm in Algorithm 1 is divided into two parts. Lines 1 to 11 form the first part, when the PREQ is received, and lines 12 to 24 form the second part, when the PREQ is transmitted. When the PREQ is received, the PIT related to the source node of the PREQ is searched, and the entry that has same index of the interface as the receiving interface is brought from the PIT. Then, the following operation is performed by comparing the direction of entry. In line 5, the NULL direction of entry means that no PREQ received or transmitted passed through this interface, so the role of the interface is assigned as Receive. In line 6, the OUT direction indicates that the PREQ received earlier was already sent using this interface; thus, the interface is assigned to IN after comparing the metric. In other words, if this interface can have a better metric when the PREQ is received, then the interface will be an IN interface. Otherwise, it becomes an OUT interface. In line 9, the IN direction shows that the corresponding interface has been receiving already, so if there

---

Active State Algorithm

(1) Receive $PREQ$ and search PIT

(2) $e^{\text{in}} \leftarrow PIT$ entry which matched with receiving interface

(3) $p^{\text{in}} \leftarrow$ received $PREQ$

(4) **switch** (e_d^{in})

(5)   **case** $NULL$: $e_{t,s,m}^{\text{in}} \leftarrow p_{t,s,m}^{\text{in}}$, $e_d^{\text{in}} \leftarrow IN$

(6)   **case** OUT: $m^{\text{in}} \leftarrow p_m^{\text{in}} - m_{\text{last}}$

(7)   **if** $(e_m^{\text{in}}$ is better than $m^{\text{in}})$ drop $PREQ$

(8)   **else** $e_{t,s,m}^{\text{in}} \leftarrow p_{t,s,m}^{\text{in}}$, $e_d^{\text{in}} \leftarrow IN$

(9)   **case** IN:

(10)   **if** $(e_{t,s,m}^{\text{in}} = p_{t,s,m}^{\text{in}})$ drop PREQ

(11)   **else** $e_{t,s,m}^{\text{in}} \leftarrow p_{t,s,m}^{\text{in}}$

(12) Process AODV routing and prepare transmitting PREQ

(13) **for** each $e \in PIT$ **where** $e \neq e^{\text{in}}$ **do**

(14) $tx[e_i] \leftarrow 1$

(15)   **switch** (e_d)

(16)   **case** $NULL$: $e_{t,s,m} \leftarrow p_{t,s,m}^{\text{out}}$, $e_d \leftarrow OUT$

(17)   **case** IN:

(18)   **if** $(e_m$ is worse than $p_m^{\text{out}})$ $e_{t,s,m} \leftarrow p_{t,s,m}^{\text{out}}$, $e_d \leftarrow OUT$

(19)   **else** $tx[e_i] \leftarrow 0$

(20)   **case** OUT: $t_{\text{out}} \leftarrow e_t \cap p_t^{\text{out}}$

(21)   **if** $(e_s < p_s^{\text{out}}$ or $(e_s = p_s^{\text{out}}, (e_m \neq p_m^{\text{out}}$ or $e_t \neq t_{\text{out}})))$

(22) $e_{s,m} \leftarrow p_{s,m}^{\text{out}}$, $e_t \leftarrow t_{\text{out}}$

(23)   **else** $tx[e_i] \leftarrow 0$

(24) **end**

(25) transmit $PREQ$ **where** $tx[\cdots] = 1$

ALGORITHM 1: The algorithm for receiving and transmitting in the active state PIT.

are no changes to the target list, SN, and metric, then the PREQ will be dropped; otherwise, transmitting PREQ will be prepared after the entry update.

When the operation of entry setting related to PREQ receiving is finished, the other entries should be configured to rebroadcast the PREQ. First, in line 12, the PREQ is prepared after the AODV routing process. At this time, the node's own address is removed from the target list of this PREQ. Unlike the operation in PREQ receiving, all entries except the received entry are examined. In line 16, the NULL direction shows that no PREQ was received or transmitted by this interface, so the interface is appointed as a transmitting interface, and the transmit array is set to 1. In line 17, the IN entry means that the interface role is already allocated as receiving, in order to receive the PREQ from neighbor nodes; thus, the interface is assigned to OUT after comparing the metric. In line 20, there are two cases for the OUT direction. The first is where the transmitting PREQ has more current information than the entry, and the second is where the PREQ has the same SN as the entry but has modified information from a neighbor node. For these cases, the entry update will be performed if the PREQ has a greater SN than the entry or if the PREQ has the same SN as the entry but the metric or target list is different.

Figure 5 shows an example of the process for the initial PIT setting of node H in Figure 3(d). When the PREQ is received from node G, the PIT is established as in the table in the left side of Figure 5, and the PREQ is sent through interfaces 0, 2, and 3. Then, when a PREQ is received from node E, the received PREQ has a better metric than the saved metric in the first entry of the PIT, so the role of the interface mapped with this entry is assigned to IN. In addition, the targets field of entry is reconfigured, and the PREQ is retransmitted through interfaces 2 and 3.

4.1.5. Computational Complexity for PREQ Prediction Scheme. Clearly, additional computation is required for the PREQ prediction scheme, so we should consider the computational cost and complexity for this scheme. Fortunately, most additional computations involve linear operation, except comparison of the targets. The computational cost of searching PIT for the source of PREQ is $O(N_{\text{src}})$, but this operation is independent of other operations. Thus, the computational cost for transmitting PREQ process, which has more operations than receiving process, is $O(N_e(N_{T(e)} + N_{T(p)}))$, where N_e indicates the number of PIT entries, which is the same as the number of interfaces of a node, $N_{T(e)}$ shows the number of targets of each entry, and $N_{T(p)}$ is the number of targets of transmitting PREQ. The maximum number of targets approaches the total number of nodes in the whole network.

4.2. PREQ Loss Recovery Scheme. The PREQ frame is propagated based on the broadcast, so the probability of loss is significantly higher than data frames. In addition, the loss of the PREQ causes a routing malfunction, and it makes the routing protocol misunderstand the availability of the path in which the loss occurs. To overcome this issue, IA-AODV

Interface	Direction	Target list	Metric
0 (E)	OUT	E, F, L	450
1 (G)	IN	E, F, L	320
2 (K)	OUT	E, F, L	463
3 (I)	OUT	E, F, L	462

Interface	Direction	Target list	Metric
0 (E)	IN	F, G, L	310
1 (G)	IN	E, F, L	320
2 (K)	OUT	F, L	453
3 (I)	OUT	F, L	452

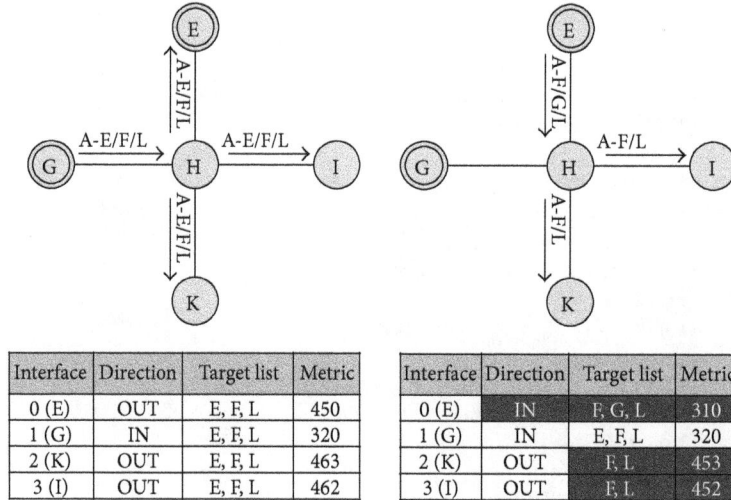

FIGURE 5: An example of the initial PIT settings in the active state.

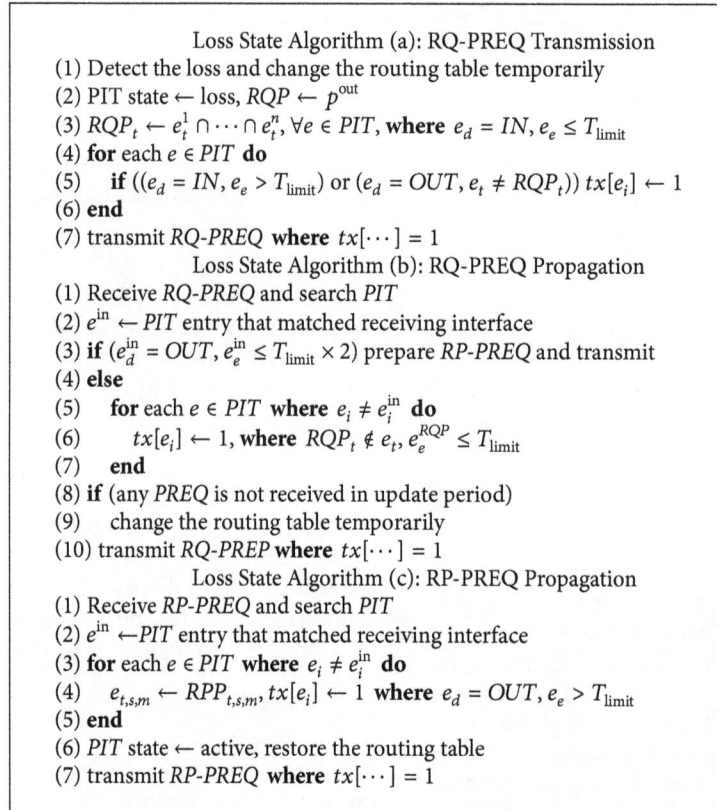

Loss State Algorithm (a): RQ-PREQ Transmission

(1) Detect the loss and change the routing table temporarily
(2) PIT state $\leftarrow$ loss, $RQP \leftarrow p^{\text{out}}$
(3) $RQP_t \leftarrow e_t^1 \cap \cdots \cap e_t^n, \forall e \in PIT$, **where** $e_d = IN, e_e \leq T_{\text{limit}}$
(4) **for** each $e \in PIT$ **do**
(5)   **if** $((e_d = IN, e_e > T_{\text{limit}})$ or $(e_d = OUT, e_t \neq RQP_t))$ $tx[e_i] \leftarrow 1$
(6) **end**
(7) transmit RQ-$PREQ$ **where** $tx[\cdots] = 1$

Loss State Algorithm (b): RQ-PREQ Propagation

(1) Receive RQ-$PREQ$ and search PIT
(2) $e^{\text{in}} \leftarrow PIT$ entry that matched receiving interface
(3) **if** $(e_d^{\text{in}} = OUT, e_e^{\text{in}} \leq T_{\text{limit}} \times 2)$ prepare RP-$PREQ$ and transmit
(4) **else**
(5)   **for** each $e \in PIT$ **where** $e_i \neq e_i^{\text{in}}$ **do**
(6) $tx[e_i] \leftarrow 1$, **where** $RQP_t \notin e_t, e_e^{RQP} \leq T_{\text{limit}}$
(7)   **end**
(8)   **if** (any $PREQ$ is not received in update period)
(9)   change the routing table temporarily
(10)   transmit RQ-$PREP$ **where** $tx[\cdots] = 1$

Loss State Algorithm (c): RP-PREQ Propagation

(1) Receive RP-$PREQ$ and search PIT
(2) $e^{\text{in}} \leftarrow PIT$ entry that matched receiving interface
(3) **for** each $e \in PIT$ **where** $e_i \neq e_i^{\text{in}}$ **do**
(4) $e_{t,s,m} \leftarrow RPP_{t,s,m}, tx[e_i] \leftarrow 1$ **where** $e_d = OUT, e_e > T_{\text{limit}}$
(5) **end**
(6) PIT state $\leftarrow$ active, restore the routing table
(7) transmit RP-$PREQ$ **where** $tx[\cdots] = 1$

ALGORITHM 2: The algorithms for a PIT in the loss state.

---

has a loss state so the PIT can respond appropriately to that problem.

The proposed IA-AODV routing protocol uses the MGMT frame as the RQ-PREQ and the RP-PREQ to resolve the PREQ loss. The RQ-PREQ and RP-PREQ contain the path information, just like a general PREQ; the path modification and the recovery request and response can be conducted at the same time using these MGMT frames. The algorithm using the RQ-PREQ and RP-PREQ is presented in Algorithm 2.

First, the condition needed to transit from the active state to the loss state is that the PREQ is not received in the specific entry of the PIT during an update period. Algorithm 2(a) indicates the process of RQ-PREQ occurrence. In lines 1 to 3

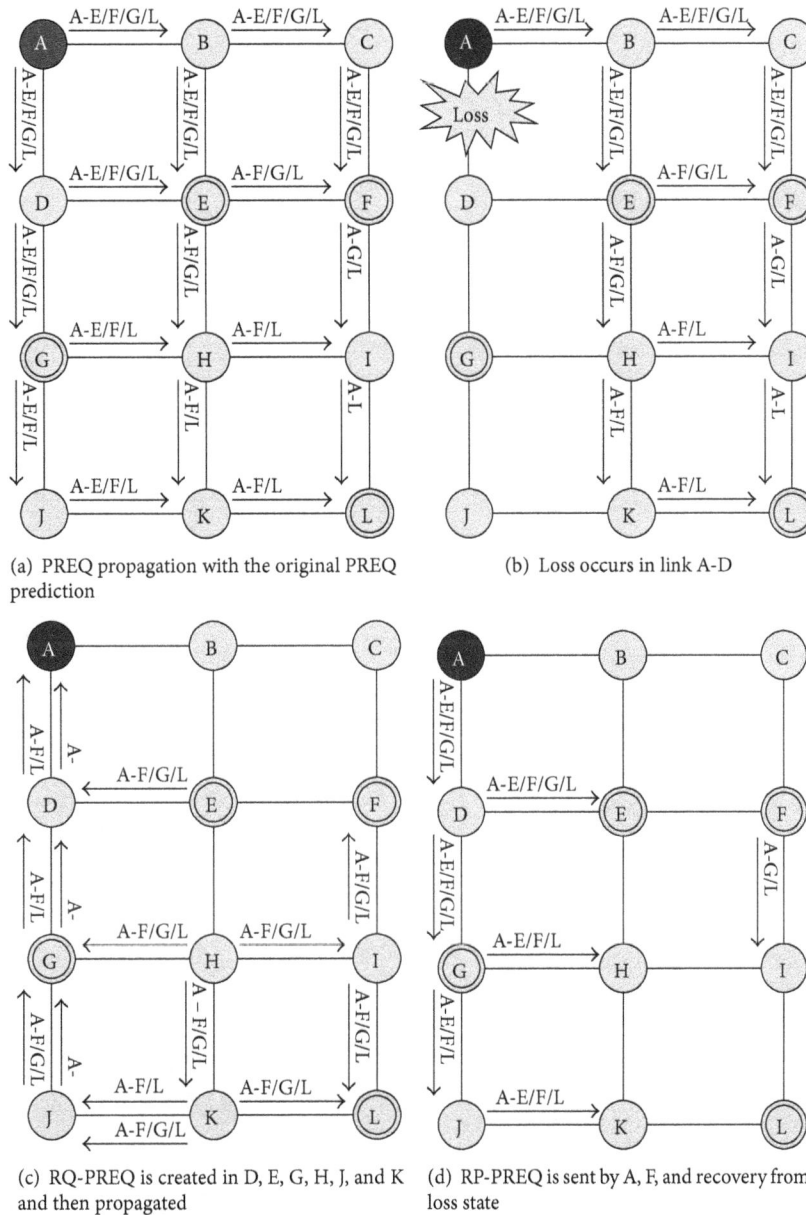

(a) PREQ propagation with the original PREQ prediction

(b) Loss occurs in link A-D

(c) RQ-PREQ is created in D, E, G, H, J, and K and then propagated

(d) RP-PREQ is sent by A, F, and recovery from loss state

FIGURE 6: An example of the PREQ loss recovery scheme.

of Algorithm 2(a), the mesh routing table will be temporarily changed if the loss is detected. Then, the RQ-PREQ is generated by using the PREQ that was recently sent, and then the target list of the RQ-PREQ is composed of the intersection of the target list of all PREQs that have been received normally. In lines 4 to 7, RQ-PREQ is sent through the IN interface that has not received a PREQ within the time limit and the OUT interface that has a different target list to the RQ-PREQ.

Algorithm 2(b) shows the process of RQ-PREQ propagation. In line 3, if the role of the interface receiving the RQ-PREQ is OUT and the elapsed time of interface is not exceeding two times of the time limit, it means that the PREQ loss occurs in the link that comprises the corresponding interface. Therefore, the mesh node generates RP-PREQ and

replies immediately. In particular, the node that has not received any PREQ within one update period temporarily replaces the routing table based on the information of the RQ-PREQ.

Finally, Algorithm 2(c) shows the process of RP-PREQ propagation. In lines 3 to 5, the RP-PREQ is propagated using the OUT interface that has not transmitted the PREQ within the time limit. When this process is done, the PIT state changes from the loss state to active state and the routing table is also restored to the state before the change. However, if the node cannot receive the RP-PREQ within a certain time then the temporary modifications of the routing table will be maintained.

Figure 6 introduces the process of the PREQ loss recovery scheme. Figure 6(a) shows PREQ propagation along with the PREQ prediction scheme, and Figure 6(b) shows the occurrence of a PREQ loss in link A-D. In this case, nodes D, G, and J cannot receive any PREQs, and the specific IN interfaces of nodes E, H, and K also are not able to receive PREQs. Therefore, these six nodes conduct propagation of the RQ-PREQ as in Figure 6(c). The PITs in nodes D, E, G, H, J, and K will transition to the loss state and will change the routing table temporarily.

Figure 6(d) shows propagation of the RP-PREQ. Nodes A and F can reply with RP-PREQ because those nodes received the RQ-PREQ through an OUT interface that sent the PREQ recently. The mesh node that received this RP-PREQ and that has a loss state PIT updates its own PIT and changes the state of the PIT to active. Also, it modifies the routing table to the state before the change and transmits the RP-PREQ via the proper interfaces. If the RP-PREQ is not received within the time limit in the mesh node that has the PIT in a loss state, the state of the PIT will be changed to inactive and all data of the PIT will be flushed. In addition, the temporary modification of the mesh routing table is maintained.

4.3. PREQ Sender Assignment Scheme. In the above sections, the MT-PREQ, the PREQ prediction scheme, and the PREQ loss recovery scheme are proposed in order to improve routing efficiency and network reliability in MIMC WMNs. In this section, we describe the scheme that reduces N_{src} in (4) when the network application that generates the many internal network paths is serviced in the MIMC WMNs. The general path discovery process of the AODV is that the node that has data sends the PREQ to the target node, and the target node replies with PREP. The initial decision about the PREQ/PREP nodes is kept until that path is not used anymore. However, if many internal network paths are generated, all nodes will handle the process of sending only one PREQ and only one PREP in the worst case scenario. Then, N_{PREQ} in (6) will increase up to the number of nodes in the network, which also increases routing complexity owing to the increase in the number of transfers required for routing. In addition, it may minimize the advantages of using MT-PREQ. Therefore, in this paper, a PREQ sender assignment scheme is proposed in order to resolve these issues.

First of all, the node that sends the PREQ frame is called the PREQ sender, and the node that replies with the PREP frame is called the PREP sender. Each node keeps the number of targets (TNUM) of the PREQ or PREP. In Figure 2, the TNUM of node A is 4, and that of node J is 3. When the new active path is required or the existing active path is disabled, the node sends the TNUM management frame, which contains the number of its own targets to all target nodes. The node receiving the TNUM frame from the target has to compare the number of its own targets. If the number of its own targets is bigger than the received TNUM, it will be a PREQ sender; otherwise, it becomes a PREP sender which waits for the PREQ. If the number of its own targets equals the received TNUM, it will be a PREQ or PREP sender by comparing the MAC address of the network interface card, which is a unique identifier.

Figure 7 shows the exchange of MGMT frames in the path discovery. Node A wants to send data to node B, so node A

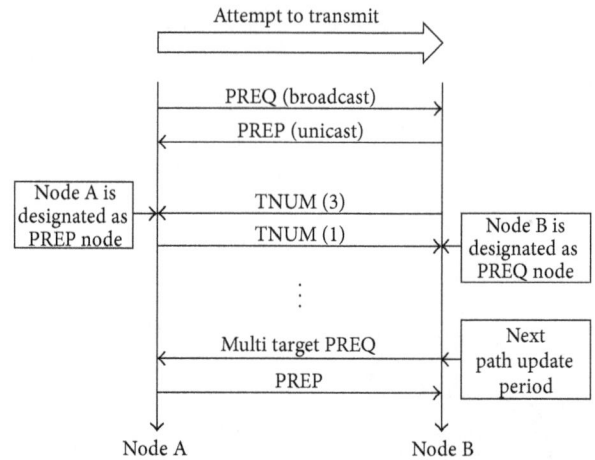

FIGURE 7: The operations by the PREQ sender assignment scheme.

transmits the PREQ, and B replies with PREP. After the path between A and B is established, nodes A and B exchange the TNUM frame containing the number of their own targets. In this process, node A will be a PREP sender and node B will be a PREQ sender. Then, node B will transmit the MT-PREQ in every path update period; and node A replies with PREP for the path update.

For this scheme, each node keeps the number of its own active paths. If the number of active paths changes, then a TNUM frame will be transferred after the exchange of PREQ/PREP in the next path update. In order to exchange the number of active paths, the node that receives the TNUM replies as the TNUM frame. If the node changes from PREP sender to PREQ sender, the state of the PIT relating to the corresponding source node will have to be changed to inactive. By this, the IA-AODV can process routing using the minimum number of PREQs, regardless of centralized traffic or distributed traffic. Thus, it achieves improvement of routing efficiency.

5. Experimental Results with a Real-World Test-Bed

5.1. Implementation of the MIMC WMNs Using IA-AODV Routing Protocol. In this section, the implementation of a real-world test-bed and the experiment scenarios are described, and the results of experiments are analyzed to evaluate the IA-AODV routing protocol in MIMC WMNs. For implementation of the mesh router in MIMC WMNs, Ubiquiti's Routerstation Pro and Mikrotik's R52Hn were utilized as the network board and the network card, respectively. Routerstation Pro is a high-performance embedded board with the Atheros AR7161 680 MHz chipset; it is suitable for the target platform of the MIMC mesh router because it has three mini-PCI slots. R52Hn uses the AR9220 chipset and supports up to 300 Mbps data rates at the PHY layer. In addition, the OpenWrt KAMIKAZE r22190 (Linux kernel 2.6.32.14) package was used for embedded Linux. The mesh engine was implemented by modifying compat-wireless-2.6.38-rc7-2, which is a driver package containing ath9k and mac80211.

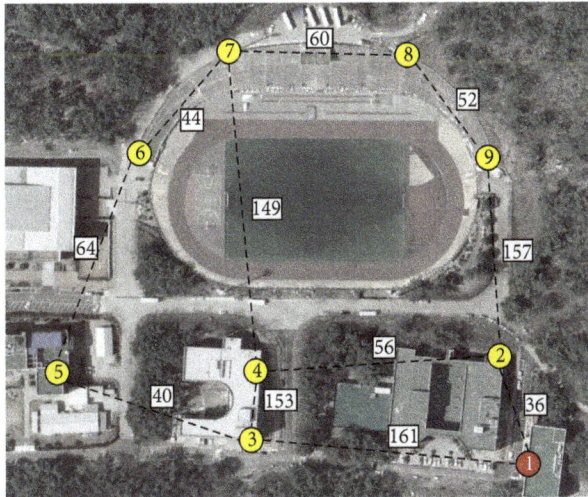

FIGURE 8: The topology of the outdoor test-bed for MIMC WMNs.

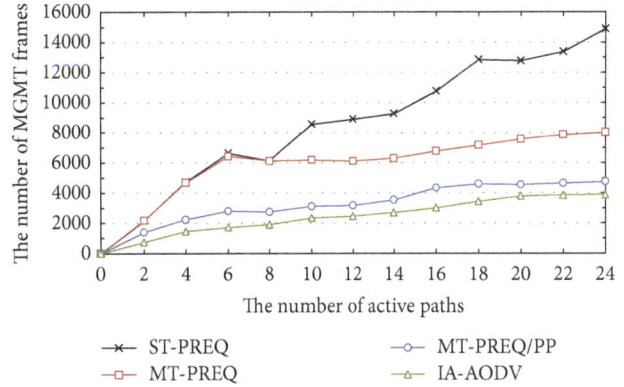

FIGURE 9: The number of MGMT frames needed for the path update process.

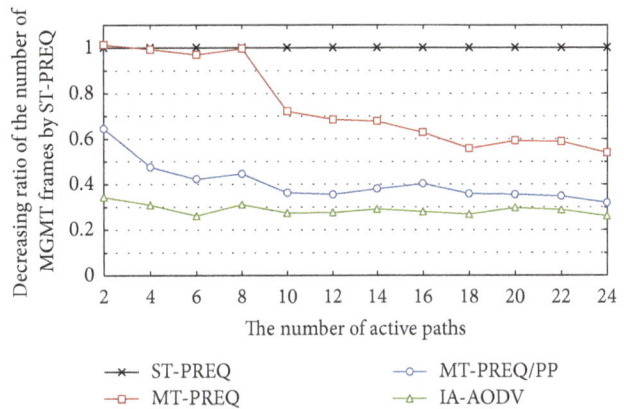

FIGURE 10: The decreasing ratio of the number of MGMT frames by ST-PREQ.

All experiments were performed on an outdoor test-bed located at Pusan National University. The test-bed was constructed on the roof of each building and the play-ground, with mesh routers using IA-AODV. The test-bed was composed of nine mesh routers as seen in Figure 8, and each mesh router had the same number of interfaces as links. Each interface runs on the 5 GHz channel and uses directional antennas. The topology is designed so that multiple multipaths can be generated.

5.2. Experiment Scenarios. The experiments used two scenarios (distributed traffic and centralized traffic) and were conducted to evaluate the efficiency of IA-AODV and the reliability of the MIMC WMNs. The number of MGMT frames, the ratio of routing malfunction, and the packet loss rate were measured in each experiment. All experiments used the iperf, which is a network performance measurement tool, and reported an average of the values that were performed 10 times; each experiment conducted UDP traffic transmission over two minutes. The source-destination pairs of UDP traffic were required to be distributed configuration for distributed traffic. In other words, nodes as many as possible were set as source or destination for the UDP traffic, so the traffic was configured to minimize the number of active paths in each node. In contrast, for the experiment using centralized traffic, the source-destination pairs were composed to be concentrated in a single node. Therefore, the traffic was organized that the active path was concentrated in a specific node.

5.3. The Evaluation and Analysis of Experimental Results

5.3.1. Experiments on Distributed Traffic. The topology in Figure 8 consists of 9 mesh nodes and 11 mesh links, and the experiment in this section makes the traffic maximize the distribution. Each mesh path becomes the active path if it is being used in communication, and it becomes inactive when communication is terminated. The experiment was set up to gradually increase the number of active paths. In the following experiments, ST-PREQ means the existing AODV routing protocol using ST-PREQ, and MT-PREQ signifies the existing AODV using MT-PREQ. Also, MT-PREQ/PP indicates the routing protocol used MT-PREQ and the PREQ prediction scheme. The IA-AODV proposed in this paper means the routing protocol that adopts full features, such as the MT-PREQ, the PREQ prediction scheme, the PREQ loss recovery scheme, and the PREQ sender assignment scheme. In addition, the number of MGMT frames of ST-PREQ, MT-PREQ, and MT-PREQ/PP means the sum of transfers of PREQ/PREP in each node. However, the number of MGMT frames of IA-AODV is the sum of transmissions of PREQ, PREP, TNUM, RQ-PREQ, and RP-PREQ.

Figure 9 shows the number of MGMT frames in two minutes in the distributed traffic. Figure 10 presents the percentage of MGMT frames of each scheme compared to ST-PREQ. Each node in the test-bed, which is used for communication, has at least two active paths in case of two to eight total active paths in the distributed traffic scenario. Therefore, each node is responsible for sending PREQ and

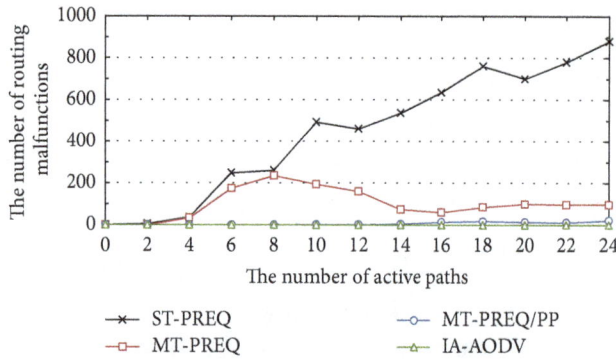

FIGURE 11: The number of routing malfunctions in 2 minutes.

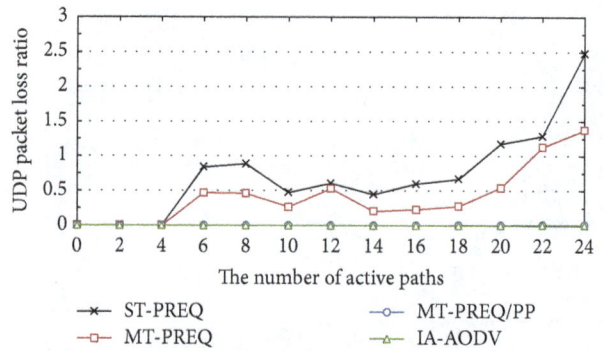

FIGURE 13: The packet loss ratio for distributed traffic.

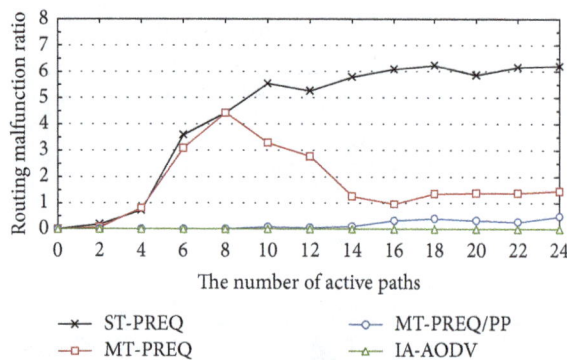

FIGURE 12: The routing malfunction ratio over distributed traffic.

PREP, so there are no differences between the number of MGMT frames using ST-PREQ and MT-PREQ in case of 2 to 8 total active paths. In terms of the number of PREQ targets, each node sends PREQs to only one target, so the MT-PREQ has no effect. However, the number of MGMT frames using the MT-PREQ is significantly reduced, when the effect of MT-PREQ occurs. Since then, even if the number of active paths increases, the MT-PREQ can maintain routing using about 74.72% of the MGMT frames' transmission compared to the ST-PREQ.

In contrast, the routing protocol using the PREQ prediction scheme can manage routing using 40.57% of the MGMT frames, and the IA-AODV routing protocol compared to ST-PREQ is able to decrease the number of MGMT frames to update the path by 28.83%. In particular, with IA-AODV, the TNUM frame, RQ-PREQ frame, and RP-PREQ frame are generated every time the active path changed. Therefore, if the active paths stay out longer, the number of MGMT frames used to perform the path update using IA-AODV can be reduced even more.

Figure 11 reports the number of routing malfunctions in 2 minutes, and Figure 12 shows the routing malfunction ratio. The routing malfunction as described in Section 3.2 means the mesh path changes to a poor path from loss of PREQ or the random receiving order of the PREQ, although the original path has no problem. The number of routing malfunctions means the sum of errors that occur in each

link, and the routing malfunction ratio is the number of malfunctions over the number of PREQs and PREPs received, which can also be defined as

Routing Malfunction Ratio

$$= \frac{\text{The number of routing malfunction}}{\text{Received PREQ + Received PREP}}. \qquad (7)$$

Similar to the number of MGMT frames, the number of routing malfunctions and the routing malfunction ratio of ST-PREQ and MT-PREQ have similar results in 2 to 8 active paths, but after that, the difference increases. The average routing malfunction ratio is 4.67% using ST-PREQ and 1.85% using MT-PREQ. The reason for the higher ratio of routing malfunctions is that a structural problem exists when ST-PREQ is applied. Although the structural problem of AODV in MIMC WMNs is removed using MT-PREQ, the routing malfunction ratio of MT-PREQ is still relatively high due to the issue of PREQ random receiving order. When MT-PREQ/PP is used, the ratio is 0.17% on average. The ratio decreases because the interface assignment of the PREQ prediction scheme can nearly remove the probability that the PREQ random receiving order problem will occur. However, it cannot handle loss of PREQs, so network reliability is not perfectly guaranteed. The IA-AODV shows a 0% routing malfunction ratio because the loss of PREQs can be resolved with the PREQ loss recovery scheme.

Figure 13 shows that the UDP packet loss ratio follows the various numbers of active paths. Overall, the pattern looks similar to the graph of the routing malfunction rate, and the average loss rate is 0.787% using ST-PREQ, 0.457% using MT-PREQ, 0.003% using MT-PREQ/PP, and 0.001% when IA-AODV is used. When the IA-AODV is adopted in MIMC WMNs, the packet loss exists due to the network congestion from increasing traffic.

In other words, when the IA-AODV proposed in this paper is used in MIMC WMNs that have distributed traffic, compared to ST-PREQ, it can use 28.83% of the MGMT frames for the path update, perfectly removing the routing malfunction, and decreasing the average loss rate by 0.0013%. Therefore, the proposed routing protocol can significantly improve not only routing efficiency but also network reliability.

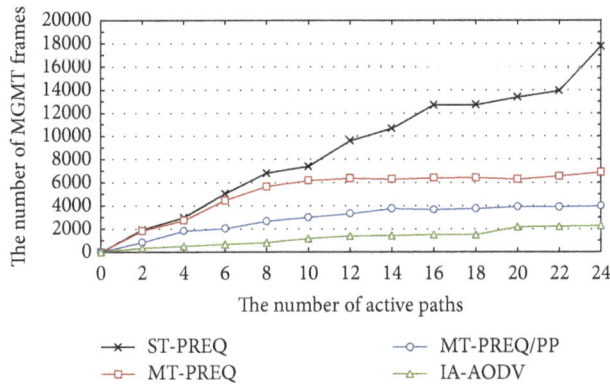

FIGURE 14: The number of MGMT frames for path update with centralized traffic.

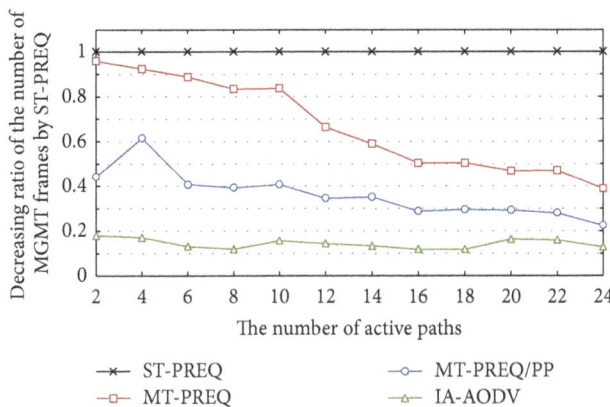

FIGURE 15: The decreasing ratio of the number of MGMT frames with the existing AODV.

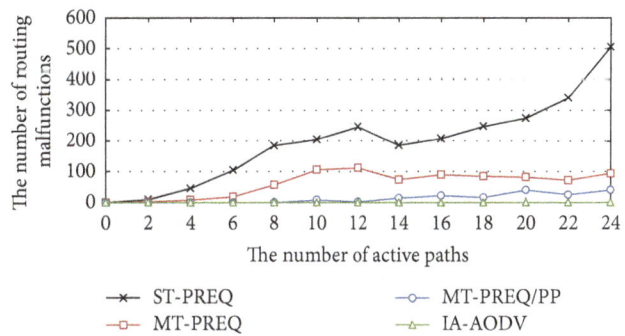

FIGURE 16: The number of routing malfunctions in 2 minutes over centralized traffic.

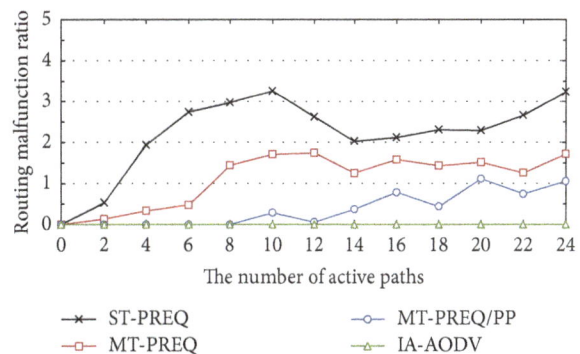

FIGURE 17: The routing malfunction ratio over centralized traffic.

5.3.2. Experiments on Centralized Traffic. For this scenario, the experiment was conducted with concentrated traffic, for this traffic form, UDP source, or destination assigned to a single node if possible. The experiment was performed with a gradual increase in the number of active paths.

Figure 14 shows the number of MGMT frames during two minutes from centralized traffics; Figure 15 shows the ratio of MGMT frames of each scheme compared to ST-PREQ. Though the UDP traffic is concentrated as much as possible, PREQ is sent by multiple nodes due to configuration so that the PREQ sender does not become one node. Thus, the efficiency of IA-AODV greatly increased compared to other routing protocols. Compared to ST-PREQ, the path update process utilized 66.91% of MGMT frames with MT-PREQ, 36.18% with MT-PREQ/PP, and 14.33% with IA-AODV. In particular, the efficiency of the PREQ sender assignment scheme will increase with centralized traffic.

When centralized traffic occurs, Figure 16 shows the number of routing malfunctions in two minutes, and Figure 17 shows the routing malfunction ratio. When using ST-PREQ the average routing malfunction ratio is 2.39%, 1.22% using MT-PREQ, 0.40% using MT-PREQ/PP, and 0% using IA-AODV, like the experiments over distributed traffic. When the experiments over centralized traffic are

compared to distributed traffic, the routing malfunction ratio was reported at a low level in every scheme.

Figure 18 shows that the UDP packet loss ratio follows the various numbers of active paths, which are formed with centralized traffic. Similar to distributed traffic, the result pattern looks like the routing malfunction ratio, and the average loss rate is 0.801% using ST-PREQ, 0.352% using MT-PREQ, 0.150% using MT-PREQ/PP, and 0.001% using the IA-AODV routing protocol. Also, the loss rate exists when using IA-AODV because of network congestion.

When the IA-AODV is adopted as a routing protocol in MIMC WMNs that have centralized traffic, similar to the experiment results over distributed traffic, it can update the path using only 14.33% of the MGMT frames, completely remove the routing malfunction, and decrease average loss rate by 0.0012%.

6. Conclusion

In this paper, we define some issues that occur when the existing AODV routing protocol has been applied to a MIMC WMN environment, and we propose IA-AODV to resolve the above problems, thereby improving routing efficiency and network reliability. When the existing AODV is adopted in MIMC WMNs, problems occur, such as the structural problem, the random receiving order of PREQs, and the PREQ loss. The IA-AODV routing protocol based on interface assignment and MT-PREQ contains a PREQ

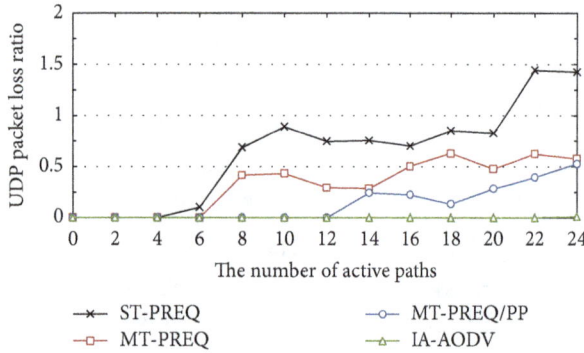

FIGURE 18: The packet loss ratio of each protocol over centralized traffic.

prediction scheme to assign the role of the interface, a PREQ loss recovery scheme to recover from PREQ loss, and a PREQ sender assignment scheme to alleviate routing complexity.

We implemented an outdoor test-bed for a MIMC WMN to evaluate IA-AODV and analyzed the proposed routing protocol from various angles. In an environment of distributed traffic, IA-AODV, compared to the existing AODV using ST-PREQ, is able to perform the path update with only 28.83% of the MGMT frames, perfectly remove the routing malfunction, and decrease average loss rate by 0.0013%. Also, with centralized traffic, IA-AODV can conduct the path update using about 14.33% of the MGMT frames, completely eliminate the routing error, and decrease the loss rate by 0.0012%. In other words, the proposed IA-AODV routing protocol has been able to significantly increase reliability and efficiency of MIMC WMNs.

One of the future directions we want to study is the effect caused by scaling up the network, such as scaling up the number of interfaces in each node and/or the number of nodes in the network. We will evaluate the computational cost for each node when the network size greatly increases.

Expression by PREQ Prediction Algorithms

PIT:	PREQ information table
e:	Entry of PIT
e^{in}:	Entry which has IN direction
P^{in}:	Received PREQ
P^{out}:	PREQ that will be sent
$x_{i,d,t,s,m,e}$:	The interface, direction, targets, SN, metric, and exp_time of entry or PREQ, respectively
m:	Metric
m_{last}:	Metric of the last hop
t:	Targets
$tx[\cdots]$:	The array that shows the interfaces used by transmission
RQP:	RQ-PREQ
RPP:	RP-PREQ
T_{limit}:	The time limit used in loss decision.

Conflict of Interests

The authors declare that there is no conflict of interests regarding the publication of this paper.

Acknowledgments

This research was supported by Basic Science Research Program through the National Research Foundation of Korea (NRF) funded by the Ministry of Education, Science and Technology (2012R1A1A2043531). This research was supported by the MSIP (Ministry of Science, ICT and Future Planning), Korea, under the ITRC (Information Technology Research Center) Support Program (NIPA-2014-H0301-14-1048) supervised by the NIPA (National IT Industry Promotion Agency).

References

[1] IEEE, "Part 11: wireless LAN medium access control (MAC) and physical layer (PHY) specifications, amendment 10: mesh networking," IEEE 802.11s, IEEE Standards Association, 2011.

[2] A. Raniwala and T. Chiueh, "Architecture and algorithms for an IEEE 802.11-based multi-channel wireless mesh network," in *Proceedings of the 24th Annual Joint Conference of the IEEE Computer and Communications Societies (INFOCOM '05)*, vol. 3, pp. 2223–2234, March 2005.

[3] A. Raniwala, K. Gopalan, and T. Chiueh, "Centralized channel assignment and routing algorithms for multi-channel wireless mesh networks," *Mobile Computing and Communications Review*, vol. 8, no. 2, pp. 50–65, 2004.

[4] P. Kyasanur and N. H. Vaidya, "Routing and interface assignment in multi-channel multi-interface wireless networks," in *Proceedings of the IEEE Wireless Communications and Networking Conference (WCNC '05)*, pp. 2051–2056, IEEE, March 2005.

[5] P. Kyasanur and N. H. Vaidya, "Routing and link-layer protocols for multi-channel multi-interface ad hoc wireless networks," *Mobile Computing and Communications Review*, vol. 10, no. 1, pp. 31–43, 2006.

[6] G. Li, L. Yang, W. S. Conner, and B. Sadeghi, "Opportunities and challenges for mesh networks using directional antennas," in *Proceedings of the 1st IEEE Workshop on Wireless Mesh Networks (WiMESH '05)*, IEEE, September 2005.

[7] U. Kumar, H. Gupta, and S. R. Das, "A topology control approach to using directional antennas in wireless mesh networks," in *Proceedings of the IEEE International Conference on Communications (ICC '06)*, pp. 4083–4088, IEEE, Istanbul, Turkey, July 2006.

[8] C. E. Perkins and E. M. Royer, "Ad-hoc on-demand distance vector routing," in *Proceedings of the 2nd IEEE Workshop on Mobile Computing Systems and Applications (WMCSA '99)*, pp. 90–100, IEEE, February 1999.

[9] B. Yu and H. Fei, "Performance impact of wireless mesh networks with mining traffic patterns," in *Proceedings of the 5th International Conference on Fuzzy Systems and Knowledge Discovery (FSKD '08)*, pp. 493–497, IEEE, October 2008.

[10] S. Waharte, R. Boutaba, Y. Iraqi, and B. Ishibashi, "Routing protocols in wireless mesh networks: challenges and design considerations," *Multimedia Tools and Applications*, vol. 29, no. 3, pp. 285–303, 2006.

[11] J.-S. Kim, S.-H. Chung, C.-W. Ahn, and W.-S. Kim, "Implementing channel-load aware routing scheme for IEEE 802.11 mesh networks," in *Proceedings of the IEEE 7th International Conference on Wireless and Mobile Computing, Networking and Communications (WiMob '2011)*, pp. 525–528, October 2011.

[12] J.-S. Kim, S.-H. Chung, Y.-S. Lee, C.-W. Ahn, W.-S. Kim, and M.-S. Jung, "Design and implementation of a WLAN mesh router based on multipath routing," in *Proceedings of the International Conference on Information Networking (ICOIN '11)*, pp. 154–159, IEEE, Barcelona, Spain, January 2011.

[13] W. S. Kim and S. H. Chung, "Design and implementation of IEEE 802.11n in multi-hop over wireless mesh networks with multi-channel multi-interface," in *Proceedings of the IEEE 9th International Conference on High Performance Computing and Communication (HPCC '12)*, pp. 707–713, IEEE, Liverpool, UK, June 2012.

[14] W.-S. Kim and S.-H. Chung, "Design of optimized AODV routing protocol for multi-interface multi-channel wireless mesh networks," in *Proceedings of the 27th IEEE International Conference on Advanced Information Networking and Applications (AINA '13)*, pp. 325–332, Barcelona, Spain, March 2013.

[15] A. A. Pirzada and M. Portmann, "High performance AODV routing protocol for hybrid wireless mesh networks," in *Proceedings of the 4th Annual International Conference on Mobile and Ubiquitous Systems: Computing, Networking and Services (MobiQuitous '07)*, pp. 1–5, IEEE, August 2007.

[16] A. A. Pirzada, M. Portmann, and J. Indulska, "Performance analysis of multi-radio AODV in hybrid wireless mesh networks," *Computer Communications*, vol. 31, no. 5, pp. 885–895, 2008.

[17] A. Valera, W. K. G. Seah, and S. V. Rao, "Cooperative packet caching and shortest multipath routing in mobile ad hoc networks," in *Proceedings of the IEEE Societies 22nd Annual Joint Conference of the IEEE Computer and Communications (INFOCOM '03)*, vol. 1, pp. 260–269, March-April 2003.

[18] I. Chakeres and E. Belding-Royer, "AODV routing protocol implementation design," in *Proceedings of the 24th International Conference on Distributed Computing Systems Workshops (ICDCSW '04)*, pp. 698–703, IEEE, March 2004.

[19] M. Singh and S. G. Lee, "Decentralized hybrid wireless mesh protocol," in *Proceedings of the 2nd International Conference on Interaction Sciences: Information Technology, Culture and Human (ICIS '09)*, pp. 824–829, ACM, November 2009.

[20] S. J. Lee and M. Gerla, "AODV-BR: backup routing in ad hoc networks," in *Proceedings of the Wireless Communications and Networking Conference (WCNC '00)*, pp. 1311–1316, IEEE, September 2000.

[21] A. P. Subramanian, M. M. Buddhikot, and S. Miller, "Interference aware routing in multi-radio wireless mesh networks," in *Proceedings of the 2nd IEEE Workshop on Wireless Mesh Networks (WiMESH '06)*, pp. 55–63, IEEE, September 2006.

[22] R. Draves, J. Padhye, and B. Zill, "Routing in multi-radio, multi-hop wireless mesh networks," in *Proceedings of the 10th annual international conference on Mobile computing and networking (MobiCom '04)*, pp. 114–128, ACM, September 2004.

[23] R. D. C. Paschoalino and E. R. M. Madeira, "A scalable link quality routing protocol for multi-radio wireless mesh networks," in *Proceedings of the 16th International Conference on Computer Communications and Networks (ICCCN '07)*, pp. 1053–1058, Honolulu, Hawaii, USA, August 2007.

[24] T. Liu and W. Liao, "Interference-aware QoS routing for multi-rate multi-radio multi-channel IEEE 802.11 wireless mesh networks," *IEEE Transactions on Wireless Communications*, vol. 8, no. 1, pp. 166–175, 2009.

[25] T. C. Tsai and S. T. Tsai, "A cross-layer routing design for multi-interface wireless mesh networks," *EURASIP Journal on Wireless Communications and Networking*, vol. 2009, Article ID 208524, 2009.

[26] T. Anker, D. Dolev, and B. Hod, "Cooperative and reliable packet-forwarding on top of AODV," in *Proceedings of the 4th International Symposium on Modeling and Optimization in Mobile, Ad Hoc and Wireless Networks (WiOpt '06)*, pp. 1–10, IEEE, April 2006.

[27] J. Boice, J. J. Garcia-Luna-Aceves, and K. Obraczka, "Combining on-demand and opportunistic routing for intermittently connected networks," *Ad Hoc Networks*, vol. 7, no. 1, pp. 201–218, 2009.

12

V-MGSM: A Multilevel and Grouping Security Virtualization Model for Mobile Internet Service

Hui Zhu, Yingfang Xue, Xiaofeng Chen, Qiang Li, and Hui Li

State Key Laboratory of Integrated Services Networks, Xidian University, No. 2 South Taibai Road, Yanta District, Xi'an 710071, China

Correspondence should be addressed to Hui Zhu; xdzhuhui@gmail.com

Academic Editor: David Taniar

With the pervasiveness of smart phones and the advance of Mobile Internet, more and more Mobile Internet services migrated to the cloud service platform for better user experience. As one of the most indispensable components of the cloud computing infrastructure, virtualization technology has attracted considerable interest recently. However, the flourish of virtualization still faces many challenges in information security. In this paper, we propose a novel architecture, called multilevel and grouping security model for virtualization (V-MGSM), for the security of resources in cloud computing platform. Specifically, to fulfill the balance between information sharing and privacy preservation, the virtual machines (VMs) are divided into diverse groups based on their corresponding entities, and each VM in the same group is assigned to different security level according to security requirements. Besides, the operation between VMs is based on mandatory access control mechanism. Detailed security analysis shows that the proposed V-MGSM can provide a secure communication mechanism for VMs and implement the synchronous updates of the borrowed data. Ultimately, we implement V-MGSM in Xen for experiments, and the results demonstrate that V-MGSM can indeed achieve data security and privacy protection efficiently for Mobile Internet service.

1. Introduction

In our information society, Mobile Internet which makes Internet connection accessible and ubiquitous is becoming increasingly adopted by ordinary consumers. With wide popularity and broad application, Mobile Internet service performs adding huge income to business communities and the Mobile Internet industry has recently started to take off [1–3]. By the end of March 2013, there were more than 81 million Mobile Internet users in China which generated beyond 20 billion dollars Mobile Internet market scale.

With tremendous advancement in mobile technology, people expect more services whenever and wherever possible. Although current mobile technologies and improvements allow shopping online, chatting online, or any other mobile applications through mobile terminals, several certain issues which hinder the communication process need to be addressed. For example, people tend to use mobile terminals such as smart phones to do anything realizable which causes series of problems including the following: the calculation load is magnified, the battery standby time is shortened, and the storage is limited. For the presented problems of mobile terminals, especially the limited computing ability and the limited storage discussed above, it is necessary to migrate Mobile Internet services to cloud platform to decrease the amount of computation and extend the standby time. A main advantage of cloud computing is to provide large storage and powerful computing ability by cloud server [4–6]. From this prospective, the cooperation between cloud computing and the Mobile Internet which is promising for cloud computing can transport applicable computing ability and storage from terminals to the cloud server. In short, the emergence of cloud computing is significant for the continued development of the Mobile Internet.

Figure 1 illustrates the necessity for traditional Mobile Internet services to be migrated to cloud service platform. Specifically, the larger the service scale is, the more the physical devices and instruments it requires to execute implementation are. In particular for Mobile Internet services, since it is constrained for the capacity of mobile terminals,

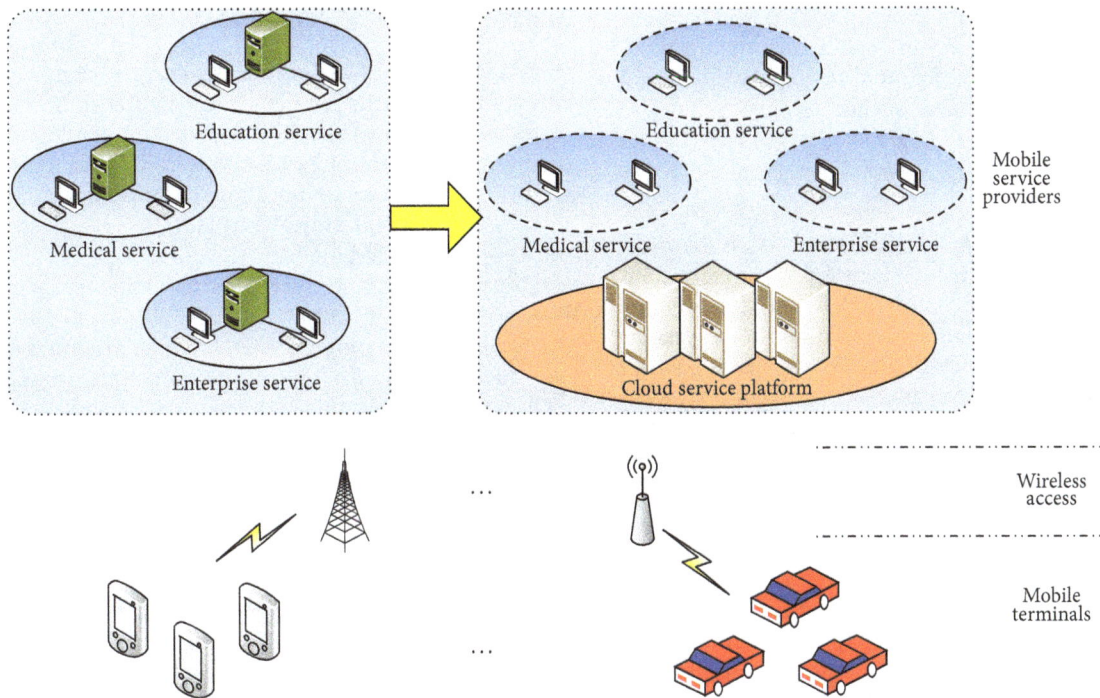

FIGURE 1: Cloud platform for Mobile Internet service.

more resources are required in the server of service provider [7, 8]. Benefiting from the increasing of computing ability and the decreasing of costs, virtualization technology brings many advantages over general technology in IT industry, such as raising utilization of hardware and promoting quality of service. For full use of infrastructure and reduction of the cost, more and more Mobile Internet service providers tend to migrate their services into a cloud service platform. However, the migrations of Mobile Internet service providers from conventional model to cloud services platform result in security issues which has become a critical concern. With virtualization, multitenant including medical service, education service, and enterprise service offers different services upon the same platform and may incur new vulnerabilities to the cloud platform, especially privacy violation between different service providers and internal access by VMs within one service provider. Even though virtualization security has attracted considerable interest in recent years and several secure solutions have been proposed [9, 10], the flourish of secure solutions for virtualization still faces many challenges in the balance between information sharing and privacy preservation. Security issues on virtualization technology have been the dominating barrier to the development and widespread use of Mobile Internet.

(i) A VM illegally accesses to another relatively important VM in a virtualization system without authorization, which may cause the leakage of confidential information.

(ii) To improve the efficiency of an organization, it is necessary to guarantee the information sharing among VMs which belong to the same organization.

(iii) Virtual machine monitor (VMM) has the highest priority in a virtualization system, and the virtualization system will be broken if VMM is accessed illegally.

In order to construct a secure communication mechanism for virtualization [11] and fulfill the balance between information sharing and privacy preservation, a new multilevel and grouping security access control model (V-MGSM) is proposed in this paper. And our main contributions are threefold.

(i) In V-MGSM model, to implement the efficient isolation for VMs corresponding to different organizations, all VMs have been divided into several groups, which can avoid the unnecessary access and provide the privacy preservation among different groups.

(ii) For reasonable and secure access control, different security levels are assigned to VMs in the same group. In particular, a low-level VM is not able to access other VMs which have higher security levels; one VM could not access other VMs which are divided into other groups; and VMs with high level could manage VMs with low level. Then, the confidential information in a high-level VM will not be obtained by some malicious low-level VMs.

(iii) In the same group, a high-level VM could borrow data from a low-level VM, and when the borrowed data is modified in the low-level VM, the corresponding data in the high-level VM will be updated synchronously.

The remainder of this paper is organized as follows. In Section 2, we survey the related works. In Section 3, we present the architecture of our proposed V-MGSM model. Then, we provide the correctness and security proof of V-MGSM model in Section 4, followed by implementation in Section 5. Finally, we draw our conclusions in Section 6.

2. Related Works

2.1. Security Mechanisms for Virtualization. With the prosperity of virtualization, the security of communications between VMs has attracted considerable interest recently. The necessity of communication between VMs in the specific application scenario was firstly pointed out in [12], which mentioned that the implementation of access control mechanism for VMs' communication was also essential. Furthermore, as a mandatory access control (MAC) mechanism, sHype [13] was one of the most well-known security architectures for VMM. Based on Xen, both the communication between VMs and the hardware resource access by VMs could be controlled well by sHype, which determine the kind of VMs that should run simultaneously according to the interest conflicts. However, the type of communication between VMs was not defined in sHype. Recently, some attention has been devoted into communication types between VMs. A Prioritized Chinese Wall (PCW) policy [14] was proposed, constructing a set of VMs which could communicate with each other dynamically. A role-based access control policy [15] was proposed later, which focused on the communication between guest VMs and VMM layer. However, they all ignored inter-VMs communication. A Virt-BLP model [16] based on BLP [17] model was proposed, which met the requirements of multilevel security for virtualization. As a security communication mechanism for virtual machine system, Virt-BLP model secured the communication between VMs, while the memory taken by access control matrix was so large.

2.2. Development of Multilevel Access Control Model. At present, there are several security models and methods used in nonvirtualized system to guarantee the access control. Bell-LaPadula Model [18], SeaView model [19], and multilevel relational (MLR) [20] model have been introduced after thirty years of the research on multilevel access control [21]. Compared with other models previously proposed, MLR was clear in semantics and perfect in function after years of research on the security access control model and has been widely used in several areas and was pretty secure as a multilevel security model. However, in a multilevel user system, due to the fact that a high-level user may want to borrow data from a low-level user when it is reasonable and necessary, a high-level user can borrow data from low-level tuple in MLR and the high-level tuple which does have this borrowed data will be updated or deleted synchronously when the borrowed data has been modified by the low-level user. Meanwhile, a deadly secure risk caused by data borrowing in MLR is that the change of a high-level user's perspective by a low-level user may result in the leakage of confidential data from high-level user if a borrowed relation exists between these two users.

3. Proposed V-MGSM Model

In this section, we redesign the architecture, elements, data explanations, security theorem, and state transition rules for better use in virtualization and propose a multilevel and grouping access control model on the basis of MLR model to construct a secure communication mechanism for VMs. We first review the architecture of virtualization, which serves as the basis of V-MGSM model.

The practical architecture of virtualization is shown as in Figure 2(a). With a software layer VMM inserted between hardware and operating systems, several VMs could run on one single physical machine at the same time. In addition, for secure communication between VMs, a mandatory access control (MAC) module is constructed to control the flow of communication between VMM and VMs. However, from different VMs' perspectives, VMM is regarded as different entities. As shown in Figure 2(b), a virtualization system could be divided into several groups according to VMs' corresponding organizations such as User System 1 and User System 2, each of which consists of VM1 and VM2, VM3 and VM4, respectively. What is more, in user system's perspective, they are managed and controlled by VMM1 and VMM2 separately, although there is only one VMM in virtualization system indeed.

3.1. Elements of V-MGSM Model. Before introducing the formal definitions of the multilevel relational model and multilevel relation for virtualization, in this part, we define the basic elements of V-MGSM model as follows.

(i) Subject and Object. Subjects represent active entities such as processes and users; objects represent passive entities such as data and files. In V-MGSM model, one VM is considered as a subject or an object based on the information flow of a communication process.

(ii) Security Level. Assuming that there are two VMs in virtual machine system, if and only if the subject and the object are in the same organization and the security level of subject is higher than or equal to that of the object, then the object can be accessed by the subject.

(iii) Data Attributes. In V-MGSM model, with virtualization system grouped on the basis of entities, in the scenario of communication between VMs, a VM could be queried by another VM only when they are corresponding to the same entity.

Definition 1 (independent multilevel relational model for virtualization). Let D denote an attribute domain, let A_i denote a data attribute from domain D, let C_i denote the security level of A_i, and let RC denote the security level of a VM. The relational instances with different levels are described as $(A_1, C_1, A_2, C_2, \ldots, A_n, C_n, RC)$. The domain of security levels could be specified by the set $\{L, \ldots, H\}$, in which all of security levels of RC and A_i from the infimum L to the supremum H are involved.

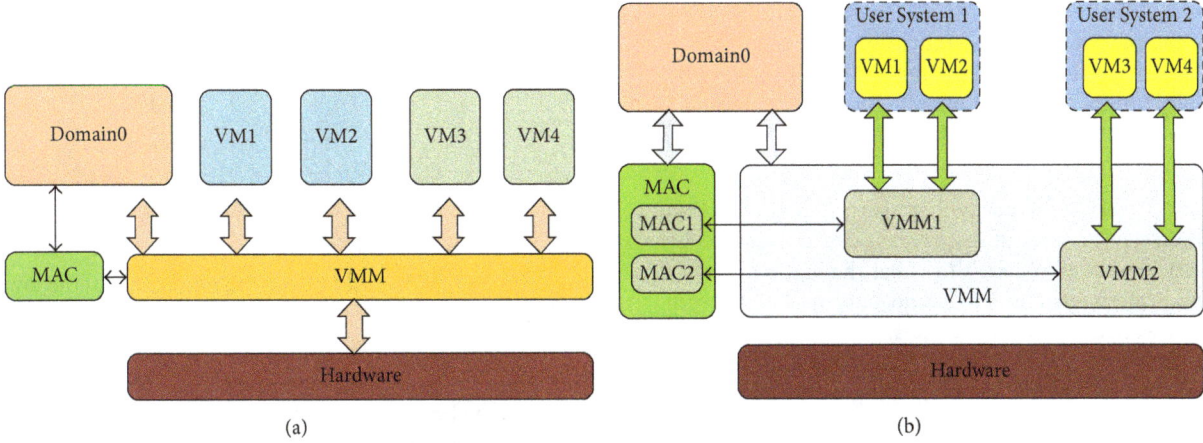

FIGURE 2: Architecture of virtualization: (a) piratical architecture; (b) architecture viewed by VMs.

Definition 2 (multilevel relation for virtualization). Let $r[rc]$ denote the security level of a VM, let $r[c_i]$ denote the security level of a_i, and the records of a VM in the multilevel relations are denoted by $(a_1, c_1, a_2, c_2, \ldots, a_n, c_n, rc)$. Then, all data of data attributes and security levels in multilevel relation meet the following expression:

$$a_i \in D_i, \qquad r[rc] \le r[c_i]. \tag{1}$$

Concretely, each relational instance $(A_1, C_1, A_2, C_2, \ldots, A_n, C_n, RC)$ consists of many records $(a_1, c_1, a_2, c_2, \ldots, a_n, c_n, rc)$, which are the records of VMs. All data and security levels in the relational instance meet the following expression:

$$a_i \in D_i \vee a_i = \text{null}, \qquad c_1, rc \in \{L, \ldots, H\}, \qquad rc \ge c_1. \tag{2}$$

3.2. Data Explanations of V-MGSM Model. To avoid the fuzziness in the querying result which may occur in accessing a VM in MLR, two important properties, Entity Multi-Instance and Record Multi-Instance, are introduced. And Table 1 is an example taken to describe the definition of Entity Multi-Instance and Record Multi-Instance. Two entities (*Gun*, *U*) and (*Gun*, *S*) are described in Table 1, which are created by *U*-subject and *S*-subject, respectively; *U* level is the lowest level, *U*-record is a basic record and could only access *U*-record, and it could only be deleted by *U*-subject as well; *TS* is higher than *U* in security level. Thus, both *TS*-record and *U*-record could be accessed by *TS*-subject; *TS*-subject has borrowed the value of attribute Range "2" which is authorized by *TS*-subject. However, the value of the attribute quantity in *TS*-subject is set by *TS*-subject as the value of attribute quantity in *U*-subject is not authorized by *TS*-subject. Then, two kinds of multi-instance including Record Multi-Instance and Entity Multi-Instance are mentioned in the example shown above.

Definition 3 (Record Multi-Instance). In a multilevel relation, the records which have the same value of A_1 and different values of RC are called Record Multi-Instance.

TABLE 1: Entity with multi-instance.

Name	C_1	Range	Quantity	RC
Gun3	S	1	Null	S
Gun3	U	2	Null	TS
Gun3	U	2	5000	U

Definition 4 (Entity Multi-Instance). In a multilevel relation, the records which have the same value of A_1 and different values of C_1 are called Entity Multi-Instance, meaning that the levels of entities are different. Then, the meaning of $r(A_1, C_1, A_2, C_2, \ldots, A_n, C_n, RC)$ can be explained as follows.

(i) Primary Key A_1 and Security Level Attribute C_1 in V-MGSM Model. Let A_1 denote the value of the primary key as well as the name of a VM, and let $r[C_1]$ denote both the identity of entity corresponding to VM and the security level of the entity in an instance r. $r[C_1] = c_1$ implies that the security level of the entity is c_1 and the entity with security level c_1 is called c_1-entity. In V-MGSM, with VMs divided into different groups according to their corresponding entities, different security levels are assigned to these entities. Given the value of C_1 at the specific moment, only one entity could be authorized by subjects who have the same value of C_1.

(ii) Security Level Attribute RC in a Record. $r[RC] = rc$ represents that record r is inserted by a VM with level rc, called rc-VM, and the data in this record is authorized by rc-VMs. If no corresponding record existed, it means that the entity is not authorized by rc-VM. The security level of a record, called rc, is used to judge whether the access is successful. Specifically, the data of VMs could be accessed by a subject who has a higher level than rc. In other words, a rc-VM could access other VMs r' if $r'[c_1] = r[c_1], r'[RC] \le rc$. The data in a record r could only be deleted or updated by its owner and a borrowed relation exists between two VMs r and r' if and only if they are divided into the same group and meet $r'[RC] \ge r[RC] \wedge r'[A_i, C_i] = r[A_i, C_i], r \in$ relation R.

3.3. Security Theorem of V-MGSM. In this part, we first elaborate the modified Simple-security-*_-property, which is selected as one of the basements for V-MGSM model and was initially proposed in BLP model. Then, four integrated properties designed for V-MGSM are introduced.

Definition 5 (Simple-security-*_-property). In V-MGSM model, the security level of a subject is not dominated by the security level of an object. VMM has the highest priority in a virtualization system; VM could only read data from other VMs with lower levels in the same group and could not read data from VMM or other VMs with higher levels, eliminating the phenomenon of reading upward. Neither VMM nor other VMs with higher security levels could update or delete any data in low-level VMs, eliminating the phenomenon of writing downward. Moreover, two basic secure theorems defined in BLP model claimed that the phenomena of reading upward and writing downward should be avoided absolutely.

Definition 6 (Entity Integrated property). Entity Integrated property can be formally represented as $r[A_1], r[C_1] \neq$ null $\wedge$ $r[RC] \geq r[C_1]$. The modified Entity Integrated property is designed to confirm that every VM is divided into a unique group. A_1 is the value of the primary key as well as the name of a VM and C_1 is the security level of an entity corresponding to VM in an instance r.

Definition 7 (Multi-Instance Integrated property). Multi-Instance Integrated property can be formally described as $A_1, C_1, RC \rightarrow A_i, 2 \leq i \leq n$. The modified Multi-Instance Integrated property is designed for the inexistence of fuzziness in the querying process, which may be caused by accessing the object by a subject in MLR. All instances with the same value of primary keys are allowed in any relation. However, given the security level, only one entity is authorized by the subject at most. For simplicity, only one entity is authorized by VMM according to the value of C_1 when processing commands at specific moment.

To demonstrate the Multi-Instance Integrated property briefly and clearly, an example shown in Table 2 is used to describe the multi-instance conflict. Two entities with the same value of primary key "Gun3" and different values of C_1 are presented in this example, which is not allowed in V-MGSM model. What is more, there is only one entity authorized to avoid the fuzziness in the querying result for a given security level. Compared with MLR, the advantage of V-MGSM model is that entity with multi-instance is permitted by crossing security levels. Meanwhile, only one entity is authorized when VM accesses other VMs.

Definition 8 (Data-Borrowing Integrated property). To all records $r \in$ relation $R, r[C_i] \leq r[RC]$, for $2 \leq i \leq n$, if $r[A_i] \neq$ null, $\exists r' \in$ relation R, and $r'[C_1] = r[C_1] \wedge r'[C_i] = r[C_i]$, then $r[A_i] = r'[A_i]$. In V-MGSM, a high-level user cares more about the borrowed relation with a low-level user, rather than the detailed value of the borrowed data. The borrowed data is vindicated by the low-level VM and the security level

TABLE 2: Multi-instance integrity conflict.

Name	C_1	Range	Quantity	RC
Gun3	S	1	Null	S
Gun3	U	2	5000	S
Gun3	U	3	5000	U

of the borrowed data should be maintained as that of the low-level VM.

(i) The Main Thought of the Improved Data Borrowing. The thought of data borrowing in MLR model is relatively unreasonable, since amount of confidential information may leak from a high-level user because a low-level user can change the view of a high-level user. Compared with MLR which is designed for nonvirtualized system, V-MGSM model is more reasonable and could be applied to a virtualization system and the information in high-level VMs will not be leaked by a low-level VM's *DELETE* or *UPDATE* operation, overcoming the disadvantage of MLR model.

(ii) Conditions for a Successful Data-Borrowing Operation. For secure inner communications between VMs within the same group, the improvement of V-MGSM mainly focuses on the aspect of data borrowing and synchronous update. In specific, we assume that the existence of the borrowed data is the basic condition for a successful data-borrowing operation by a high-level user; then, a low-level VM owns the value of the borrowed attribute; finally, the security level of borrowed data in the high-level VM is still maintained as the original security level, which ensures that the borrowed information is still vindicated by its creator. If none of the conditions is met, the borrowed data should be set null.

(iii) How to Perform Synchronous Update. The security level of the borrowed data is maintained as the original level in high-level VM to perform synchronous update of the borrowed data when it has been modified or deleted by low-level VM.

Therefore, the borrowed data from low-level VMs is set null or updated synchronously in high-level VMs when deleted or modified by the creator, avoiding losing important information.

Definition 9 (Nonborrowed Attribute Integrated property). Nonborrowed Attribute Integrated property can be formally described as $(A_i, A_j \in$ NBA$)$, $R[C_i] = R[C_j]$. NBA denotes nonborrowed attributes in relation R. For all data in VMs, one instance of multirelation R meets Nonborrowed Attribute Integrated property when the security levels of A_i belonging to nonborrowed attributes are the same.

3.4. State Transition Rules to V-MGSM Model. In this part, four orders (*INSERT*, *DELETE*, *UPDATE*, and *SELECT*) and their corresponding grammars are defined as the state transition rules of V-MGSM model.

Definition 10 (*INSERT* manipulation language). Grammatical form of *INSERT* manipulation language is described as follows:

$$INTERT\ INTO\ R\left[\left(A_1,\left[A_2\right],\ldots\right)\right]$$

$$VALUES\ \left(a_1,\left[a_2\right],\ldots\right) \qquad (3)$$

$$a_1 \in D_1, a_2 \in D_2, \ldots.$$

Semantics. R denotes the name of a relation, A_1 denotes the name of VM, and [ ] denotes the alternative items. Every *INSERT* operation could create one record with security level c into relation R by a c-subject at most. For all $2 \leq i \leq n$, if A_i is in the list of "into" attribute, then $r[A_i] = a_i \wedge r[C_i] = r[RC]$. Another case, $r[A_i] = null \wedge r[C_i] = r[RC] = c$, should be taken into account if the value of A_i is not inserted into record r. *INSERT* operation is permitted only when none of records $r' \in$ relation R meets $r'[A_1] = r[A_1] \wedge r'[RC] = c$. One record could be inserted successfully if the generated state of virtualization satisfies Entity Integrated (EI) property, Multi-Instance Integrated property, and Nonborrowed Attribute Integrated (NBAI) property simultaneously.

Definition 11 (*DELETE* manipulation language). Grammatical form of *DELETE* manipulation language is described as follows:

$$DELETE\ FROM\ R$$

$$\left[WHERE\ p\right]. \qquad (4)$$

Semantics. R denotes the name of a relation, p denotes a check expression which could be security attribute expression or data expression, and the data of any VM could only be deleted by its creator. *DELETE* operation is permitted if and only if $r[RC] = c$ and meets p and then deletes record r. In V-MGSM model, another situation shown as follows is taken into account because of data borrowing. For all records $r \in$ relation R, we can obtain $r'[A_i] = null$, if $\exists r' \in$ relation R, and $r'[C_1] = r[C_1] \wedge r'[RC] > r[RC] \wedge r'[A_i, C_i] = r[A_i, C_i]$.

Definition 12 (*UPDATE* manipulation language). Grammatical form of *UPDATE* manipulation can be described as follows:

$$UPDATE\ SET\ A_2 = s_2\left[, A_3 = s_3\right]$$

$$\left[WHERE\ p\right] \qquad (5)$$

$$2 \leq i \leq n.$$

Semantics. For *UPDATE* manipulation language, the data of any VM could only be updated by its creator. In V-MGSM model, another situation, synchronous update of the borrowed information, is taken into consideration if high-level VM has borrowed data from low-level VM. The value of A_1 is not allowed to be modified in *UPDATE* operation. For all records r in relation R, if $p \wedge r[RC] = c$, then $r[A_i, C_i] = (s_i, c)$. The value of the updated attribute could only be updated by its creator, so the security level of updated attribute is maintained as that of its owner whether it has

been borrowed or not. The value of updated attribute in both of the low-level VM and high-level VM is updated synchronously when borrowed relation exists between these two VMs. Therefore, if $\exists r' \in$ relation R, $r'[A_i, C_i] = r[A_i, C_i] \wedge r'[RC] > c$, and $r'[C_i] = c$, then $r'[A_i] = s_i$.

Definition 13 (*SELECT* manipulation language). Grammatical form of *SELECT* manipulation language is described as follows:

$$SELECT\ A_1\left[, A_2\right]$$

$$FROM\ R_1\left[, R_2\right] \qquad (6)$$

$$WHERE\ p.$$

Semantics. Let $*$ denote selecting all data from relation R, and let p denote a check expression which could be security attribute expression or data expression. Data of VMs in relation $R_1[, R_2]$ will be calculated in p, and record r will be accessed by a higher-level subject in the same group. If $r[C_1] = r'[C_1] \wedge r[RC] \leq c'$, then the information of VM could be queried by c'-subject.

4. Correctness and Security Analysis

4.1. Correctness of V-MGSM Model. To prove the correctness of V-MGSM model unambiguously, two steps are necessary. One is the proof of all records r in a relation R that meet four integrated properties; the other one is the proof that any sequence of operations including *INSERT*, *DELETE*, and *UPDATE* transforms an arbitrary legal state into another legal state.

Definition 14. V-MGSM model is correctly equal to the fact that all records in a relation R in a legal state in virtualization system meet four integrated properties defined in Section 3.

Definition 15. A model is correctly equal to the fact that any legal state can be transformed into another legal state by any sequence of *INSERT*, *DELETE*, and *UPDATE* which are previously defined.

It is necessary to stress that *INSERT*, *DELETE*, and *UPDATE* operations change an arbitrary legal state in virtualization system into another legal state. Evidently, any *SELECT* operation does not change any state because no data is inserted, updated, or deleted. Then, we assume that record r has been operated in the following steps.

(i) A generated data by an *INSERT* operation does not break Entity Integrated property, Data-Borrowing Integrated property, or Nonprimary Key Integrated property according to the semantics of *INSERT* operation. Therefore, we just need to prove that an *INSERT* operation does not break Multiple Instance Integrated property. Based on the manipulation regulation of *INSERT* operation, one record r is inserted successfully only when r' which meets $r'[A_1] = r[A_1] \wedge r'[RC] = c$ does not exist in relation R. In a word, the generated state by an *INSERT* operation does not break four previously defined integrated properties.

(ii) The reason why any *DELETE* operation does not break four integrated properties is explained in detail. Entity Integrated property, Multiple Instance Integrated property, and Nonprimary Key Integrated property are met because no record r' in the original relation R has been inserted or updated after a *DELETE* operation. For $r \in$ relation R, if $\exists r' \in$ relation R and $r'[C_1] = r[C_1] \wedge r'[RC] > r[RC] \wedge r'[A_i, C_i] = r[A_i, C_i]$, then we can obtain $r'[A_i] =$ null, and Data-Borrowing Integrated property is met mandatorily in relation R. Therefore, *DELETE* operation does not break four defined integrated properties.

(iii) Any *UPDATE* operation does not break four integrated properties. Actually, all integrated properties have been met mandatorily based on the semantics of *UPDATE* operation. Therefore, if the initial state is correct, any state transformed from any correct state by an *UPDATE* operation is correct or legal as we defined in Definition 15.

Hence, all related operations can transform a legal state into another legal state without breaking four integrated properties. Eventually, we can prove that the proposed V-MGSM model is correct.

4.2. Security Analysis. To prove the security of V-MGSM model, we define the secure state and explain the reason why the state is still secure after executing series of operations mentioned in the earlier part.

Definition 16. Secure state in V-MGSM model means that no downward information flow exists during communication process between VMs.

The output of any subject s_2 belonging to $SV(c)$ is not influenced by deleting any input of subject s_1 belonging to $SH(c)$. Let S be the set of all subjects including all VMs, and let R be all records with different levels in a virtual machine system. Then, we define $SV(c)$ as a set of all subjects whose security levels are lower than or equal to c, $RV(c)$ denote a set of all records whose levels are lower than or equal to c, and $SH(c)$, $RH(c)$: S-$SV(c)$ are $RH(c)$, R-$RV(c)$, respectively. For any access level c, we have obtained the following equations with great ease according to the meaning of each notation:

$$S = SV(c) \cup SH(c) \wedge SV(c) \cap SH(c) = \emptyset,$$
$$R = RV(c) \cup RH(c) \wedge RV(c) \cap RH(c) = \emptyset. \tag{7}$$

To prove the security of V-MGSM model, we take sequence of operations including *INSERT, DELETE, SELECT,* and *UPDATE* by VMs as input in V-MGSM and output the result which includes two parts: (1) a set of records or result of failed access is returned to VM by any *SELECT* operation; (2) the successful or failed information returns to the subject after *INSERT, DELETE,* or *UPDATE* operation.

Upon analyzing the possible access results returned to the subject, the security proof of V-MGSM consists of two situations.

(i) For any access level c, the output to any subject $s \in SV(c)$ is not influenced by changing $RH(c)$ and four cases in the following need to be taken into consideration.

(a) Due to the semantics of SELECT operation, when executing a SELECT manipulation by c'-subject and c' is less than or equal to c, no record in the set $RH(c')$ will be returned to subject s. Therefore, the output to the subject $s \in SV(c)$ is not influenced by changing $RH(c)$ as $RH(c') \supseteq RH(c)$.

(b) The *INSERT* operated by a c'-subject is declined if record $r'' \in$ relation R, and $r''[A_1] = r[A_1] \wedge r''[RC] = c'$. Go a further step; the *INSERT* operation will also be declined whether the generated record breaks EI property, Multi-Instance Integrated property, NBAI property, or Data-Borrowing Integrated property.

(c) The *DELETE* operated by a subject s is declined only when the deleted record r is not created by the subject s, which means that the subject s does not have the right to execute any operation on this record as the subject does not have the ownership of this record.

(d) The *UPDATE* operated by a subject s is declined whether the update operation results in breaking EI property, Multi-Instance Integrated property, and NBAI property or it is not executed by its creator.

Wherein r is the record created by a c-subject and r' is the record created by a c'-subject, only records r and r' are involved and should be taken into account. The successful or failed information, which is delivered to the subject $s \in SV(c)$, is not influenced by changing $RH(c)$ because $r, r' \in RH(c') \supseteq RH(c)$. Then, we can draw a conclusion that $RV(c)$ is not changed by deleting any input of a subject $s \in SH(c)$.

(ii) The state in V-MGSM model is modified only by *INSERT, DELETE,* or *UPDATE* operation and we assume that c' is greater than c in following situations.

(a) c'-VM could generate a c'-record by *INSERT* operation.

(b) c'-VM could only delete c'-record.

(c) c'-VM could update c'-record.

Since c' is greater than c, $r' \notin RV(c)$.

Finally, we draw a conclusion that no downward information flow exists during communication process. The results returned to any subject s_2 belonging to set $SV(c)$ which is not influenced by deleting the input of any subject s_1 belonging to set $SH(c)$. Thus, any state of virtualization transformed from a secure initial state is maintained securely after new state transition rules are carried out.

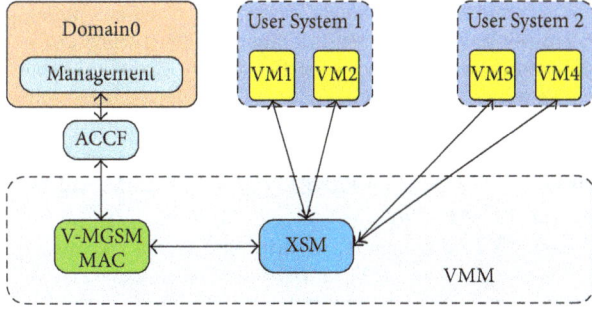

FIGURE 3: The implementation of MAC framework in Xen.

5. Implementation of V-MGSM Model

As shown in Figure 3, we construct a MAC framework which is composed of VMM, Management, XSM, V-MGSM MAC, ACCF, and VMs. Specifically, ACCF denotes access control configuration file, Management is used to manage the access control over other VMs in Domain0, and V-MGSM MAC is a security module which is implemented as a security hook function based on XSM, provided by Xen, and manages communications between VMs according to the state transition rules in V-MGSM model. When a VM makes a request to access another VM with access attribute *SELECT, UPDATE,* or *DELETE,* the XSM intercepts the request; and XSM transfers this access request to V-MGSM MAC; then, ACCF stores the access authority and relation between subjects and objects, which is used to help V-MGSM MAC to determine whether the access request is declined or not; finally, V-MGSM MAC returns the decision. If the returned decision is "Yes," then XSM authorizes the subject to access the object with the access attribute; otherwise, the access is declined.

The implementation is built in a workstation with four quad-core Intel Core i7, 2.5 GHz CPUs, 16 GB memory, and 1 T storage. In specific, the version of Xen is 3.3.1, the operating system in Domain0 is CentOS 5.4, and each VM covered in the test operates on Ubuntu 8.10. A web service is provided by Apache in Domain0, through which each VM could log in and query information of shared memory; in addition, a software used to apply to sharing memory and set the value in shared memory, called Virmem, is executed in every VM.

Step 1 (the initialization of the virtualization in Xen). Five VMs are created in Xen and denoted by VM1, VM2, VM3, VM4, and VM5, respectively, and the security levels of them are $H3, H2, H1, H3$, and $H2$, where $H3 > H2 > H1$. For efficient privacy preservation between different organizations, VM1, VM2, and VM3 are grouped into User System 1, and VM4 and VM5 are in User System 2; the unique identifications of User System 1 and System 2 are L1 and L2, respectively. Then, the Virmem in VM3, VM4, and VM5 is applied for sharing the memory and sets the first byte of each shared memory by "e," "k," and "m," respectively. And VM2 is applied for sharing the memory to VM1 and sets the first byte's value by "b."

TABLE 3: The test's results of V-MGSM.

(a) The result of VM1's first query

VM name	C_1	Data (first byte)	C_2	RC
VM1	L1	b	H2	H3
VM2	L1	b	H2	H2
VM3	L1	e	H1	H1

(b) The result of VM4's query

VM name	C_1	Data (first byte)	C_2	RC
VM4	L2	k	H3	H3
VM5	L2	m	H2	H2

(c) The result of VM2's query

VM name	C_1	Data (first byte)	C_2	RC
VM2	L1	f	H2	H2
VM3	L1	e	H1	H1

(d) The result of VM1's second query

VM name	C_1	Data (first byte)	C_2	RC
VM1	L1	f	H2	H3
VM2	L1	f	H2	H2
VM3	L1	e	H1	H1

Step 2 (VM1 in User System 1 and VM4 in User System 2 query all shared memory in Domain0 by web service, resp.). Then, the results are shown in Tables 3(a) and 3(b). As VM1 owns the highest security level, all data in User System 1 but not User System 2 could be queried by VM1. And as shown in Table 3(a), the values of C_1 in these records are the same which means that all VMs are corresponding to the same group User System 1. As shown in Table 3(b), when VM4 queries all shared memory in User System 2, the value of C_1 is also the same. Therefore, the result shows that our proposed V-MGSM model is correctly and securely applied to the privacy preservation for virtualization.

Step 3 (the value of shared memory is updated by VM2). VM2 in User System 1 updates the first byte's value of shared memory to "f" by Virmem. Then, VM2 queries shared memory by web service in Domain0, and the result is shown as in Table 3(c).

Step 4 (VM1 queries shared memory again). As shown in Table 3(d), the value of data in both first and second records is modified to "f." Hence, the value of borrowed shared memory in a high-level VM could be updated synchronously and successfully if it is updated by its owner, a low-level VM.

The experimental results show that multilevel security access control for VMs and efficient isolation for VMs corresponding to different groups and synchronous update have been implemented in a virtualization system and perform a secure communication process between VMs.

6. Conclusions

In this paper, based on multilevel relation, a new multilevel and grouping security model, called V-MGSM, is proposed for virtualization. In specific, on the basis of the security requirements for different VMs, we redesign the elements, integrated properties, and data manipulation languages of V-MGSM model, and all VMs are divided into several groups based on their corresponding entities, which can avoid the unnecessary access and provide the privacy protection among different groups. Besides, different security levels are assigned to VMs in the same group, and the confidential information in a high-level VM will not be obtained by some malicious low-level VM. Then, when the borrowed data is modified in the low-level VM, the corresponding data in the high-level VM will be updated synchronously. In addition, detailed security analysis shows that the proposed V-MGSM can provide a secure communication mechanism for VMs and achieve the synchronous updates of the borrowed data. Eventually, we implement V-MGSM in Xen for experiments, and the results demonstrate that V-MGSM can indeed achieve efficiency for Mobile Internet service.

Disclosure

The abstract of this paper has been presented in the 5th International Conference on Intelligent Networking and Collaborative Systems, pp. 9–16, 2013 [1].

Conflict of Interests

The authors declare that there is no conflict of interests regarding the publication of this paper.

Acknowledgments

This work was supported by National Natural Science Foundation of China (nos. 61303218 and 61272455); China 111 Project (B08038); and the Fundamental Research Funds for the Central Universities of China (no. K5051301017).

References

[1] H. Zhu, Y. Xue, Y. Zhang, X. Chen, H. Li, and X. Liu, "V-MLR: a multilevel security model for virtualization," in *Proceedings of the 5th International Conference on Intelligent Networking and Collaborative Systems (INCoS '13)*, pp. 9–16, IEEE, September 2013.

[2] A. Flahive, D. Taniar, and W. Rahayu, "Ontology as a Service (OaaS): a case for sub-ontology merging on the cloud," *The Journal of Supercomputing*, vol. 65, no. 1, pp. 185–216, 2013.

[3] D. Taniar and S. Goel, "Concurrency control issues in Grid databases," *Future Generation Computer Systems*, vol. 23, no. 1, pp. 154–162, 2007.

[4] S. Goel, H. Sharda, and D. Taniar, "Replica synchronisation in grid databases," *InternationalJournal of Web and Grid Services*, vol. 1, no. 1, pp. 87–112, 2005.

[5] D. Taniar, C. H. Leung, W. Rahayu, and S. Goel, *High Performance Parallel Database Processing and Grid Databases*, vol. 67, John Wiley & Sons, 2008.

[6] A. Flahive, D. Taniar, and W. Rahayu, "Ontology as a Service (OaaS): extracting and replacing sub-ontologies on the cloud," *Cluster Computing*, vol. 16, no. 4, pp. 947–960, 2013.

[7] S. Shukla and R. Kumar Singh, "Security of cloud computing system using object oriented technique," in *Proceedings of the 3rd International Conference on Computing, Communication and Networking Technologies (ICCCNT '12)*, 9, p. 1, IEEE, July 2012.

[8] H. Zhu, T. Liu, G. Wei, and H. Li, "PPAS: privacy protection authentication scheme for VANET," *Cluster Computing*, vol. 16, no. 4, pp. 873–886, 2013.

[9] F. Sabahi, "Cloud computing security threats and responses," in *Proceedings of the IEEE 3rd International Conference on Communication Software and Networks (ICCSN '11)*, pp. 245–249, 2011.

[10] C. Li, A. Raghunathan, and N. K. Jha, "A trusted virtual machine in an untrusted management environment," *IEEE Transactions on Services Computing*, vol. 5, no. 4, pp. 472–483, 2012.

[11] S. M. Bellovin, "Virtual machines, virtual security?" *Communications of the ACM*, vol. 49, no. 10, p. 104, 2006.

[12] J. E. Smith and R. Nair, "The architecture of virtual machines," *Computer*, vol. 38, no. 5, pp. 32–38, 2005.

[13] R. Sailer, T. Jaeger, E. Valdez et al., "Building a MAC-based security architecture for the Xen open-source hypervisor," in *Proceedings of the 21st Annual Computer Security Applications Conference (ACSAC '05)*, pp. 276–285, December 2005.

[14] G. Cheng, H. Jin, D. Zou, A. K. Ohoussou, and F. Zhao, "A prioritized chinese wall model for managing the covert information flows in virtual machine systems," in *Proceedings of the 9th International Conference for Young Computer Scientists (ICYCS '08)*, pp. 1481–1487, November 2008.

[15] M. Hirano, T. Shinagawa, H. Eiraku et al., "Introducing role-based access control to a secure virtual machine monitor: security policy enforcement mechanism for distributed computers," in *Proceedings of the IEEE Asia-Pacific Services Computing Conference (APSCC '08)*, pp. 1225–1230, IEEE, 2008.

[16] Q. Liu, G. Wang, C. Weng, Y. Luo, and M. Li, "A mandatory access control framework in virtual machine system with respect to multi-level security I: theory," *China Communications*, vol. 7, no. 4, pp. 137–143, 2010.

[17] D. E. Bell and L. J. La Padula, "Secure computer system: unified exposition and multics interpretation," DTIC Document, 1976.

[18] D. E. Bell and L. J. LaPadula, "Secure computer systems: mathematical foundations," Tech. Rep., DTIC, 1973.

[19] T. F. Lunt, D. E. Denning, R. R. Schell, M. Heckman, and W. R. Shockley, "SeaView security model," *IEEE Transactions on Software Engineering*, vol. 16, no. 6, pp. 593–607, 1990.

[20] R. Sandhu and F. Chen, "The multilevel relational (MLR) data model," *ACM Transactions on Information and System Security*, vol. 1, no. 1, pp. 93–132, 1998.

[21] X. Chen, J. Li, and W. Susilo, "Efficient fair conditional payments for outsourcing computations," *IEEE Transactions on Information Forensics and Security*, vol. 7, no. 6, pp. 1687–1694, 2012.

13

A Pervasive Promotion Model for Personalized Promotion Systems on Using WLAN Localization and NFC Techniques

Kam-Yiu Lam,[1] Joseph Kee Yin Ng,[2] Jiantao Wang,[2]
Calvin Ho Chuen Kam,[2] and Nelson Wai-Hung Tsang[1]

[1]*Department of Computer Science, City University of Hong Kong, Kowloon Tong, Hong Kong*
[2]*Department of Computer Science, Hong Kong Baptist University, Kowloon Tong, Hong Kong*

Correspondence should be addressed to Joseph Kee Yin Ng; jng@comp.hkbu.edu.hk

Academic Editor: David Taniar

In this paper, we propose a novel *pervasive business model* for sales promotion in retail chain stores utilizing WLAN localization and *near field communication (NFC)* technologies. The objectives of the model are to increase the customers' *flow* of the stores and their *incentives* in purchasing. In the proposed model, the NFC technology is used as the first mean to motivate customers to come to the stores. Then, with the use of WLAN, the movements of the customers, who are carrying smartphones, within the stores are captured and maintained in the *movement database*. By interpreting the movements of customers as indicators of their interests to the displayed items, *personalized* promotion strategies can be formulated to increase their incentives for purchasing future items. Various issues in the application of the adopted localization scheme for locating customers in a store are discussed. To facilitate the item management and space utilization in displaying the items, we propose an *enhanced R-tree* for indexing the data items maintained in the movement database. Experimental results have demonstrated the effectiveness of the adopted localization scheme in supporting the proposed model.

1. Introduction

With the advances in electronic and wireless communication technologies, the design and development of novel mobile and pervasive computing applications have attracted great interests in both the industry and academia. Nowadays, people can easily use their smartphones as minicomputers to play computer games and retrieve various useful information, for example, shopping information, through cellular or wireless networks. It can be anticipated that pervasive computing applications will become one of the core computing applications in the coming future [1], and more and more smart electronic devices will be equipped with powerful computing capabilities to support novel and advanced mobile services to the users.

One of the typical pervasive computing applications that is attracting tremendous attention in the business sector is the *location-based services* [2]. For location-based services, the services to be delivered to mobile clients depend on their current locations. For example, in a museum, different descriptions about the exhibits will be provided to the visitors depending on the current locations of the visitors. In general, location-based services can be classified into two main categories. Firstly, they are the *smart services* for mobile users such as mobile control and navigation services, for example, Google map and smart cars. Secondly, it is the mobile information services. The business sector is exploring the locations and movements information of mobile users aiming at improving the business operations and the increase of sales volume, for example, *mobile advertisement* [3].

A fundamental technology in providing effective location-based services is to have a good localization system to determine the current locations of mobile users accurately. Although GPS is a very successful technique for localization, it is mainly effective for outdoor localization. Even with the help of mobile phone technologies in localization, its accuracy may not be good enough for many indoor location-based applications which require an accuracy within a range of 1 to 2 m.

Another common technology for localization is to use the *wireless local area network (WLAN)* which has been shown to be more suitable for localization in indoor environment [4, 5]. In addition, the coverage of WLAN is getting larger and larger in many big cities and various efficient techniques have been proposed to improve its accuracy and deployment [5]. Experimentally, it has been shown that the accuracy achieved from a WLAN in localization can be within one metre with good calibration [4].

Another type of wireless communication technology that is receiving growing interests in recent years is the *near field communication (NFC)*. Functionally, NFC is not a communication device but an identification technique similar to RFID and smart cards. Currently, various studies are conducted to look into the feasibility of NFC for payment systems, object tracking, and logistic applications. Since more and more smartphones are equipped with NFC, its applications are growing rapidly. Unlike WLAN, NFC is for very short range communication, that is, within several centimetres. Although the close range of communication of NFC limits its communication capability, it can be used as a reliable and easy-to-use location indicator if the location of one of the NFC devices is known, that is, the presence of an NFC device next to another NFC device which has a known location.

Although various novel location-based services are emerging, for example, counting people in a tour group and taking roll calls for class attendance in universities, the application potentials of the location information obtained from WLAN for location-based business applications still have a lot of room for further exploration especially when they are combined with NFC and other location-based technologies. According to behavioral studies in pervasive promotion systems [1, 6, 7], the information for the movements of customers within a store could be very useful for marketing and for planning promotion strategies in retail stores. In this paper, by exploring the advanced techniques in pervasive computing and localization using both WLAN and NFC, we formulate a *pervasive promotion model* for the development of a *personalized promotion system* for sales promotion in retail chain stores. Various technical issues in the design of the system are discussed in this paper.

The remainder of the paper is organized as follows. In Section 2, we briefly review the related work on location-based services and the different technologies for localization. In Section 3, we define the proposed pervasive promotion model. In Section 4, we discuss the details of the design of the personalized promotion system according to the proposed model. An illustrative example is presented in Section 5. In Section 6, we report the performance of the localization scheme in supporting the proposed system. We conclude the paper and briefly discuss the future works in Section 7.

2. Related Work and Background

In this section, we review the previous work on applying location-based services in business advertisement and customer behavioral prediction. Then, we present the latest technologies for localization using WLAN and the basic principles and performance characteristics of NFC.

2.1. Location-Based Business Services. With the popularity of smartphones, business companies are developing various novel location-based services for catching new business opportunities. One of the popular location-based services is mobile advertisement [2, 8]. In [8], the iMAS system was proposed. It uses GPS to provide location-based intelligent advertising services such that different advertising information is provided to clients at different locations. Similarly, in [2], a location-based mobile advertisement publishing system for vendors was proposed. The system allows the vendors to edit advertisements and publish advertisements to customers through a wireless infrastructure such as 3G networks or WiMAX. In addition, the advertisement data desired by the customers can also be viewed through QR code readers.

In addition to mobile advertisements, some researchers studied how to provide user-adaptive advertisement. In [9], how to publish advertisements in a supermarket that can be adapted to the concerning information of a particular client from his own profile is presented. Then, specific advertising hints can be delivered as a navigation service to the client within the supermarket while he is shopping around.

As demonstrated in [6], economic incentive is a very effective way for shaping consumer behavior. Yamabe et al. [6] studied various incentive mechanisms and how they alter the consumer behavior. They proposed four micropricing models to alter the behavior of consumers. In the micropricing models, different pricing systems are adopted for different services to be provided to a consumer. Similar to [6], Kanda et al. [10] studied the behavior of customers in particular their movements in a public area such as in a train station to identify the real customers. It uses a humanoid robot to identify the real customers based on the collected movement behavior information, for example, the movement speeds and paths of the real customers, in the monitored region. In [7], product involvement was applied for mobile advertising to study the effect of involvements on consumers' purchase intention. It was shown that impulse buying tendency influences the degree that a customer may be engaged in a product purchase. You et al. [1] studied the relationship between shopping time and purchasing. It divides the time that a customer spends within a store to shopping time and entertainment time. Then, it proposed a phone-based system to sense physical shopping activities while tracking the shopping time of customers in a store.

Similar to the previous work, for example, [1, 6, 10], in this paper, we explore the movements of customers to formulate personalized promotion strategies with the purpose to increase the incentives of customers for purchasing. However, unlike those mentioned above, we apply the latest localization method using WLAN to capture the movements of customers in a store to determine their interests in an item/items displaying in the store [4] and then formulate promotion strategies. Different from the work in [1], we combine NFC and WLAN in the personalized promotion system to increase the flow volume of customers into the stores. To the best of our knowledge, there is a lack of effective personalized promotion systems for retail sales. In addition, we combine the movement database with an enhanced R-tree index such that the data items maintained in the movement database can be retrieved efficiently and various spatial queries can be supported for enhancing the space management in a store. Although R-tree [11] has been shown to be effective for managing spatial objects and location information of moving objects, to the best of our knowledge, there is still lack of any research work on applying R-tree for tracking the movement behavior of mobile clients in business promotion and space management in retail stores.

2.2. Localization Using Wireless Communication

2.2.1. Localization Using WLAN. The basic principle in localization is to use a set of sensors with fixed and known positions to collect radio signals from a mobile device carried by the mobile user [12–15]. With the obtained data, they perform location estimation such as using triangulation. Various estimation techniques have been proposed to improve the accuracy of the estimation such as the sequential monte carlo localization (MCL) method [16] and the sequential greedy optimization (SGO) method [17].

According to the distance or a distance dependent parameter parameter between the sensors and the mobile device, localization techniques can be classified into range-based and range-free methods [18]. Although the range-free methods offer a straightforward and simple localization option for many wireless localization applications since they do not require any specific ranging devices, their accuracies are normally limited.

Compared with the range-free methods, the range-based methods, in general, can achieve a higher accuracy for localization. They explore the measured values in wireless communication to estimate the location of a mobile user. Some of the commonly used physical measures are the time-of-arrival (TOA), time-difference-of-arrival (TDOA), angle-of-arrival (AOA), and received signal strength indicator (RSSI) [5, 16, 19, 20]. Amongst these measures, the RSSI-based localization techniques provide a class of low-cost and easy-implemented range-based localization schemes [13–15, 19]. Some popular techniques based on RSSI are the fingerprinting, aggregated signal layout, and center of gravity (CG) [4]. To further enhance the performance in using RSSI, the signal aggregated method was proposed in [21]. In [5], an efficient approach was proposed to reduce the manual calibration and to optimize the accuracy in indoor positioning using RSSI. In this paper, we also adopt RSSI together with the signal aggregated method as the basic measures for localization using WLAN to determine the movements of customers within a store.

2.2.2. Localization Using NFC. Near field communication (NFC) [22–24] is a very short range of communication technology using radio frequency, that is, 13.56 MHz. When two NFC devices are brought within a close range, that is, less than 10 cm, they can determine each other automatically and establish a communication channel for data exchange. Basically, there are two types of NFC: active and passive. In passive NFC, a passive sensor/tag is activated when an active NFC device is within a close range of communication of the passive tag. Since we put the passive tag at a fixed location

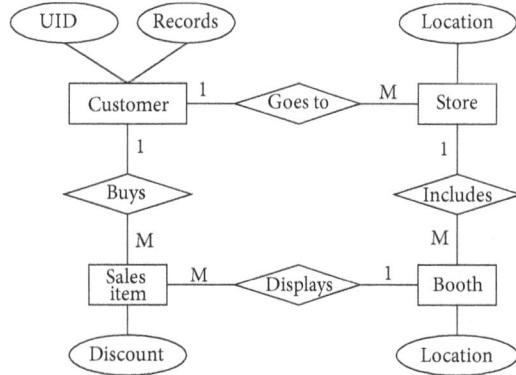

FIGURE 1: The ER model of the main entities.

and it has a unique identifier, the location information of the passive tag and identifier can be transferred to the active NFC device. Similar to passive RFID, another benefit of a passive NFC is that the passive NFC tag can obtain power from the radio frequency generated from the active NFC device. Each active NFC device is self-powered and generates its own oscillating magnetic field for communication. In the proposed promotion system, we use passive NFC tags to issue some discount policies to attract customers to enter into the chain stores for promotion purpose.

The emergency of NFC-enabled smartphones opens up a lot of new business opportunities to the smartphone operators [23, 24]. Nowadays, NFC can be used to replace smart cards for various identity checking purposes such as in taking attendances and even be used for e-payment to replace the Octopus system, the most popular e-payment system in Hong Kong. When the NFC works with other networks such as cellular networks and WLAN, its applications could be much wider since NFC and the networks supplement the limitations of each other. WLAN and cellular networks require a more complex system infrastructure and a higher security measure. On the other hand, although NFC has limited communication range, it has a simple "touch" interface to facilitate its operation. The close range of communication of NFC can also enhance the security of the communication and makes the transmission more difficult to be captured and attacked by malicious devices.

3. Pervasive Promotion Model

In this section, we introduce a *pervasive promotion model* for promoting *selected* items in retail sales to registered customers. The pervasive promotion model is for retail chain stores which have many stores distributed in a city such as the Park'N Shop Supermarket and the convenient store Mannings Company Limited in Hong Kong. The main entities in the system are *customer, store, booth,* and *item* as shown in Figure 1. The model is *pervasive* in the sense that it integrates the promotion of items with the shopping movement of customers while they are shopping around in any one of the chain stores. In addition, the customers can also exchange promotion information with each other

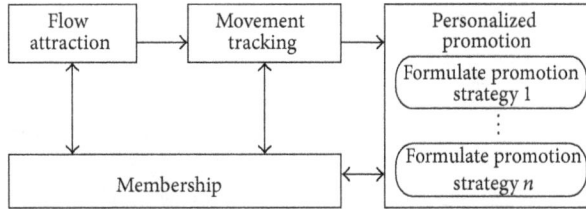

FIGURE 2: The pervasive promotion model for promoting selected items in retail chain stores.

whenever they meet each other independent of their current locations or remotely through the internet.

In retail sales, the first objective in promotion is to increase the *flow volume* of customers of the stores. Then, the second objective is to increase the *incentives* of the customers for purchasing. In the following sections, we will discuss how the proposed pervasive promotion model can achieve the two objectives effectively.

As shown in Figure 2, the pervasive promotion model consists of four main components:

(i) *membership,*

(ii) *flow attraction,*

(iii) *movement tracking,*

(iv) *personalized promotion.*

3.1. Membership. Membership is commonly adopted in retail stores to provide discount and rebate to registered customers for improving the relationships between customers and the stores. However, most of them are *not* personalized to the customers such that the relationship is just defined between a group of customers and a store or a chain of stores; that is, the same promotion strategy is applied to the same group of registered customers as shown in Figure 3(b).

Personalized relationships are not easy to be formulated as they require the identification of the items that the customers may be interested to purchase. A wrong guess could be annoying and generate negative impacts to the image of the stores and companies, for example, receiving a large number of uninterested promotion messages. In this proposed model, by exploring the movements of customers within the stores, we formulate the relationships between a customer and the items which may be interesting to the customer. Then, different promotion strategies can be applied to different customers based on the defined relationships specifically as shown in Figure 3(a). This not only can improve the relationships between the customers and the stores as the customers are treated specifically by the stores. This can also provide a higher flexibility in promoting the items; for example, different discount rates may be applied to different customers in promoting the same item.

In addition to tracking the movements of customers within the stores, the purchase histories of the customers as well as their responses to the promotion messages are also maintained; that is, the promotion of an item is successful and the customer has taken the promotion and purchased the item. This information will also be considered in formulating

the promotion strategies to improve the effectiveness of the promotion system.

3.2. Flow Attraction. The second component is *flow attraction*. It is how to increase the flow volume of customers of the stores. If more customers are walking around within a store, for either window shopping or real shopping, the sales volume of the store has a higher potential to be increased [1]. In the proposed pervasive promotion model, we make use of the close range communication of NFC to motivate the customers to visit the stores.

Once a registered customer carrying an NFC-enabled smartphone, which is running *apps* specially designed for the personalized promotion system, comes to a close range of a passive NFC tag in any one of the chain stores, the tag will be activated and its unique ID code and its location information will be transferred to the *apps*.

In order to increase the incentive of a customer to go to check with the NFC tags installed in the stores, a simple *discount promotion strategy* may be applied. For example, each checking with an NFC tag will give a bonus mark to the customer and a customer can accumulate up to a maximum mark in each day from the same store.

To make the promotion scheme more attractive and to increase the number of customers to visit the stores, the accumulated bonus marks of a customer may be *transferred* to another customer if both of them have NFC-enabled smartphones and are the registered customers of the chain stores. In this way, they may be able to maximize the discount and purchase benefits that they can obtain according to the received promotion messages.

3.3. Movement Tracking. The third component of the pervasive promotion model is tracking the *movements* of customers within a store. This is a localization problem of the registered customers within a store. In our system, we adopt the WLAN for localization. A number of WLAN access points (APs) are installed at the ceiling of each store. They are used to determine which booth a customer is currently locating at. In a store, a set of nonoverlapping rectangular booths are conceptually defined as shown in Figure 4. The size of a booth may range from one to several square metres. Each booth is assigned a unique booth ID (B_{ID}) and a set of coordinates (e.g., x_1, x_2, x_3, and x_4) to represent the identity of the booth and its location in the store, respectively. The rectangular shape of a booth can facilitate the booth management using the *R-tree index*. This will be discussed in Section 4.3. Note that it is not necessary to have a WLAN access point for each booth for localization. The important issue in localization is to distinguish which booth a customer is currently locating at or he is locating in a nonbooth area. Within each booth, items are being displayed. It is assumed that the items displaying under a booth are of similar types, for example, watches or baby food. Some items are selected to be promotion items (I_P) while the others are nonpromotion items (I_{NP}). I_P are those items that are popular and promotion strategies have been defined for increasing their sales volumes.

In location tracking, it is not necessary to continuously monitor the location of a customer for every step. Otherwise,

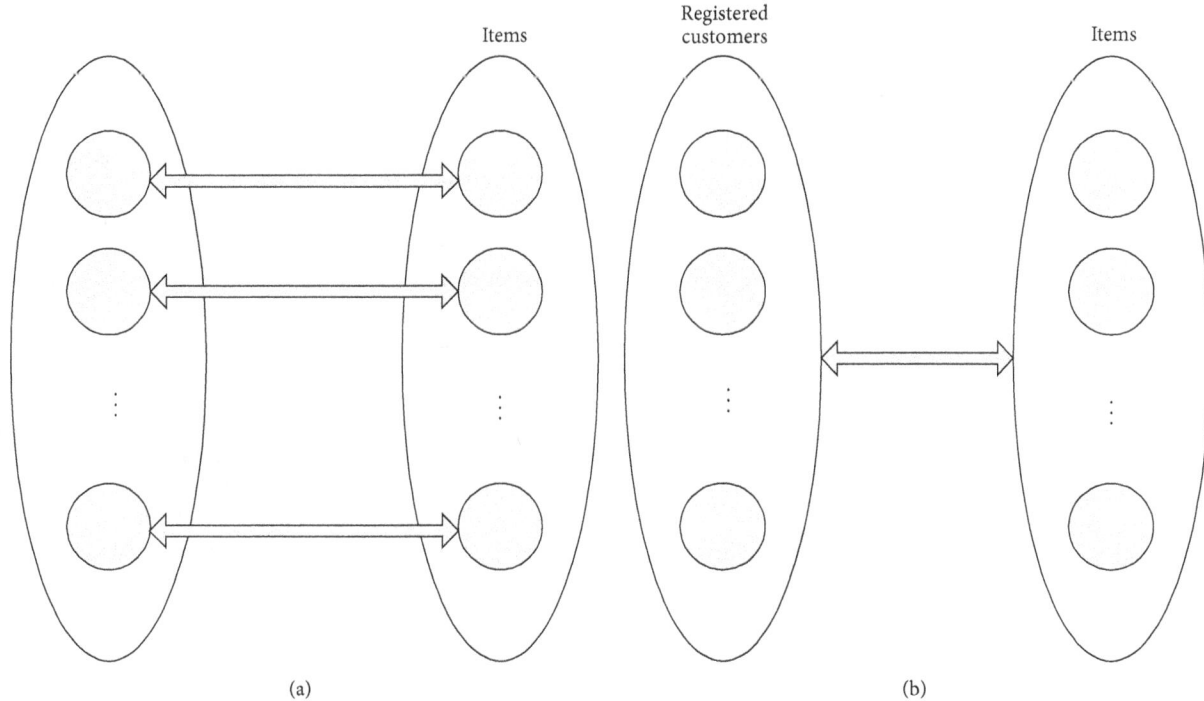

FIGURE 3: Personalized promotion versus nonpersonalized promotion.

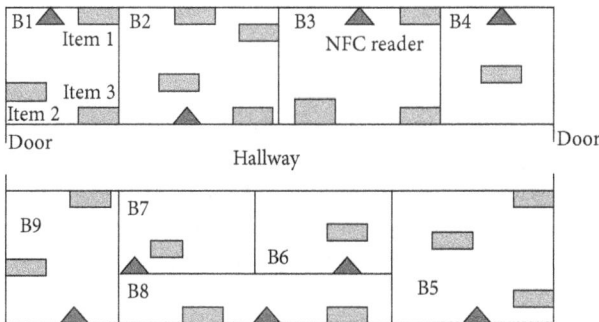

FIGURE 4: The booths in a store.

the localization cost in terms of processing cost and communication cost of the network could be expensive. Instead, we use an *adaptive periodic scheme* (to be discussed in next section) for checking the current location of a customer in a store.

In localization, we determine the booth ID (B_{ID}) of the customer in the store. If the $B_{ID}(t)$ of customer C at time t is different from the previous $B_{ID}(t-s)$ and the total duration of the customer in a booth is larger than a predefined threshold T_{update}, a time-stamped record called *customer-item record* will be generated for each promotion item, I_p, in the booth. The generated customer-item records will be forwarded and inserted into the *movement database* which contains the information of the "customers' interests" to the items centrally maintained for all the stores. Each customer-item record is an indicator to show that the customer may be interested in the item.

3.4. Personalized Promotion.

The last component of the pervasive promotion model is the *personalized promotion*. Once the *app* running in the smartphone of a registered customer obtains the ID of an NFC tag, it will be activated and the personalized promotion component will be triggered. The interested items of the customer maintained in the *movement database* based on his previous visits to the stores will be searched using the customer ID obtained from his smartphone. If the store has the same items, promotion strategies will be formulated to determine how to promote the items such as the discount rates of the items to the customer. Then, time-stamped *promotion coupons* will be forwarded and displayed in his smartphone through the installed mobile *apps* for the customer to consider. The time-stamp of the promotion coupon indicates the validity period of the coupon. It could be within the same day or week depending on the formulated promotion strategies. Therefore, by controlling the number of promotion items and the definitions of the conditions for formulating the promotion rules, the number of promotion messages can be controlled so that a customer will not be annoyed by a large number of uninterested promotion messages. Since this paper is concentrated on the business model and the design of the personalized promotion system, the discussion on the discount rules is out of the scope of this paper.

It is expected that, through the pervasive promotion model, a mutual benefit can be achieved to both the stores and the customers; that is, the customers can purchase the items they find intersting using lower prices or with more benefits, for example, bonus items, while the sales volume of the stores can be increased. Furthermore, different promotion

TABLE 1: Details of the databases.

	Database	Description of the database
Frontend	Store database	Store record contains the information about the stores such as store ID, booths' location, booth ID, and sale items ID.
Backend	Item database	Item records contain information about the items such as the location of an item in a booth in each store and its price.
	Movement database	Booth records contain the information of the booth IDs and the associated list of customer-item records.
	Customer database	Customer records contain the information of registered customers such as customer ID, the name, and telephone number.
	Promotion database	Promotion records contain the information about the promotion rules for each promotion item.

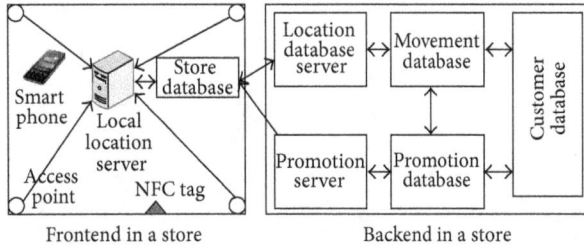

FIGURE 5: System architecture of the personalized promotion system.

strategies can be applied to different customers according to their loyalties to the stores and their previous purchase histories (e.g., total sales purchased in a month). In addition, the personalized promotion can make the customers have a good feeling of being important as they are being treated specifically. Increasing the loyalty of customers to the stores may motivate further purchases from them.

4. System Architecture and Technical Issues

In this section, according to the pervasive promotion model introduced in Section 3, we present the design of the *personalized promotion system*.

4.1. System Architecture and Operations. Figure 5 shows the architecture of the personalized promotion system. It consists of two subsystems: frontend and backend. Table 1 summarizes the set of databases maintained in the frontend and backend systems.

4.1.1. Frontend System. At the frontend system, each store has a passive NFC tag with a fixed location, a set of NFC-enabled smartphones carrying by the registered customers visiting the stores, *apps* for the personalized promotion system running on the customers' smartphones, a set of fixed WLAN access points, and a local location server which maintains the *store database*. The store database maintains information about the store such as store ID, store address, booth IDs, booth locations, and sales items in the store.

The local location server connects to a set of WLAN access points (APs) installed at the ceiling of the store. The

apps running in an NFC-enabled smartphone communicate with the NFC tag when they are within a close range, that is, next to each other. Then, the *apps* get the tag ID and communicate with the WLAN access points installed in the store to establish a connection with one of them to acquire an IP address for the smartphone. Note that although a wireless connection can be established between a smartphone and the WLAN once the smartphone is within the store, in our system the connection is established "only" after a customer has checked with an NFC tag. The reason is that these are registered customers. Once a customer uses his smartphone to check with the NFC tag, it is an "agreement" between store and the customer that he is allowed to be connected and monitored in the store.

The WLAN access points measure the *received signal strength indicator (RSSI)* from the smartphone of the customer periodically and pass the measured signal strengths to the local location server. The server then performs location estimation to determine the booth that the customer is currently staying at and to determine the period of time that the customer has been staying for in the booth by comparing with the previous location record of the customer. The details of the localization estimation method will be discussed in Section 4.2. Since the number of registered customers in a store could be large, we adopt an adaptive periodic scheme for setting the period for checking the locations of a customer such that the sampling period $P_{i,j}$ for customer i in booth j is adaptive:

$$P_{i,j} = \max\left((x + n * m), T_{\text{check}}\right), \tag{1}$$

where n is the number of consecutive times that the customer remains in the same booth while x and m are predefined tuning parameters to control the estimation period. T_{check} is the maximum period for localization with T_{check} smaller than T_{update} which is the time threshold for generating a location update. Larger values for m and x increase the estimation period and reduce the estimation cost at the local location server.

If there is a change in location, that is, the latest B_{ID} being different from the previous one, and the period of time of the customer staying in the booth, T_{period}, is larger than T_{update}, a *customer-item record* will be generated for each *promotion item*, I_p, displaying in the booth. Please be reminded that it is assumed that the items in a booth are assumed to be the

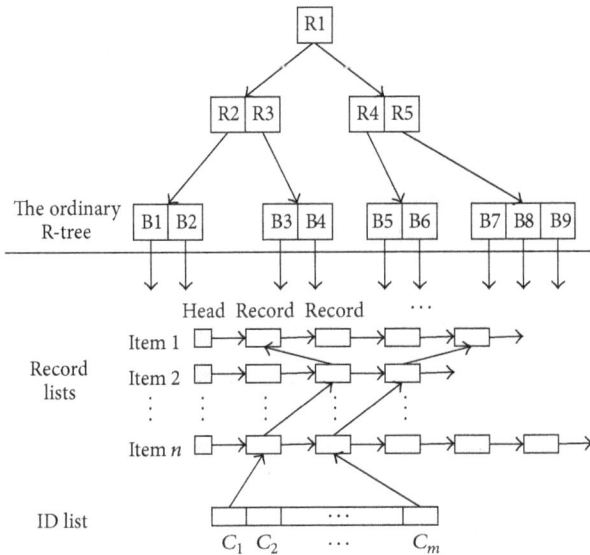

FIGURE 6: Enhanced R-tree for a store in Figure 4.

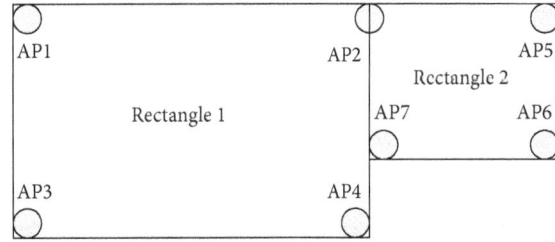

FIGURE 7: The floor plan of a store.

same types. Therefore, the customer may be interested in all of them. Each record consists of the customer ID, booth ID, item ID, the date, the start time, and the period of time that the customer has stayed in the booth.

4.1.2. Backend System.
The backend system consists of two servers: the location database server and the promotion server. The backend system connects to the frontend system through a high-speed wired network. The location database server maintains the movement database for all the stores. It is indexed by an enhanced R-tree [11]. Each data item, called a *booth record*, maintains the information for a booth such as its identity and physical coordinates (e.g., x_1, x_2, y_1, and y_2) in the store. The details of the enhanced R-tree will be discussed in Section 4.3.

Under each booth record, there is a list of items displaying in the booth sorted according to their IDs. In the movement database, an item in a booth may be pointed by a customer ID if the customer is interested at the item defined according to a customer-item record as shown in Figure 6. In addition to the R-tree index, a key is maintained to sort the list of customer IDs. Note that a customer ID may point to a list of items as he may be interested in several items distributed in different booths. If there is a move of an item from a booth to another booth, the movement database and the item database will be updated accordingly. The item database contains the details of the items and the distribution of items in each booth and each store.

Since the item lists pointed by a customer ID are linked together, we may use the customer ID to obtain the set of items the customer is interested in when he accumulates marks in a store using his smartphone. After identifying his items of interest (obtaining the list of items pointed by the customer ID), the promotion server searches the *promotion database* which defines the promotion rules for each promotion item. Following the rules, the promotion server formulates the promotion strategies for the selected items and then promotion coupons will be generated and displayed through the *apps* in the customer's smartphone.

4.2. Localization Using WLAN.
From our previous studies on localization using WLAN, the fingerprinting approach is good and accurate in providing room-level or, in this case, booth-level localization services with an average accuracy of about 1.6 m [25]. However, the disadvantage is the difficulties in setup and heavy maintenance cost of the fingerprint database. Fortunately, this cost can be reduced by the aggregated signal layout technique where the accuracy is roughly maintained [21].

Figure 7 shows an example floor plan of a store. It is assumed that it is rectangular or consists of several connected rectangles. Access points are installed strategically within each rectangle and most of the time at the corner of each rectangle. These access points form a WLAN to support the communication with the smartphones as well as for location estimation of the customers within the booths.

Since we adopt using the aggregated signal layout approach [21] for location estimation, we have to find out the signal-distance relationship for the target operation environment. For this real-time calibration, access points act as sensors to capture the communication signals between the access points and the WiFi-enabled smartphone as well as among all the access points themselves. As these access points are installed at strategic points within the operation environment in such a way that the access points receive signals among each other from different distances, a real-time signal to distance relationship can then be obtained.

With the average received signal strength of each communication between a pair of access points being detected, a (*signal, distance*) data point for each pair of access points can be obtained. Then we can use a linear equation $y = a + bx$ that best fits the data points, with the y-offsets defined as

$$R^2 = \sum \left[y_i - f(a, b) \right]^2, \tag{2}$$

where y_i is the distance in a (*signal, distance*) data point and $f(a, b) = a + bx$. a and b with minimum R^2 are obtained by the linear least squares fitting. Then, the solutions to b and a are

$$b = \frac{ss_{xy}}{ss_{xx}}, \qquad a = \overline{y} + b\overline{x}, \tag{3}$$

where

$$ss_{xx} = \sum \left(x_i - \overline{x}\right)^2, \qquad ss_{xy} = \sum \left(x_i - \overline{x}\right)\left(y_i - \overline{y}\right). \quad (4)$$

With a real-time derivation of the signal-distance relationship, a signal map can be obtained for each access point. When a WiFi-enabled smartphone enters the operation environment, the WiFi signals can be picked up by the nearby access points and vice versa. If the area is marked by a grid of a certain size, each access point that detects the user's WiFi signal can calculate the possible positions within the grid where the smartphone is currently locating. Then the aggregated signal layout method can perform location estimation by overlapping or stacking up the signal layouts of these access points. Considering a single access point, given the received signal strength from a WiFi-enabled smartphone, the distance between the phone and the access point could be obtained from the sensor's signal layout; the interdistance draws the possible locations of the smartphone in a circular perimeter. By aggregating the signal layouts of all the access points, essentially locating the common points of the perimeters, the estimated location of the smartphone can be obtained.

Since it is common that the received signal strength does not conform to an access point signal layout, we allow some errors to the distance estimation. For instance, we allow a search window of ±2 dbm to our aggregated signal layout method, such that when an access point receives signal strength of −68 dbm, the algorithm searches the sensor's signal layout for position marker distances within the grid that falls between signal strengths −66 and −70 dbm. This search window we called signal spread is used to ensure more overlaps between layers of the signal layouts of the access points. With such a localization scheme, experimental results have shown that we can have a booth-level location estimation ranging from 0.9 m to 1.4 m. Please refer to [21] for the details of the aggregated signal layout method for localization within a wireless network. Since the accuracy may have error in estimation, the problem will be more serious if the booth size is small. In order to improve the location estimation accuracy, we use an adaptive rate measure (ARM) scheme to estimate the current booth ID of a customer. Under ARM, if the dimension of a booth is smaller (e.g., the length of edges of a rectangle), we repeat with more number of estimations to obtain the average value in localization for a customer in order to improve the accuracy in estimating the booth that the customer is currently locating.

One of the problems in determining the booth that a customer is currently locating using the above location estimation method is the bing-bong effect. This is a customer locating at the connection line of two booths and moving back and forth between them. For this case, similar to ARM, we increase the number of estimations as the problem may be due to estimation error and then take the average as a damping factor to determine the booth that the customer is currently locating. If the customer is always located between the two booths, that is, the average location being close by a threshold distance to both booths, two sets of customer-item records will be generated.

4.3. The Enhanced R-Tree Index.

The R-tree index was proposed for indexing spatial data [11]. In an R-tree index, each node, except the root node, has M maximum number of entries and m minimum number of entries with $m \leq M/2$. The root node has at least two entries. All the leaf nodes are at the same level; that is, it is a balanced tree. Each node in the R-tree contains a set of entries of the form (ptr, rect), where, for leaf node, ptr is a pointer to a record maintained in the database and rect is the minimum bounding rectangle (MBR) of the data item as shown in Figure 6. The MBR is used to represent the physical location and the size of a booth. For inner node, ptr is the pointer to a child node at the next lower level of the tree and rect is the MBR that bounds all the MBRs of the child node under it.

An enhanced R-tree is proposed to index the booth records of the stores. They are maintained in the movement database. The enhanced R-tree can facilitate the management of the items and space utilization of the booths by supporting various types of spatial queries. As shown in Figure 6, the index for the movement database consists of two parts. At the upper level, an R-tree is used to index the spaces covered by each booth in the stores. For example, the R-tree with fanout 3 in Figure 6 indexes the stores in Figure 4. The inner nodes cover the areas indexed by the booths pointed by their leaf nodes. Each leaf node points to the items under the booth. Under each item, if it is a promotion item, a list of customer-item records (called item list) is maintained according to the start time (e.g., time when the customer starts to stay in the booth) of the records. In addition, a set of array lists, $C_1, C_i, \ldots, C_m$ (called customer list) sorted according to customer ID, is maintained for searching the items that are of interest to customer i according to the received customer-item records. Since the number of customers could be large, a hash table is used to search the customer list to obtain the list of items that are of interest to the customer. A newly arrived customer-item record is inserted into customer ID list according to the customer ID and its duration. In addition, it is inserted into the item list according to their start time and item ID. Since the customer-item records under each item list are time-stamped according to the start time of the records, old customer-item records may be removed periodically by searching the item lists.

5. Illustrative Example

In this section, we give an illustrative example to show how the personalized promotion system retrieves the items that a customer may be interested in and how the system can help to increase the sales volume of the retail chain stores. Table 2 presents the promotion rule of some items. For example, if a customer has accumulated 15 marks, he can get 10% discount for item 1 in booth B2.

Once a registered customer enters a store for accumulating mark using his NFC-enabled smartphone, his smartphone will acquire an IP address from the WLAN installed in the store. Then, the location database server will retrieve the customer's customer-item records by invoking RetrieveCustRecord. If the corresponding sales items in the promotion database have discount rules defined at that time,

TABLE 2: Example promotion rules for the sales items in booths B1 and B2 of Figure 4.

Booths	Items	Marks			
		5	15	30	45
B1	1	0.05	0.1	0.12	0.2
	2	0.05	0.1	0.15	0.2
	3	0.05	0.05	0.1	0.15
B2	1	0.05	0.1	0.1	0.15
	2	0.05	0.1	0.15	0.15
	3	0.05	0.1	0.1	0.15
	4	0.05	0.1	0.15	0.20

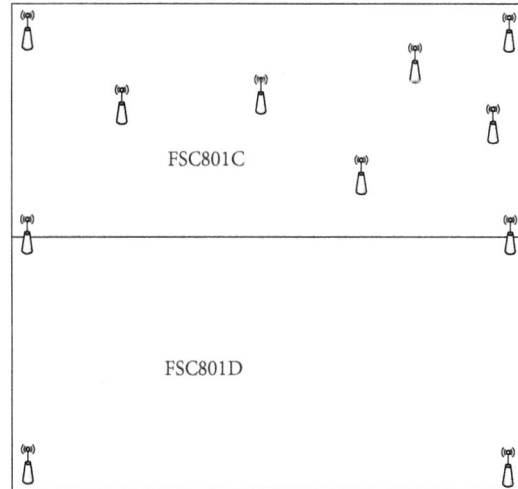

FIGURE 8: The locations of the access points in the test site.

the corresponding promotion information will be transferred to the customer by invoking IssueInfoToCustomer. At the same time, his identity and movement will be monitored periodically for further customer-item records generation. Meanwhile, if a customer does not know the location of a particular sales item that he wants to check with, he can submit a query RetrieveItem to obtain the location information from the *Item database.*

The generated customer-item records maintained by the local location server will be forwarded to the location database server by AddCustRecord periodically. After inserting all the customer-item records, the location database server will exploit the received customer-item records. If a sales item appears in many customer-item records, the SetPromotion will be invoked to set the discount rate for that item according to the *discount rule* defined by the chain stores. Meanwhile, the chain stores can retrieve the booth that contains the most booth records by invoking GetHottestBooth to identify which items are most popular to the customers. Other spatial queries may also be submitted for how to place the items in the booths so as to attract the customers' attention. Table 3 shows some example queries of the promotion systems.

6. Experimental Results

In this section, we report the experimental results of our study on the *effectiveness* of the adopted localization scheme in identifying the booth in which the user device (a WiFi-enabled smartphone) is locating. We constructed and conducted experiments in the network labs FSC301C and FSC301D at Hong Kong Baptist University (HKBU). The two labs are rectangular in shape and adjacent to each other.

6.1. Experimental Setup. Figure 8 shows the floor plan and the layout of the access points in the two labs FSC801C and FSC801D. All the access points were mounted at the ceiling forming a WLAN as well as a sensing network. The access points were programmable routers, Linksys WRT54G, which were burned with an open-source custom-made firmware, acting as the WiFi sensors to obtain information of the data packets transmitted within the WLAN. All the access points continuously listen to the packets sent and received

by the user device and also among the access points themselves. Packet information such as the service set identifier (SSID), extended service set identifier (ESSID), received signal strength indicator (RSSI), noise level, traffic rate, and traffic frequency was measured by a custom-made wireless data acquisition application written for the Linksys WRT54G WLAN router. For transmission efficiency and the reduction of data traffic, this useful information was selected (filter), grouped, time-stamped, and sent via TCP sockets through the router to be stored into the database maintained in the location database server. Then, the location estimation process would perform the estimation to find out which booth the user was currently locating at.

As shown in Figure 8, the access points in FSC801C were placed at the four corners first and the remaining five access points were called signal calibrators. Each signal calibrator was placed at a position such that the distances between it and the access points at each corner were different. Similarly, in FSC801D, two access points were placed at the far end corners from FSC801C. Furthermore, as shown in Figure 9, we divided FSC801C into 42 marker positions (each with a marker number) with horizontal and vertical dimensions of 1.9 m and 1.4 m, respectively. They were the basic unit for location estimation.

6.2. Experimental Results and Discussions. In the first set of experiments, we studied the estimation accuracy of the adopted localization scheme when the size of test site was varied. We placed the user device at different arbitrary marker positions (e.g., marker numbers of 9, 13, 23, and 25) in lab FSC801C and turned off all the calibrators except one of them. As shown in Table 4, the estimated accuracy of correctly identifying the marker number where the user device was locating was in a range of 0.584 to 0.832. It is interesting to see that the estimation accuracy is not significantly affected by the size of the test site and the number of access points used for localization. The average accuracies of using FSC801C and both FSC801C and FSC801D are similar, for

TABLE 3: Example queries in the personalized promoted system.

Queries	Descriptions
SetPromotion(*item, marks, discnt*)	Set the discount rate as *discnt* for item *item* with the accumulated *marks*.
AddCustRecord(*UID, BID*)	Add the customer-item record with the customer ID *UID* to the booth record with *BID* to the movement database.
RetrieveCustRecord(*UID, n*)	Retrieve the top *n* items of the customer-item records that the customer *UID* is interested from the movement database.
IssueInfoToCustomer(*UID, item*)	Issue discount information of data item *item* to the customer with identity number *UID*.
RetrieveItem(*item*)	Retrieve the location of the given item *item*.
GetHottestBooth(*date, BID*)	Get the booth ID with the maximum number of customer-item records on a particular day *date*.
GetColdestBooth(*date, BID*)	Get the booth ID with the minimum number of customer-item records on a particular day *date*.
GetRecords(*area, S_s, S_e*)	Get the number of customer-item records in the given area *area* for a given time period $[S_s, S_e]$.
GetHighTimeofBooth(*BID, hour, record*)	Given a booth ID, get the largest total stay period of the customer-item records for each hour.

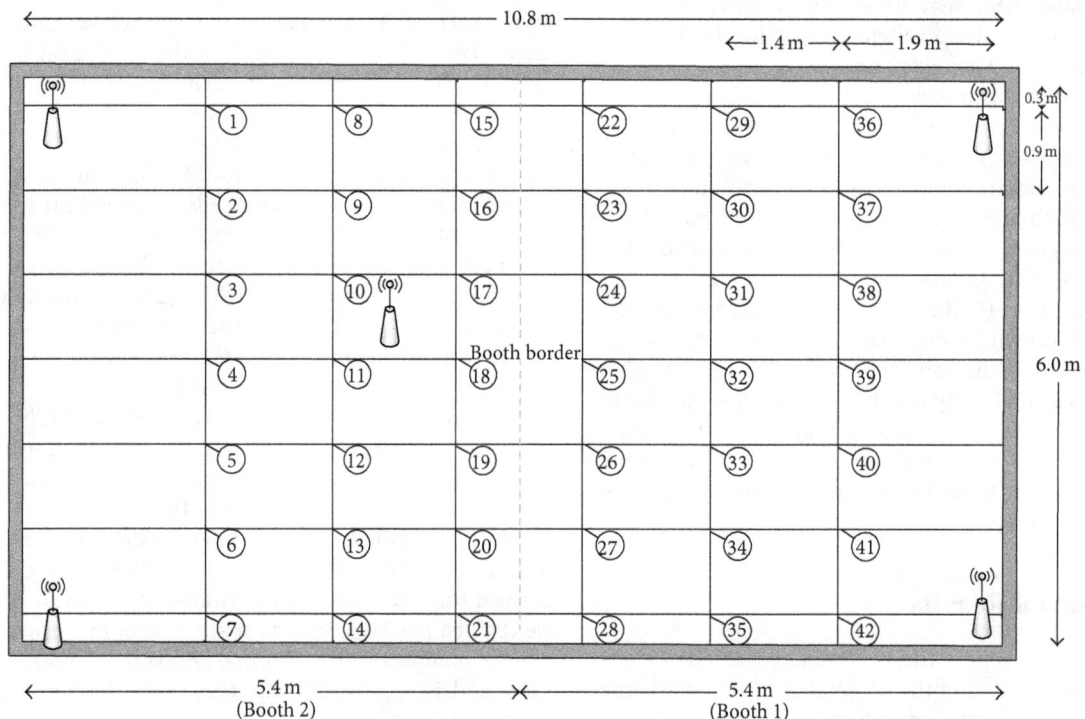

FIGURE 9: The floor plan and the layout of the access points in the test site (FSC801C).

example, 0.693 and 0.702, respectively. Therefore, to simplify the experimental setup, in the remaining experiments, we adopted the settings in lab FSC801C for testing and turned off all the access points in lab FSC801D.

In the second set of experiments, different numbers of access points (e.g., 3, 5, 7, and 9) were used. The access points were first placed at the corners and the remaining access points were calibrators. The results, as shown in Table 5, indicate that increasing the number of access points does not have serious impact on the estimation accuracy. Using more access points (or calibrators) does not always give better

estimation accuracy. Using four access points at the corners and one signal calibrator gives an estimation accuracy similar to the performance of different numbers of access points.

In the third set of experiments, we looked at the size of a booth on the effectiveness of the adopted localization scheme in localization, that is, the effectiveness to identify the current booth ID when the booth size was varied. Following the settings in the second set of experiments, we used five access points with one of them as the calibrator in FSC801C. In this set of experiments, we assumed that the booths were squares with equal size and their dimension

TABLE 4: Probabilities of the user location in a specific marker position estimated with different test sites.

Marker position	Test site	
	FSC801C	FSC801C + FSC801D
9	0.832	0.766
13	0.652	0.584
23	0.646	0.764
25	0.64	0.694
Average	0.692	0.702

TABLE 5: Probabilities of the user location in a specific marker position estimated with different number of APs.

Marker position	Number of APs			
	3	5	7	9
9	0.658	0.815	0.628	0.641
13	0.591	0.575	0.575	0.583
23	0.741	0.84	0.901	0.783
25	0.725	0.445	0.442	0.527
Average	0.678	0.668	0.636	0.6335

TABLE 6: The accuracy of the location estimation.

Within a radius of	500 samples	50 samples	25 samples	10 samples
0.83 m	0.973	0.738	0.503	0.258
2.15 m	0.995	0.954	0.875	0.62
2.50 m	1	0.987	0.977	0.908

was described by the radius of the circle that could include the square. Furthermore, we assumed that the user was at the center of the booth (center of the circle). Then we measured the probability that the user was estimated to be in that booth when the booth size was varied. Since we took signal strength data samples for location estimation, the more the samples, the better the accuracy (i.e., the adaptive rate measure (ARM)). Since we had 5 access points in the test site and each access point could generate 5–10 samples per second, it roughly took 10 to 20 seconds for collecting 500 samples for achieving a high accuracy for identifying the current booth ID. As shown in Table 6, when the size of the booth is within a radius of 0.83 m, we need to collect more samples to have a higher probability to identify the booth ID. As the booth size increases to a circle of radius of 2.15 m, there is a high probability (0.875) to identify the booth ID correctly by collecting only 25 data samples, that is, around 0.5 to 1 second to complete the localization estimation. Our experimental result shows that if the booth size is 2.5 m or above in radius, we can correctly identify which booth the user is currently locating at.

In the last set of experiments, we considered lab FSC801C consisting of Booth 1 and Booth 2 with the common border at the middle of the room. We moved the user within a booth to the common edge of the two booths instead of putting him at the center of the booth to test the localization scheme when

the pingpong effect was serious. Table 7 shows the results of the second set of experiments. It indicates that the probability of each estimated marker position being identified is the correct booth ID. That is, referring to Figure 9 and Table 7, for Marker 1 at the lowest left corner, the probability of Marker 1 estimated to be in Booth 2 is 0.952. Similarly, for Marker 22 which is nearby the booth border, the accuracy is only about 0.483 of guessing it right at Booth 1.

As shown in Table 7, indicated by the average probability for each row of markers, the accuracy of the estimation of getting the correct booth ID is lower if the user is closer to the booth border. On the contrary, if the user is far from the booth border, for example, 3.5 m away, the probability of correctly identifying the booth ID is around 0.809 to 0.896. We use a signal strength graphical icon of 5 ratings to show visually the pattern of correctly identifying the booth ID within our test site. Because of the positioning of access points installed, there may be some discrepancies among different markers. However, in general, if one is not too close to the border, that is, within 0.7 m, we can obtain the correct booth ID in most of the cases.

7. Conclusions and Future Work

Location-based services are receiving tremendous interests in the business sector and various novel location-based services are developing accordingly. However, most of them just use simple location technologies to get the current location of a mobile client to determine the services to be provided to the client. There is still lack of a good business model to show how the technologies can be integrated into the design of practical business systems. In this paper, we apply WLAN for localization and combine it with the NFC technology to propose a *pervasive promotion model* for the design of an effective personalized promotion system for retail chain stores. The localization technique using WLAN is adopted to capture the movements of customers within the stores by assuming that the movements within a store are effective indicators of the customers' interest in sales items. The architecture and operations of the personalized promotion system are discussed and various technical issues in localization using WLAN are studied for improving the accuracies in identifying the booth that a customer is currently locating. In addition, an enhanced R-tree index is proposed to support different types of spatial queries for better management of the items and booths in the stores. Experimental results have demonstrated the effectiveness of the adopted localization scheme in tracking the movements of customers in different booths. An important future work is to study the effectiveness of the proposed model and promotion system in retail sales in practices. A survey and questionnaire may be conducted to study how the customers respond to the system and the promotion strategies adopted in the model. The model may be applied to formulate promotion solutions for specific retail sale systems, for example, the convenient stores. Another important future work is to study the impacts of different placements of access points on the effectiveness of the adopted localization scheme.

TABLE 7: Probabilities of the user location estimated within the correct booth area.

	Probabilities of estimated marker position falling in the correct booth							Average	Distance from border
Booth 1	0.789	0.827	0.871	0.772	0.82	0.861	0.724	0.809	3.5 m
	0.619	0.741	0.588	0.888	0.857	0.854	0.952	0.786	2.1 m
	0.483	0.425	0.357	0.459	0.997	0.571	0.588	0.554	0.7 m
Booth 2	0.799	0.898	0.673	0.929	0.718	0.697	0.408	0.732	0.7 m
	0.884	0.983	0.721	0.966	0.895	0.939	0.701	0.87	2.1 m
	0.952	0.823	0.956	0.963	0.769	0.867	0.942	0.896	3.5 m

Conflict of Interests

The authors declare that there is no conflict of interests regarding the publication of this paper.

Acknowledgments

The work reported here was supported in part by the HKBU Research Centre for Ubiquitous Computing (RCUC), the Institute of Computational and Theoretical Studies (ICTS), and the HKBU Strategic Development Fund under the Grant no. HKBU SDF 10-0526-P08. The work was supported in part by a Strategic Research Grant (Project no. 7004044) from City University of Hong Kong.

References

[1] C.-W. You, C.-C. Wei, Y.-L. Chen, H.-H. Chu, and M.-S. Chen, "Using mobile phones to monitor shopping time at physical stores," *IEEE Pervasive Computing*, vol. 10, no. 2, pp. 37–43, 2011.

[2] C.-R. Dow, Y.-H. Lee, J. Liao, H.-W. Yang, and W.-L. Koo, "A location-based mobile advertisement publishing system for vendors," in *Proceedings of the 8th International Conference on Information Technology: New Generations (ITNG '11)*, pp. 24–29, Las Vegas, Nev, USA, April 2011.

[3] M. Strohbach and M. Martin, "Toward a platform for pervasive display applications in retail environments," *IEEE Pervasive Computing*, vol. 10, no. 2, pp. 19–27, 2011.

[4] Q. J. Cheng, J. K.-Y. Ng, and K. C. Y. Shum, "A wireless LAN location estimation system using center of gravity as an algorithm selector for enhancing location estimation," in *Proceedings of the 26th IEEE International Conference on Advanced Information Networking and Applications (AINA '12)*, pp. 261–268, IEEE, Fukuoka, Japan, March 2012.

[5] M. Ficco, C. Esposito, and A. Napolitano, "Calibrating indoor positioning systems with low efforts," *IEEE Transactions on Mobile Computing*, vol. 13, no. 4, pp. 737–751, 2013.

[6] T. Yamabe, V. Lehdonvirta, H. Ito, H. Soma, H. Kimura, and T. Nakajima, "Applying pervasive technologies to create economic incentives that alter consumer behavior," in *Proceedings of the 11th International Conference on Ubiquitous Computing (UbiComp '09)*, pp. 175–184, 2009.

[7] D. Drossos and K. Fouskas, "Mobile advertising: product involvement and its effect on intention to purchase," in *Proceedings of the 9th International Conference on Mobile Business and 9th Global Mobility Roundtable (ICMB-GMR '10)*, pp. 183–189, June 2010.

[8] C. Evans, P. Moore, and A. Thomas, "An intelligent mobile advertising system (iMAS): Location-based advertising to individuals and business," in *Proceedings of the 6th International Conference on Complex, Intelligent, and Software Intensive Systems (CISIS '12)*, pp. 959–964, July 2012.

[9] L. Spassova, G. Kahl, and A. Krüger, "User-adaptive advertisement in retail environments," in *Workshop on Pervasive Advertising and Shopping*, pp. 1–9, 2010.

[10] T. Kanda, D. F. Glas, M. Shiomi, H. Ishiguro, and N. Hagita, "Who will be the customer?: a social robot that anticipates people's behavior from their trajectories," in *Proceedings of the 10th International Conference on Ubiquitous Computing (UbiComp '08)*, pp. 380–389, September 2008.

[11] A. Guttman, "R-trees: a dynamic index structure for spatial searching," in *SIGMOD Conference*, vol. 14, pp. 47–57, 1984.

[12] H. Liu, H. Darabi, P. Banerjee, and J. Liu, "Survey of wireless indoor positioning techniques and systems," *IEEE Transactions on Systems, Man and Cybernetics Part C: Applications and Reviews*, vol. 37, no. 6, pp. 1067–1080, 2007.

[13] R. W. Ouyang, A. K.-S. Wong, and C.-T. Lea, "Received signal strength-based wireless localization via semidefinite programming: noncooperative and cooperative schemes," *IEEE Transactions on Vehicular Technology*, vol. 59, no. 3, pp. 1307–1318, 2010.

[14] Z. Guo, Y. Guo, F. Hong et al., "Perpendicular intersection: locating wireless sensors with mobile beacon," *IEEE Transactions on Vehicular Technology*, vol. 59, no. 7, pp. 3501–3509, 2010.

[15] L. Jun, H. V. Shukla, and J.-P. Hubaux, "Non-interactive location surveying for sensor networks with mobility-differentiated toA," in *Proceedings of the 25th IEEE International Conference on Computer Communications (INFOCOM '06)*, pp. 1241–1252, April 2006.

[16] L. Hu and D. Evans, "Localization for mobile sensor networks," in *Proceedings of the 10th Annual International Conference on Mobile Computing and Networking (MobiCom 2004)*, pp. 45–57, October 2004.

[17] Q. Shi, C. He, H. Chen, and L. Jiang, "Distributed wireless sensor network localization via sequential greedy optimization algorithm," *IEEE Transactions on Signal Processing*, vol. 58, no. 6, pp. 3328–3340, 2010.

[18] T. He, C. Huang, B. Blum, J. Stankovic, and T. Abdelzaher, "Range-free localization schemes for large scale sensor networks," in *Proceedings of the 9th annual international conference on Mobile computing and networking (MobiCom '03)*, pp. 81–95, San Diego, Calif, USA, September 2003.

[19] G. Wang and K. Yang, "A new approach to sensor node localization using rss measurements in wireless sensor networks," *IEEE Transactions on Wireless Communications*, vol. 10, no. 5, pp. 1389–1395, 2011.

[20] X. Li, "Collaborative localization with received-signal strength in wireless sensor networks," *IEEE Transactions on Vehicular Technology*, vol. 56, no. 6, pp. 3807–3817, 2007.

[21] K. C. Y. Shum, Q. J. Cheng, J. K. Y. Ng, and D. Ng, "A signal strength based location estimation algorithm within a wireless network," in *Proceedings of the 25th IEEE International Conference on Advanced Information Networking and Applications (AINA '11)*, pp. 509–516, March 2011.

[22] C. A. Opperman and G. P. Hancke, "A generic NFC-enabled measurement system for remote monitoring and control of client-side equipment," in *Proceedings of the 3rd International Workshop on Near Field Communication*, pp. 44–49, February 2011.

[23] C. Opperman and G. Hancke, "Using NFC-enabled phones for remote data acquisition and digital control," in *Proceedings of the IEEE AFRICON*, pp. 1–6, 2011.

[24] A. Juntunen, S. Luukkainen, and V. K. Tuunainen, "Deploying NFC technology for mobile ticketing services—identification of critical business model issues," in *Proceedings of the 9th International Conference on Mobile Business and 9th Global Mobility Roundtable (ICMB-GMR '10)*, pp. 82–90, IEEE, Athens, Greece, June 2010.

[25] W. Yeung, J. Zhou, and J. Ng, "Enhanced fingerprint-based location estimation system in wireless LAN environment," in *Emerging Directions in Embedded and Ubiquitous Computing*, vol. 4809 of *Lecture Notes in Computer Science*, pp. 273–284, Springer, Berlin, Germany, 2007.

14

Secure Mobile Agent from Leakage-Resilient Proxy Signatures

Fei Tang,[1,2] Hongda Li,[1,2] Qihua Niu,[1,2] and Bei Liang[1,2]

[1] The Data Assurance and Communication Security Research Center, Chinese Academy of Sciences, No. 89 Minzhuang Road, Haidian District, Beijing 100093, China
[2] State Key Laboratory of Information Security, Institute of Information Engineering, Chinese Academy of Sciences, No. 89 Minzhuang Road, Haidian District, Beijing 100093, China

Correspondence should be addressed to Fei Tang; tangfei127@163.com

Academic Editor: David Taniar

A mobile agent can sign a message in a remote server on behalf of a customer without exposing its secret key; it can be used not only to search for special products or services, but also to make a contract with a remote server. Hence a mobile agent system can be used for electronic commerce as an important key technology. In order to realize such a system, Lee et al. showed that a secure mobile agent can be constructed using proxy signatures. Intuitively, a proxy signature permits an entity (delegator) to delegate its signing right to another entity (proxy) to sign some specified messages on behalf of the delegator. However, the proxy signatures are often used in scenarios where the signing is done in an insecure environment, for example, the remote server of a mobile agent system. In such setting, an adversary could launch side-channel attacks to exploit some leakage information about the proxy key or even other secret states. The proxy signatures which are secure in the traditional security models obviously cannot provide such security. Based on this consideration, in this paper, we design a leakage-resilient proxy signature scheme for the secure mobile agent systems.

1. Introduction

Mobile agents [1–3] are designed as some autonomous software entities which are able to sign some messages in a remote server on behalf of a customer without exposing its secret key. Therefore, a mobile agent system can be used for electronic commerce in many ways such as negotiating something with other entities, searching and buying special products or services on behalf of a customer, and selling products on behalf of a shopping server. As shown by previous works, a mobile agent system can be constructed using some proxy signature schemes; for example, Lee et al. [4] used a strong nondesignated proxy signature scheme; they also provided an RSA-based and Schnorr-based constructions of secure mobile agent.

Proxy Signatures. This notion was first introduced by Mambo et al. [5] in 1996. In a proxy signature scheme, an entity called delegator may delegate its signing right to another entity called proxy who can then sign some specified messages on

behalf of the delegator; we call such signatures as proxy signatures. Finally, the verifier can be convinced from the proxy signatures that the original signer's agreement on the signed message and such proxy signatures must be computed by the proxy rather than the delegator. Obviously, proxy signatures are very useful in many application scenarios, for example, mobile agents [3, 6–9] and mobile communications [10, 11]. In the existing proxy signature schemes, the model of delegation by warrant [5] (a signed warrant, e.g., $W := ID_{\text{proxy}} \parallel \mathcal{M} \parallel$ indate $\parallel \cdots$, used to describe the validity of the delegation) has received the most attention. Kim et al. [12] suggested that a proxy key should be generated from such warrant. After Mambo et al.'s seminal work, many variants or improved schemes have been proposed (e.g., see [4, 11, 13–17]).

BPW Transformation. Boldyreva et al. [13] (henceforth called BPW) have given a secure generic construction of proxy schemes in the model of delegation by warrant from any secure ordinary signature scheme. Informally, to generate a proxy key, the original signer first signs a concatenation of the proxy's public key and a warrant with a specific way to obtain a delegation certificate. Then the proxy could set up the proxy key by himself using this delegation certificate. Finally, the proxy could sign some messages that are described in the warrant on behalf of the original signer (cf. Section 4 of [13] for detailed description).

Multilevel Proxy Model. Malkin et al. [14] extended the general proxy signatures to the scenario of multilevel proxy, where the proxy can also delegate the proxy signing right to another proxy (in such setting the former proxy also is a delegator); similarly, the second proxy also can delegate its proxy signing right to another, and so on. We call the identities that the original signer and all proxies construct a delegation chain, that is, (original signer)-(1th proxy)-(2th proxy)-$\cdots$-(jth proxy)-$\cdots$.

Security Models for Proxy Signatures. Due to the additional property of the proxy signatures, how to define the security for the proxy signatures is more complicated than the standard signatures [18]. In [19], Mambo et al. introduced several security notions (then enhanced by Lee et al. [4]) for the proxy signatures (here we omit them; please refer to [4, 19] for detailed description). These notions provide some intuitive security requirements for the proxy signatures, but corresponding security definitions are unclear (i.e., lacking of formal definitions), so many constructions were shown to be insecure and then fixed and finally to be shown insecure again (e.g., [4, 19, 20]). Subsequently, Boldyreva et al. [13] first presented a well-defined security model for the proxy signatures. In their model, the adversary is allowed to corrupt an arbitrary number of users and learn their secret keys. Moreover, the adversary can also register some public keys on behalf of new users. Then, the adversary interacts with honest users playing the role of a delegator or a proxy and it can see the transcripts of all executions of the delegation protocol between the honest users. It is a rather strong security model. Malkin et al. [14] later extended this model to allow multilevel proxy signatures; they also showed that proxy signatures are equivalent to key-insulated signatures [21]. The models of [13, 14] both are registered key models, which means that it is required that the adversary submits the secret and public keys of all users used in the model except a single challenging user. Schuldt et al. [15] got rid of this requirement and gave a new security model, existential unforgeability under adaptive chosen message attack with proxy key exposure (EU-CMA-PKE). In this model, adversary directly controls all user's secret keys of the delegation chain except the challenging user; furthermore, the adversary can corrupt some user to obtain the proxy keys (see Section 4 of [15] for more detailed description).

Black-Box Assumption versus Reality. In the security model of cryptographic schemes, traditionally, it is assumed that the secret internal state (secret key, randomness, etc.) of the schemes is completely hidden to the adversary, and hence the adversary in the traditional black-box model only can access an oracle to learn the input and output behaviors about the scheme. Unfortunately, many cryptographic engineers

have shown that this assumption is not true in real world applications. They have designed a large class of realistic attacks, called side-channel attacks, to detect some leakage information about the secret state, for example, timing attacks [22], power consumption [23], and fault attacks [24, 25]. Therefore, if we implement a mobile agent system from a secure proxy signature that is in the traditional security model, it may be also insecure if the device of mobile agent encounters the side-channel attacks.

Leakage-Resilient Cryptography. To resist such side-channel attacks, cryptographers have proposed many countermeasures in the past few years. Leakage-resilient cryptography is one of them, which means that a cryptosystem is also secure; even the adversary obtains some bounded (even arbitrary) leakage information about the secret internal state.

To model the security of cryptographic schemes in the leakage-resilient cryptography setting with a formal way, considering an adversary attacks a scheme besides the ordinary queries (as in the black-box model), it also can adaptively choose arbitrary polynomial time computable functions (named leakage functions) $f_i : \{0,1\}^* \to \{0,1\}^\lambda$ to obtain some information about the secret internal state. The restrictions of the input and output for such leakage functions depend on the leakage models. Here, we briefly present some of them.

(i) Only computation leaks model, introduced by Micali and Reyzin [26]: in this model, leakage is assumed to only occur on values that are currently accessed during the computation. Therefore, the input of the leakage function f_i is confined to the *active* part of the internal secret state, while the passive part of the secret state is not taken as input to the leakage function.

(ii) Bounded leakage model: the overall amount of the leakage should be bounded on a prespecified value λ.

(iii) Continual-leakage model, introduced by Brakerski et al. [27] and Dodis et al. [28], independently: in this model, the secret key is allowed to be refreshed, while the corresponding public key remains fixed. Then the amount of the leakage is bounded only in between any two successive key refreshes and the overall amount can be unbounded.

Many cryptographic schemes have been proposed in the leakage-resilient cryptography setting based on different leakage models, for example, leakage-resilient stream ciphers [29], leakage-resilient zero knowledge [30], leakage-resilient PKE [31, 32], leakage-resilient IBE [33, 34], and leakage-resilient signatures [35–40].

Leakage-Resilient Signatures. In this paper, we focus on the construction of leakage-resilient signature schemes. Alwen et al. [35] gave a construction of leakage-resilient signature scheme in the random oracle model which may tolerate leakage of up to half the secret key. Then Katz and Vaikuntanathan [38] constructed a bounded leakage-resilient signature scheme in the standard model which can tolerate leakage of up to $\ell - \ell^\epsilon$ (ℓ denotes the bit-length of the secret key) bits of information about the secret key. In the same paper, they also introduced the notion of fully leakage-resilient signatures which means that it is EU-CMA secure even the adversary may obtain leakage information on all internal state values that are used throughout the lifetime of the scheme. Boyle et al. [36] then improved their scheme to a full one which can be resilient to any leakage of length $(1 - o(1))\ell$ bits. Faust et al. [37] constructed a tree-based leakage-resilient signature scheme (in the model of "only computation leaks") which can be instantiated with any 3-time bounded leakage-resilient signature. Their scheme resilient to $\lambda = \lambda'/3$ bits per signing process, where λ' is size of the underlying 3-time signature scheme, can leak in total.

Our Contribution. Proxy signatures are often proposed for use in applications where signing is done in a potentially hostile environment; for example, if we use a proxy signature to realize a mobile agent system, then the proxy key is stored in a laptop, or even an IC card, which might become infected by malware. In such setting, an adversary who launches side-channel attacks can detect some leakage information about the proxy key or even other internal states. Based on this consideration, we construct a proxy signature scheme in the setting of leakage-resilient cryptography, the leakage-resilient proxy signature (LRPS), for the first time. The proposed LRPS scheme maintains the properties of these two primitives, leakage-resilient cryptography and proxy signatures.

To define the security notion to the LRPS scheme, we combine the existing security models of proxy signatures and leakage-resilient cryptography to put forward the security model of existential unforgeability against the adaptive chosen message and leakage attacks (EU-CMLA (We also introduce the notion of EU-CMLA-PKE which is extended from EU-CMA-PKE in [15] for the full construction of the LRPS in Appendices.)). Furthermore, we also construct a concrete LRPS scheme under the delegation by warrant and multilevel proxy models, it can be regarded as a concrete implementation of the BPW transformation in the setting of leakage-resilient cryptography. We use a tree-based signature scheme to construct the proxy signature scheme, which is different than the method that [13, 15] adopted; they both adopted an aggregate signature [41]. Hence our construction provides an alternative method to the construction of the proxy signatures. The concrete construction of the LRPS scheme is based on Faust et al.'s [37] (henceforth called FKPR, in TCC 2010) leakage-resilient signature scheme.

2. Definitions

In this section, we present some basic definitions for this paper: the notion of the stateful signatures and its security in the black-box model and in the presence of leakage, respectively.

2.1. Notations. 1^k denotes the string of k ones for $k \in N$. $|x|$ denotes the length of the bit string x if x is a bit string; $|S|$ denotes the number of the entries in the set S. $s \xleftarrow{\$} S$ means

randomly choosing an element s from the set S. We write $y \leftarrow A(x)$ to indicate that running the algorithm A with input x and then outputs y and $y \xleftarrow{\$} A(x)$ has the same indication except that A is a probabilistic algorithm. We use the notation $s_1 \parallel s_2$ to denote the concatenation of the bit strings s_1 and s_2; if they are not strings, we assume that they will be encoded as a string before the concatenation takes place. Lastly we write PPT for the probabilistic polynomial time.

2.2. Stateful Signatures. A signature scheme SIG consists of three algorithms, key generation, signing, and verification denoted by Kg, Sign, and Vfy, respectively. We say that a signature scheme is stateful if the Sign algorithm is stateful, which means that the secret key will be refreshed after (or before) each signing process, while its corresponding public key remains fixed. That is to say, SIG = (Kg, Sign, Vfy) is a stateful signature scheme if it satisfies the following.

(i) Kg is a PPT algorithm that takes as input a security parameter k and then outputs the signer's initial secret key SK_0 and public key PK. We write it $(SK_0, PK) \xleftarrow{\$} Kg(1^k)$.

(ii) Sign is a PPT algorithm run by the signer who takes as input its stateful secret key SK_{i-1} and a message m_i and then outputs a signature Σ_i and the next stateful secret key SK_i. We write it $(\Sigma_i, SK_i) \xleftarrow{\$} Sign(SK_{i-1}, m_i)$.

(iii) Vfy is a deterministic algorithm run by the verifier who takes as input the signer's public key PK, the signed message m_i, and the corresponding signature Σ_i and then outputs 1 if it is valid; else it outputs 0. We write it $1/0 \leftarrow Vfy(PK, m_i, \Sigma_i)$.

2.3. Security of Stateful Signatures in the Black-Box Model. The definition of existential unforgeability against adaptive chosen message attack (EU-CMA) for the stateful signatures is defined by the following experiment $\mathbf{Exp}_{SIG,\mathscr{A}}^{eu\text{-}cma}$ which is played by a EU-CMA adversary $\mathscr{A}$ and a challenger $\mathscr{B}$.

(i) $\mathscr{B}$ runs $(SK_0^*, PK^*) \xleftarrow{\$} Kg(1^k)$ and gives PK^* to $\mathscr{A}$.

(ii) $\mathscr{A}$ can adaptively ask $\mathscr{B}$ for the following:

signing query $\mathscr{S}\mathscr{Q}$: m_i

$\mathscr{B}$ runs $(\Sigma_i, SK_i^*) \xleftarrow{\$} Sign(SK_{i-1}^*, m_i)$ and returns Σ_i to $\mathscr{A}$.

(iii) At some point, $\mathscr{A}$ outputs (m^*, Σ^*).

We say that $\mathscr{A}$ wins the above experiment $\mathbf{Exp}_{SIG,\mathscr{A}}^{eu\text{-}cma}$ if $1 \leftarrow Vfy(PK^*, m^*, \Sigma^*)$ and m^* was not submitted to the signing query. We denote the probability of $\mathscr{A}$ succeeded by $\mathbf{Adv}_{SIG,\mathscr{A}}^{eu\text{-}cma}$. We say SIG is EU-CMA secure if $\mathbf{Adv}_{SIG,\mathscr{A}}^{eu\text{-}cma}$ is negligible for every PPT adversary $\mathscr{A}$.

2.4. Security of Stateful Signatures in the Presence of Leakage. In the setting of the leakage-resilient cryptography, adversary $\mathscr{A}$ can obtain λ bits of leakage information with every signing query. With the ith signing query, the adversary $\mathcal{A}$ adaptively chooses any computable leakage function $f_i : \{0,1\}^* \rightarrow \{0,1\}^\lambda$ to the leakage query and then obtains the output Λ_i of f_i which takes as input the active part SK_{i-1}^{*+} of the stateful secret key and the randomness r_i used in the signing phase. Formally, the model of existential unforgeability against adaptive chosen message and leakage attacks (EU-CMLA) is defined by the following experiment $\mathbf{Exp}_{SIG,\mathcal{A}}^{eu\text{-}cmla}$ which is played by a EU-CMLA adversary $\mathcal{A}$ and a challenger $\mathcal{B}$.

(i) $\mathcal{B}$ runs $(SK_0^*, PK^*) \stackrel{\$}{\leftarrow} Kg(1^k)$ and gives PK^* to $\mathcal{A}$.

(ii) $\mathcal{A}$ can adaptively ask $\mathcal{B}$ for the following:

    (a) signing query $\mathcal{SQ}$: m_i

        $\mathcal{B}$ runs $(\Sigma_i, SK_i^*) \stackrel{\$}{\leftarrow} Sign(SK_{i-1}^*, m_i, r_i)$ and returns Σ_i to $\mathcal{A}$;

    (b) leakage query $\mathcal{LQ}$: f_i

        $\mathcal{B}$ runs $\Lambda_i \leftarrow f_i(SK_{i-1}^{*+}, r_i)$ and if $|\Lambda_i| \neq \lambda$ then it returns $\bot$; else it returns Λ_i to $\mathcal{A}$.

(iii) At some point, $\mathcal{A}$ outputs (m^*, Σ^*).

We say that $\mathcal{A}$ wins the above experiment $\mathbf{Exp}_{SIG,\mathcal{A}}^{eu\text{-}cmla}$ if $1 \leftarrow Vfy(PK, m^*, \Sigma^*)$ and m^* was not submitted to the signing query. We denote the probability of $\mathcal{A}$ succeeded by $\mathbf{Adv}_{SIG,\mathcal{A}}^{eu\text{-}cmla}$. We say SIG is EU-CMA secure if $\mathbf{Adv}_{SIG,\mathcal{A}}^{eu\text{-}cmla}$ is negligible for every PPT adversary $\mathcal{A}$.

3. Leakage-Resilient Proxy Signatures

As outlined in the Introduction, there exists three entities in a proxy signature scheme: an original signer, a (or multi) proxy signer, and a verifier. A delegator, whether it is the original signer or a proxy signer, wants to delegate its signing right, whether original signing is right (i.e., the delegator is the original signer) or proxy signing is right (i.e., the delegator is a proxy signer) to a proxy. Finally, the verifier can be convinced with the original signer's agreement on the signed message and the identities of the proxy signers from the proxy signatures.

In the multilevel proxy model, a delegation chain, (ori ginal signer)-(1th proxy)-(2th proxy)-·····-(jth proxy)-····, consists of an original signer and j (or more) proxy signers. To identify them, we require a list $\mathcal{PK}$ of their public keys in the proxy signatures.

In the BPW transformation, the delegator will sign its proxy's public key and corresponding warrant to obtain a certificate to generate the proxy key. Therefore, to verify the validity of the delegation, it is also required that the proxy signatures contain a list $\mathcal{W}$ of the warrants and $\mathcal{C}$ of the certificates of the delegations.

3.1. Syntax. Formally, we define the stateful proxy signatures (under the BPW transformation) as follows. That is to say, $SIG^* = (Kg^*, Sign^*, Vfy^*, \langle Del^*, PKg^* \rangle, PSign^*, PVfy^*)$ is a stateful proxy signature scheme if the first three algorithms are defined as Kg, Sign, and Vfy of the scheme SIG, respectively, and the latter three algorithms satisfy the following.

(i) $\langle Del^*, PKg^* \rangle$ is a pair of interactive PPT delegation protocol which means that the delegator D whose stateful key is $(SK_{D(i-1)}, PK_D)$ delegates its signing right to a proxy P who has a stateful key pair $(SK_{P(i'-1)}, PK_P)$.

    (a) Del* is run by the delegator with input $(SK_{D(i-1)}, PK_P, \mathcal{PK}, \mathcal{W}, \mathcal{C}, j, W_j)$, where $\mathcal{PK}, \mathcal{W}$, and $\mathcal{C}$ are the lists of public keys, warrants, and delegation certificates of the previous delegators, respectively, j describes the current proxy is the jth proxy in the delegation chain ($j = 0$ means that the delegator is the original signer), and W_j is the warrant for the current delegation.

    (b) PKg* is run by the proxy with input $(SK_{P(i'-1)}, PK_P, PK_D)$ to generate its proxy key.

As a result of this interactive algorithm, the algorithm Del* has no local output except that the delegator's next stateful key SK_{Di}. The local output of PKg* is the delegation information $(\mathcal{PK}', \mathcal{W}', \mathcal{C}', j, SK_{P(i'-1)})$, where $\mathcal{PK}'$, $\mathcal{W}'$, and $\mathcal{C}'$ are the lists of public keys, warrants, and certificates in the delegation chain extended with the public key of the proxy and warrant and certificate of the current delegation, respectively. We write it $(SK_{Di}, \mathcal{PK}', \mathcal{W}', \mathcal{C}', j, SK_{P(i'-1)}) \stackrel{\$}{\leftarrow} \langle Del^*(SK_{D(i-1)}, PK_P, \mathcal{PK}, \mathcal{W}, \mathcal{C}, j, W_j), PKg^*(SK_{P(i'-1)}, PK_P, PK_D) \rangle$.

(ii) PSign* is a PPT algorithm run by a proxy that takes as input its delegation information $(\mathcal{PK}, \mathcal{W}, \mathcal{C}, j, SK_{P(i'-1)})$ and a message m_i and then outputs a proxy signature $(\mathcal{PK}, \mathcal{W}, \mathcal{C}, j, P\Sigma_i)$ on behalf of the delegator and its next stateful key $SK_{Pi'}$. We write it $(\mathcal{PK}, \mathcal{W}, \mathcal{C}, j, P\Sigma_i, SK_{Pi'}) \stackrel{\$}{\leftarrow} PSign^*(\mathcal{PK}, \mathcal{W}, \mathcal{C}, j, SK_{P(i'-1)}, m_i)$.

(iii) PVfy* is a deterministic algorithm run by the verifier who takes as input $(\mathcal{PK}, \mathcal{W}, \mathcal{C}, j, m_i, P\Sigma_i)$ and then outputs 1 if it is valid; else it outputs 0. We write it $1/0 \leftarrow PVfy^*(\mathcal{PK}, \mathcal{W}, \mathcal{C}, j, m_i, P\Sigma_i)$.

In the real world applications, user's long-term secret key should be stored in a secure way and thus to guarantee that no information about the long-term key is leaked while the proxy key is exposed, it is better to generate a proxy key independent of the long-term key. We call such construction a full construction. There exists a simple method to the full construction from any BPW transformed proxy signature (cf. Section 5 of [15]).

(i) After obtaining the delegation information $(\mathcal{PK}, \mathcal{W}, \mathcal{C}, j, SK_{P(i'-1)})$, the proxy first generates a fresh proxy key pair $(SK_{P0}', PK_P') \stackrel{\$}{\leftarrow} Kg^*(1^k)$.

(ii) Compute $(cert', SK_{Pi'}) \stackrel{\$}{\leftarrow} Sign^*(SK_{P(i'-1)}, 00 \parallel PK_P' \parallel 0 \parallel cert)$, where $cert \in \mathcal{C}$ is the delegation certificate from the delegator.

(iii) The new delegation information is $(\mathcal{PK}', \mathcal{W}, \mathcal{C}', j', SK_{P0}')$, where $PK' \in \mathcal{PK}'$ and $cert' \in \mathcal{C}'$.

The concrete full construction of such proxy signature scheme and corresponding security analysis are presented in Appendices.

3.2. Implement Secure Mobile Agent from Proxy Signature Scheme.

When we realize a mobile agent system construction by using a secure proxy signature scheme let the clients be the delegators and let the mobile agent be the proxy. Then the clients and the agent together run the interactive delegation protocol to delegate the client's signing right to the agent. Finally, the agent can sign some specified messages on behalf of the client. A secure proxy signature scheme implies a secure mobile agent system; similarly, a leakage-resilient proxy signature scheme means that the corresponding mobile agent system can be resilient to some bounded information leakage.

3.3. Security of the Leakage-Resilient Proxy Signatures.

We put forward the security model of existential unforgeability against adaptive chosen message and leakage attacks (EU-CMLA) for the proxy signatures in the presence of leakage. It defined by the following experiment $\mathbf{Exp}^{\text{eu-cmla}}_{\text{SIG}^*,\mathscr{A}}$ which is played by a challenger $\mathscr{B}$ and a EU-CMLA adversary $\mathscr{A}$ who controls all user's secret keys except the challenging user.

(i) $\mathscr{B}$ runs $(\text{SK}^*_0, \text{PK}^*) \xleftarrow{\$} \text{Kg}^*(1^k)$ and gives PK^* to $\mathscr{A}$.

(ii) $\mathscr{A}$ can adaptively ask $\mathscr{B}$ for the following:

(a) delegation to SK^*_{i-1}: PK_D

$\mathscr{B}$ interacts with $\mathscr{A}$ through the delegation protocol by running algorithm $\text{PKg}^*(\text{SK}^*_{i-1}, \text{PK}^*, \text{PK}_\text{D})$. When it is finished, $\mathscr{B}$ will obtain the delegation information $(\mathscr{PK}', \mathscr{W}', \mathscr{C}', j, \text{SK}^*_{i-1})$.

(b) delegation of SK^*_{i-1}: $(\text{PK}_\text{P}, W_j)$

$\mathscr{B}$ interacts with $\mathscr{A}$ through the delegation protocol to generate a proxy key to PK_P; $\mathscr{B}$ runs $\text{Del}^*(\text{SK}^*_{i-1}, \text{PK}_\text{P}, \mathscr{PK}, \mathscr{W}, \mathscr{C}, j, W_j)$. When it is finished, $\mathscr{B}$ returns the transcript of the delegation to $\mathscr{A}$;

(c) self-delegation of SK^*_{i-1}: W

$\mathscr{B}$ first runs $(\text{SK}'_0, \text{PK}') \xleftarrow{\$} \text{Kg}^*$ and then runs the delegation protocol to generate a proxy key to the challenging user itself, $(\text{SK}^*_i, \mathscr{PK}', \mathscr{W}', \mathscr{C}', j', \text{SK}'_0) \xleftarrow{\$} \langle \text{Del}^*(\text{SK}^*_{i-1}, \text{PK}', \mathscr{PK}, \mathscr{W}, \mathscr{C}, j, W), \text{PKg}^*(\text{SK}'_0, \text{PK}', \text{PK}^*) \rangle$. When it is finished, $\mathscr{B}$ will obtain the delegation information $(\mathscr{PK}', \mathscr{W}', \mathscr{C}', j', \text{SK}'_0)$ and send the transcript of the delegation to $\mathscr{A}$;

(d) ordinary signing queries of SK^*_{i-1}: m_i

$\mathscr{B}$ runs $(\Sigma_i, \text{SK}^*_i) \xleftarrow{\$} \text{Sign}^*(\text{SK}^*_{i-1}, m_i)$ and returns Σ_i to $\mathscr{A}$;

(e) proxy signing queries of SK^*_{i-1}: $(\mathscr{PK}, \mathscr{W}, \mathscr{C}, j, m_i)$

$\mathscr{B}$ runs $(\mathscr{PK}, \mathscr{W}, \mathscr{C}, j, P\Sigma_i, \text{SK}^*_i) \xleftarrow{\$} \text{PSign}^*(\mathscr{PK}, \mathscr{W}, \mathscr{C}, j, \text{SK}^*_{i-1}, m_i)$ and returns $(\mathscr{PK}, \mathscr{W}, \mathscr{C}, j, P\Sigma_i)$ to $\mathscr{A}^*$;

(f) leakage queries: f_i

$\mathscr{A}$ may adaptively launches leakage query after each query to the delegation protocol, ordinary signing, or proxy signing oracle; that is, these algorithms have taken as input the secret key SK^*_{i-1}. $\mathscr{B}$ runs $\Lambda_i \leftarrow f_i(\text{SK}^{*+}_{i-1}, r_i)$ and if $|\Lambda_i| \neq \lambda$ then it returns $\bot$; else it returns Λ_i to $\mathscr{A}$.

(iii) At some point, $\mathscr{A}$ outputs a forgery which must be one of the following cases.

(1) Ordinary signature of PK^*: (m^*, Σ^*)

if $1 \leftarrow \text{Vrf}^*(\text{PK}^*, m^*, \Sigma^*)$ and m^* has not been submitted to the ordinary signing queries, then output 1; else output 0.

(2) Proxy signature of PK^*: $(m^*, (\mathscr{PK}, \mathscr{W}, \mathscr{C}, j, P\Sigma^*))$, PK^* is the last entry in $\mathscr{PK}$

if $1 \leftarrow \text{PVrf}^*(\mathscr{PK}, \mathscr{W}, \mathscr{C}, j, m^*, P\Sigma^*)$ and $(\mathscr{PK}, \mathscr{W}, \mathscr{C}, j, m^*)$ has not submitted to the proxy signing queries, then output 1; else output 0.

(3) Proxy signature on behalf of PK^*: $(m^*, (\mathscr{PK}, \mathscr{W}, \mathscr{C}, j, P\Sigma^*))$, PK^* is the nth entry in $\mathscr{PK}$.

If $1 \leftarrow \text{PVrf}^*(\mathscr{PK}, \mathscr{W}, \mathscr{C}, j, m^*, P\Sigma^*)$ and $\mathscr{A}$ has not queried the delegation of SK^*_{i-1} oracle on inputs $(\text{PK}_{n+1}, W_{n+1})$, that is, the $(n+1)$-th entry in the set $\mathscr{PK}$), then output 1 else output 0.

We say that $\mathscr{A}$ wins the above experiment $\mathbf{Exp}^{\text{eu-cmla}}_{\text{SIG}^*,\mathscr{A}}$ if it outputs a valid forgery. We denote the probability of $\mathscr{A}$ succeeded by $\mathbf{Adv}^{\text{eu-cmla}}_{\text{SIG}^*,\mathscr{A}}$. We say SIG^* is EU-CMLA secure if $\mathbf{Adv}^{\text{eu-cmla}}_{\text{SIG}^*,\mathscr{A}}$ is negligible for every PPT adversary $\mathscr{A}$.

Remark. In the model of EU-CMA-PKE, $\mathscr{A}$ is allowed to query a redelegation of a user's proxy key. However, we define the LRPS under the BPW transformation model (i.e., the user's proxy key is exactly its secret key), so in the model of EU-CMLA, $\mathscr{A}$ can run the redelegation by itself except that the redelegation of SK^*_{i-1} which can be obtained from the query of delegation of SK^*_{i-1} in such setting. Similarly, $\mathscr{A}$ has no need to query the proxy key exposure queries.

4. Construction of Leakage-Resilient Proxy Signatures

In this section, we present a concrete construction of the LRPS scheme SIG^* based on FKPR signature scheme which can be instantiated with any EU-CMTLA (existential unforgeability against chosen message and total leakage attacks) 3-time signature scheme $\text{sig} = (\text{kg}, \text{sign}, \text{vfy})$.

Before giving the detailed description of the SIG^*, we first introduce some notations relative to the tree-based (with depth $d \in N$) signature. We denote the all bit strings of length at most d (including the empty string ε) with $\{0,1\}^{\leq d} = \bigcup_{i=1}^{d} \{0,1\}^i \cup \varepsilon$ (size $2^{d+1} - 1$). The left and right child of an internal node (or root) $w \in \{0,1\}^{\leq d-1}$ are denoted by $w \parallel 0$ and $w \parallel 1$, respectively, and $\text{par}(w)$ denotes the node w's parent node. Depth-first traversal algorithm can be used to traverse and label the tree. For a node $w \in \{0,1\}^{\leq d} \setminus 1^d$, we define algorithm $\mathsf{DF}(w)$ as the node traversed after w in the depth-first traversal; that is,

$$\mathsf{DF}(w)$$
$$:= \begin{cases} w \parallel 0, & \text{if } |w| < d \\ & (w \text{ is the root or an internal node}) \quad (1) \\ w' \parallel 1, & \text{if } |w| = d, \\ & \text{where } w = w' \parallel 0 \parallel 1^j \ (w \text{ is a leaf}). \end{cases}$$

When the depth-first algorithm traverses the binary tree, each node w is associated with a secret-public key pair $(\mathrm{sk}_w, \mathrm{pk}_w)$ by invoking the kg algorithm of the underlying signature scheme sig. The following notations will be used in the latter part of this paper. Let $w = w_1 w_2 \cdots w_t$ be a bit string with length t.

(i) $\Gamma_w := \{(\mathrm{pk}_w, \phi_w), \ldots, (\mathrm{pk}_{w_1 w_2}, \phi_{w_1 w_2}), (\mathrm{pk}_{w_1}, \phi_{w_1})\}$ is a "signature path" from w to the root; $\phi_{w'}$ is a signature of $010 \parallel \mathrm{pk}_{w'}$ with its parent's key $\mathrm{sk}_{\mathrm{par}(w')}$; that is,

$$\phi_{w'} \xleftarrow{\$} \mathrm{sign}(\mathrm{sk}_{\mathrm{par}(w')}, 010 \parallel \mathrm{pk}_{w'}).$$

(ii) $S_w := \{\mathrm{sk}_{w_1 w_2 \cdots w_i} \mid w_{i+1} = 0\}$ is a subset of the secret keys on the path from the root ε to node w. $\mathrm{sk}_{w'} \in S_w$ if and only if the path goes to the left child $w' \parallel 0$ at the node w'. (The reason is that, in this case, the node w''s right child $w' \parallel 1$ will be traversed after node w under the depth-first traversal. Consequently, we need the secret key $\mathrm{sk}_{w'}$ of node w' to sign its right child $w' \parallel 1$'s public key $\mathrm{pk}_{w' \parallel 1}$.)

The stateful secret key of the scheme SIG^* will have the form (w, S_w, Γ_w) (i.e., using stacks S_w and Γ_w to keep track of the state, or node w). For a stack S, define the following three algorithms:

(1) $\mathsf{push}(S, a)$: putting element a on the stack S;

(2) $a \leftarrow \mathsf{pop}(S)$: removing the topmost element from the stack S and assigning it to a;

(3) $\mathsf{trash}(S)$: removing the topmost element from the stack S.

4.1. Construction. To avoid trivial attacks against this scheme, we use the idea of Boldyreva et al. [13], attach a 3-bit string as the prefix of the text that will be signed, that is, $111\parallel$(text which will be to compute ordinary signatures), $010\parallel$(text which will be to compute signature paths), $100\parallel$(text which will be to compute delegation certificates), and $101\parallel$(text which will be to compute proxy signatures), respectively. The LRPS scheme SIG^* is constructed as follows.

(i) $\mathsf{Kg}^*(1^k)$:

$(\mathrm{sk}_\varepsilon, \mathrm{pk}_\varepsilon) \xleftarrow{\$} \mathsf{Kg}(1^k), S_\varepsilon := \mathrm{sk}_\varepsilon, \Gamma_\varepsilon := \emptyset, \mathrm{SK}_\varepsilon := (w_\varepsilon, S_\varepsilon, \Gamma_\varepsilon), \mathrm{PK} := \mathrm{pk}_\varepsilon$; return $(\mathrm{SK}_\varepsilon, \mathrm{PK})$.

(ii) $\mathsf{Sign}^*(\mathrm{SK}_w, m)$: (to ease exposition, the signing process of the root ε (i.e., $\sigma \xleftarrow{\$} \mathrm{sign}(\mathrm{sk}_\varepsilon, 111 \parallel m)$) is not contained in this formalizing description)

parse $\mathrm{SK}_w := (w, S_w, \Gamma_w)$; if $w = 1^d$ return $\bot$; $\widehat{w} \leftarrow \mathsf{DF}(w), (\mathrm{sk}_{\widehat{w}}, \mathrm{pk}_{\widehat{w}}) \xleftarrow{\$} \mathsf{Kg}(1^k)$

$\sigma \xleftarrow{\$} \mathrm{sign}(\mathrm{sk}_{\widehat{w}}, 111 \parallel m); \mathrm{sk}_{\mathrm{par}(\widehat{w})} \leftarrow \mathsf{pop}(S_w); \phi_{\widehat{w}} \xleftarrow{\$} \mathrm{sign}(\mathrm{sk}_{\mathrm{par}(\widehat{w})}, 010 \parallel \mathrm{pk}_{\widehat{w}})$

$$\begin{aligned} &\text{if } \widehat{w}_{|\widehat{w}|} = 0, \quad S_w \longleftarrow \mathsf{push}\left(S_w, \mathrm{sk}_{\mathrm{par}(\widehat{w})}\right) \\ &\quad \text{if } |\widehat{w}| < d, \quad S_{\widehat{w}} \longleftarrow \mathsf{push}\left(S_w, \mathrm{sk}_{\widehat{w}}\right) \\ &\quad\quad \text{if } |w| = d, \quad w = w' 01^j, \quad (2) \\ &\quad\quad \text{for } i = 1, \ldots, j+1, \text{ do } \mathsf{trash}\left(\Gamma_w\right) \end{aligned}$$

$\Gamma_{\widehat{w}} \leftarrow \mathsf{push}(\Gamma_w, (\mathrm{pk}_{\widehat{w}}, \phi_{\widehat{w}})); \Sigma := (\sigma, \Gamma_{\widehat{w}}); \mathrm{SK}_{\widehat{w}} := (\widehat{w}, S_{\widehat{w}}, \Gamma_{\widehat{w}})$; return $(\Sigma, \mathrm{SK}_{\widehat{w}})$.

(iii) $\mathsf{Vfy}^*(\mathrm{PK}, m, \Sigma)$:

parse $\Sigma := (\sigma, \Gamma_{\widehat{w}_1 \widehat{w}_2 \cdots \widehat{w}_{|\widehat{w}|}}), \mathrm{pk}_\varepsilon := \mathrm{PK}$; for $i = 1, \ldots, |\widehat{w}|$ do

if $0 \leftarrow \mathrm{vfy}(\mathrm{pk}_{\widehat{w}_1 \cdots \widehat{w}_{i-1}}, 010 \parallel \mathrm{pk}_{\widehat{w}_1 \cdots \widehat{w}_i}, \phi_{\widehat{w}_1 \cdots \widehat{w}_i})$ return 0;

else return $\mathrm{vfy}(\mathrm{pk}_{\widehat{w}_1 \widehat{w}_2 \cdots \widehat{w}_{|\widehat{w}|}}, 111 \parallel m, \sigma)$.

(iv) $\mathsf{Del}^*(\mathrm{SK}_{\mathrm{D}(i-1)}, \mathrm{PK}_\mathrm{P}, \mathscr{PK}, \mathscr{W}, \mathscr{C}, j, W_j)$:

D runs $(\mathrm{cert}_j, \mathrm{SK}_{\mathrm{D}i}) \xleftarrow{\$} \mathsf{Sign}^*(\mathrm{SK}_{\mathrm{D}(i-1)}, 100 \parallel \mathrm{PK}_\mathrm{P} \parallel j \parallel W_j)$ and

then sends $(\mathscr{PK}, \mathscr{W}, \mathscr{C}, j, W_j, \mathrm{cert}_j)$ to P.

(v) $\mathsf{PKg}^*(\mathrm{SK}_{\mathrm{P}(i'-1)}, \mathrm{PK}_\mathrm{P}, \mathrm{PK}_\mathrm{D})$:

P first checks the validity of the delegation certificates, for $k = 1, \ldots, j$ does

if $0 \leftarrow \mathsf{Vfy}^*(\mathrm{PK}_{k-1}, 100 \parallel \mathrm{PK}_k \parallel k \parallel W_k, \mathrm{cert}_k)$, it returns $\bot$ and rejects this delegation;

otherwise, run $\mathscr{PK} \leftarrow \mathsf{push}(\mathscr{PK}, \mathrm{PK}_\mathrm{P}), \mathscr{W} \leftarrow \mathsf{push}(\mathscr{W}, W_j), \mathscr{C} \leftarrow \mathsf{push}(\mathscr{C}, \mathrm{cert}_j)$;

finally, set the delegation information as $(\mathscr{PK}, \mathscr{W}, \mathscr{C}, j, \mathrm{SK}_{\mathrm{P}(i'-1)})$.

If someone, whose key pair is $(\mathrm{SK}_{\mathrm{SD}(i-1)}, \mathrm{PK}_{\mathrm{SD}})$, wants to designate itself as a proxy it runs $(\mathrm{SK}'_{\mathrm{P}0}, \mathrm{PK}'_\mathrm{P}) \xleftarrow{\$} \mathsf{Kg}^*(1^k)$ to generate a fresh key pair as the proxy key and creates a certificate $(\mathrm{cert}', \mathrm{SK}_{\mathrm{SD}i}) \xleftarrow{\$} \mathsf{Sign}^*(\mathrm{SK}_{\mathrm{SD}(i-1)}, 100 \parallel \mathrm{PK}'_\mathrm{P} \parallel 0 \parallel W')$, then does

$$\begin{aligned} \mathscr{PK} &\longleftarrow \mathsf{push}\left(\mathscr{PK}, \mathrm{PK}'_\mathrm{P}\right), \\ \mathscr{W} &\longleftarrow \mathsf{push}\left(\mathscr{W}, W'\right), \quad (3) \\ \mathscr{C} &\longleftarrow \mathsf{push}\left(\mathscr{C}, \mathrm{cert}'\right); \end{aligned}$$

finally, it sets the delegation information as $(\mathscr{PK}, \mathscr{W}, \mathscr{C}, j, \mathrm{SK}'_{\mathrm{P}0})$.

(vi) $\mathsf{PSign}^*(\mathscr{PK}, \mathscr{W}, \mathscr{C}, j, \mathrm{SK}_{\mathrm{P}(i-1)}, m)$:

$(\Sigma, \mathrm{SK}_{\mathrm{P}i}) \overset{\$}{\leftarrow} \mathsf{Sign}^*(\mathrm{SK}_{\mathrm{P}(i-1)}, 101 \parallel m)$ and output the proxy signature $(\mathscr{PK}, \mathscr{W}, \mathscr{C}, j, P\Sigma := \Sigma)$.

(vii) $\mathsf{PVfy}^*(\mathscr{PK}, \mathscr{W}, \mathscr{C}, j, m, P\Sigma)$:

V first checks the validity of the delegation certificates, for $k = 1, \ldots, j$ does

if $0 \leftarrow \mathsf{Vfy}^*(\mathrm{PK}_{k-1}, 100 \parallel \mathrm{PK}_k \parallel k \parallel W_k, \mathsf{cert}_k)$ returns 0;

else it returns $\mathsf{Vfy}^*(\mathrm{PK}_j, 101 \parallel m, P\Sigma)$.

Upper Bound of the Number of the Messages Can Be Signed. For a fixed signing key, in both of the schemes FKPR and SIG*, the upper bound of the number of the message that can be signed is $q = 2^{d+1} - 2$. We can see that, from the above construction, each internal node is used only one time to the signing algorithm. However, the key (with respect to the scheme **sig**) of any leaf can be signed three times. Hence, the upper bound of the number of the message can be signed and could be increased to $2^{d+2} - 4$ that is double the number of the previous upper bound, as well as the FKPR scheme.

We should stress here that there is a disadvantage to our scheme which is based on tree-based signature compared to that constructed based on aggregate signature [13, 15]; that is, in those schemes, the verification of the delegation certificates can be executed at a time due to the property of aggregability of the aggregate signatures [41].

4.2. Security. We now analyze the security of the proposed LRPS scheme.

Theorem 1. *If the FKPR scheme (denoted by* **SIG***) is EU-CMLA secure, then the proxy signature scheme* **SIG*** *also is EU-CMLA secure.*

Our proof line is similar to that of Boldyreva et al.'s [13]. If there exists a EU-CMLA adversary and $\mathscr{A}$ can break the security of the scheme SIG*, then we can construct a challenger $\mathscr{B}$ to break the security of the FKPR scheme SIG.

(i) Initially, $\mathscr{B}$ will be given a challenging public key PK' and can adaptively make signing query ($\mathscr{SQ}$) and leakage query ($\mathscr{LQ}$) in the experiment $\mathbf{Exp}^{\mathrm{eu\text{-}cmla}}_{\mathrm{SIG}, \mathscr{B}}$. $\mathscr{B}$ first sets $\mathrm{PK}^* := \mathrm{PK}'$ as the challenging public key of the experiment $\mathbf{Exp}^{\mathrm{eu\text{-}cmla}}_{\mathrm{SIG}^*, \mathscr{A}}$ and sends it to $\mathscr{A}$. Then it plays the experiment with $\mathscr{A}$.

(ii) $\mathscr{A}$ may adaptively ask $\mathscr{B}$ for the following.

(a) Delegation to SK^*_{i-1}: PK_{D}

$\mathscr{B}$ interacts with $\mathscr{A}$ through the delegation protocol by running $\mathsf{PKg}^*(*, \mathrm{PK}^*, \mathrm{PK}_{\mathrm{D}})$. When it is finished, $\mathscr{B}$ will obtain the delegation information $(\mathscr{PK}', \mathscr{W}', \mathscr{C}', j, *)$. $\mathscr{B}$ can run the

PKg^* algorithm even if it has no idea about the SK^*_{i-1}, because SK^*_{i-1} will be set as the proxy key of the challenging user, so upon completion, $\mathscr{B}$ does not know the corresponding proxy key.

(b) Delegation from SK^*_{i-1}: $(\mathrm{PK}_{\mathrm{P}}, W_j)$

$\mathscr{B}$ interacts with $\mathscr{A}$ through the delegation protocol to generate a proxy key to PK_{P}. $\mathscr{B}$ makes the signing query $\mathscr{SQ}$ with input $00 \parallel \mathrm{PK}_{\mathrm{P}} \parallel j \parallel W_j$; then it will be returned Σ. After the delegation protocol is finished, $\mathscr{A}$ will obtain the delegation information $(\mathscr{PK}', \mathscr{W}', \mathscr{C}', j, *)$, where $\mathrm{PK}_{\mathrm{P}} \in \mathscr{PK}'$, $W_j \in \mathscr{W}'$, and $\mathsf{cert}_j := \Sigma \in \mathscr{C}'$.

(c) Self-delegation of SK^*_{i-1}: W

$\mathscr{B}$ runs the delegation protocol to generate a proxy key of PK^* to itself. $\mathscr{B}$ first runs $(\mathrm{SK}'_0, \mathrm{PK}') \overset{\$}{\leftarrow} \mathsf{Kg}^*$ and then makes the signing query $\mathscr{SQ}$ with input $00 \parallel \mathrm{PK}' \parallel 0 \parallel W$; then it will be returned to Σ. Finally, $\mathscr{B}$ will return the delegation information $(\mathscr{PK}', \mathscr{W}', \mathscr{C}', 0, \mathrm{SK}'_0)$ and sends the delegation transcripts to $\mathscr{A}$, where $\mathrm{PK}' \in \mathscr{PK}'$, $W \in \mathscr{W}'$, and $\mathsf{cert}' := \Sigma \in \mathscr{C}'$.

(d) Ordinary signing queries of SK^*_{i-1}: m_i

$\mathscr{B}$ makes the signing query $\mathscr{SQ}$ with input $11 \parallel m_i$; then it will be returned to signature Σ. Finally, $\mathscr{B}$ returns Σ to $\mathscr{A}$.

(e) Proxy signing queries of SK^*_{i-1}: $(\mathscr{PK}, \mathscr{W}, \mathscr{C}, j, m_i)$

$\mathscr{B}$ makes the signing query $\mathscr{SQ}$ with input $01 \parallel m_i$; then it will be returned to signature Σ. Finally, $\mathscr{B}$ returns $(\mathscr{PK}, \mathscr{W}, \mathscr{C}, j, P\Sigma := \Sigma)$ to $\mathscr{A}$.

(f) Leakage queries: f_i

$\mathscr{A}$ may make query f_i for the leakage information after each delegation protocol, ordinary signing, or proxy signing query. To answer it, $\mathscr{B}$ makes the same query to $\mathscr{LQ}$; it will be returned as a valid leakage information Λ_i or $\perp$ if f_i is illegal. Finally, $\mathscr{B}$ returns it to $\mathscr{A}$.

Remark. In the construction of scheme SIG*, except for the Sign^* algorithm, there are also two algorithms using the signing or proxy signing key, the Del^* and PSign^*. Actually, however, they are also a signing algorithm just with different input of text, so the leakage information answered by $\mathscr{B}$ (from $\mathscr{LQ}$) is indistinguishable to what $\mathscr{A}$ obtains in the real interaction in the experiment $\mathbf{Exp}^{\mathrm{eu\text{-}cmla}}_{\mathrm{SIG}^*, \mathscr{A}}$.

(iii) Finally, according to the assumption, $\mathscr{A}$ outputs a forgery for the challenging public key PK^* with respect to scheme SIG*. It must be one of the following cases. We now show the challenger $\mathscr{B}$ how to translate $\mathscr{A}$'s forgery as a forgery with respect to the FKPR scheme SIG.

(1) Ordinary signature of PK^*: (m^*, Σ^*)

If $\mathscr{A}$ outputs an ordinary signature (m^*, Σ^*) of PK^*, then $\mathscr{B}$ outputs $(11 \parallel m^*, \Sigma^*)$.

(2) Proxy signature of PK^*: $(m^*, (\mathscr{PK}, \mathscr{W}, \mathscr{C}, j, P\Sigma^*))$, PK^* is the last entry in $\mathscr{PK}$.

If $\mathscr{A}$ outputs a proxy signature $(m^*, (\mathscr{PK}, \mathscr{W}, \mathscr{C}, j, P\Sigma^*))$ of PK^*, $\mathscr{B}$ outputs $(01 \parallel m^*, \Sigma^*)$.

(3) Proxy signature on behalf of PK^*: $(m^*, (\mathscr{PK}, \mathscr{W}, \mathscr{C}, j, P\Sigma^*))$, PK^* is the nth entry in the list $\mathscr{PK}$.

If $\mathscr{A}$ outputs a proxy signature $(m^*, (\mathscr{PK}, \mathscr{W}, \mathscr{C}, j, P\Sigma^*))$ on behalf of PK^*, then $\mathscr{B}$ outputs $(00 \parallel \text{PK}_{n+1} \parallel n + 1 \parallel W_n, \text{cert}_{n+1})$.

Analysis of $\mathscr{B}$. It is clear that the view of $\mathscr{A}$ which is answered by $\mathscr{B}$ in the above experiment is identical to what $\mathscr{A}$ obtains in the real interaction in the experiment $\text{Exp}_{\text{SIG}^*, \mathscr{A}}^{\text{eu-cmla}}$. We now show that any valid output of the adversary $\mathscr{A}$ can be translated to a valid forgery with respect to the FKPR scheme SIG.

(1) If $\mathscr{A}$ outputs an ordinary signature (m^*, Σ^*), $1 \leftarrow \text{Vrf}^*(\text{PK}^*, m^*, \Sigma^*)$, and m^* has not been submitted to the ordinary signing queries, so $\mathscr{B}$ does not make the signing query $\mathscr{SQ}$ with input $11 \parallel m^*$. Therefore, $(11 \parallel m^*, \Sigma^*)$ is a valid forgery with respect to the scheme SIG.

(2) If $\mathscr{A}$ outputs a proxy signature $(m^*, (\mathscr{PK}, \mathscr{W}, \mathscr{C}, j, P\Sigma^*))$, $1 \leftarrow \text{PVrf}^*(\mathscr{PK}, \mathscr{W}, \mathscr{C}, j, m^*, P\Sigma^*)$, and $(\mathscr{PK}, \mathscr{W}, \mathscr{C}, j, m^*)$ has not submitted to the proxy signing queries, so $\mathscr{B}$ does not make the signing query $\mathscr{SQ}$ with input $01 \parallel m^*$. Therefore, $(01 \parallel m^*, P\Sigma^*)$ is a valid forgery with respect to the scheme SIG.

(3) If $\mathscr{A}$ outputs a proxy signature on behalf of PK^*: $(m^*, (\mathscr{PK}, \mathscr{W}, \mathscr{C}, j, P\Sigma^*))$, where PK^* is the nth entry in $\mathscr{PK}$, $1 \leftarrow \text{PVrf}^*(\mathscr{PK}, \mathscr{W}, \mathscr{C}, j, m^*, P\Sigma^*)$ and $\mathscr{A}$ does not make the query of delegation from SK_{i-1}^* with input $(\text{PK}_{n+1}, W_{n+1})$ $((n + 1)$th entry in $\mathscr{PK}$), so $\mathscr{B}$ does not make the signing query $\mathscr{SQ}$ with input $00 \parallel \text{PK}_{n+1} \parallel n + 1 \parallel W_n$. Therefore, $(00 \parallel \text{PK}_{n+1} \parallel n + 1 \parallel W_n, \text{cert}_{n+1})$ is a valid forgery with respect to the scheme SIG.

From the above analysis, we can see that the challenger $\mathscr{B}$'s output of forgery is contradictory to the security of the FKPR scheme SIG (cf. Theorem 1 of [37]) and thus proves the security of the LRPS scheme SIG^*.

5. Conclusion

In this paper, we design a leakage-resilient proxy signature scheme, the LRPS. To model the security of such schemes, we adapt the existing models of the proxy signature schemes which are proposed by Schuldt et al. (in PKC 2008) [15] and Boldyreva et al. (in Jour. Crypto. 2012) [13] to the leakage-resilient cryptography setting and give an extended model, EU-CMLA, for the LRPS schemes. Furthermore, we present a concrete construction based on Faust et al.'s (in TCC 2010) [37] LR signature scheme. This construction is provably secure under the given security model.

Appendices

Now we show that their proposed proxy signature scheme SIG^* in Section 4 which is based on the BPW transformation can be used to produce a secure full construction (denoted by SIG^{**}) of the proxy signature scheme.

A. Construction

As said before, to guarantee that no information about the user's long-term secret key is leaked if its proxy keys are exposed, we had better let a proxy generate fresh and independent keys (PK, SK) in a delegation, create a certificate for PK, and keep the SK as the proxy secret key; to record the proxy public keys of the proxies maintain a separate list $\mathscr{FK}$ to store them. The construction of the scheme $\text{SIG}^{**} = (\text{Kg}^{**}, \text{Sign}^{**}, \text{Vfy}^{**}, \langle \text{Del}^{**}, \text{PKg}^{**} \rangle, \text{PSign}^{**}, \text{PVfy}^{**})$ is as follows, where the algorithms $\text{Kg}^{**}, \text{Sign}^{**}, \text{Vfy}^{**}$ are the same as the algorithms $\text{Kg}^*, \text{Sign}^*, \text{Vfy}^*$ of the scheme SIG^*, respectively. Here we should stress that the following construction is based on Schuldt et al.'s [15] idea, while their scheme is based on sequential aggregate signature, but ours is based on tree-based signature and we focus on the realization of the leakage-resilient proxy signature.

In the scheme SIG^*, the proxy's proxy key is in fact exactly its long-term secret key and hence it delegates its own signing right or proxy's signing right to the next proxy, it takes as input its secret key to run the delegation algorithm Del^*. However, when we consider the full construction of the proxy signature scheme, proxy's secret key and proxy's key are different and independent, and thus when it delegates its own signing right to a proxy it takes as input its secret key; when it delegates its proxy signing right to the next proxy, then it takes as input the proxy key. To uniformly describe these two cases, we use sk to denote the input to the Del^{**} algorithm run by the delegator in the scheme SIG^{**}. For ease of description, here we describe the stateful signing algorithm Sign^{**} as a nonstateful formalization.

(i) $\text{Del}^{**}(\text{sk}, \text{PK}_\text{p}, \mathscr{PK}, \mathscr{FK}, \mathscr{W}, \mathscr{C}, W)$: it is divided into the following two cases depending on $(\mathscr{PK}, \mathscr{W})$

(a) If $\mathscr{PK}$ and $\mathscr{W}$ are empty (i.e., sk is an long-term secret key), the delegator constructs lists $\mathscr{PK} = \{\text{PK}_\text{D}, \text{PK}_\text{p}\}$, $\mathscr{FK} = \emptyset$, and $\mathscr{W} = \{W\}$. Then compute cert $\overset{\$}{\leftarrow} \text{Sign}^{**}(\text{sk}, 100 \parallel \mathscr{PK} \parallel \mathscr{FK} \parallel \mathscr{W})$ and send the delegation information $(\mathscr{PK}, \mathscr{FK}, \mathscr{W}, \text{cert})$ to the proxy.

(b) If $\mathscr{PK}$ and $\mathscr{W}$ are not empty (i.e., sk is a proxy key), the delegator constructs lists $\mathscr{PK} \leftarrow \text{push}(\mathscr{PK}, \text{PK}_\text{p})$ and $\mathscr{W} \leftarrow \text{push}(\mathscr{W}, W)$. Then compute cert $\overset{\$}{\leftarrow} \text{Sign}^{**}(\text{sk}, 100 \parallel \mathscr{PK} \parallel \mathscr{FK} \parallel \mathscr{W})$ and send the delegation information $(\mathscr{PK}, \mathscr{FK}, \mathscr{W}, \mathscr{C}, \text{cert})$ to the proxy.

(ii) $\mathsf{PKg}^{**}(SK_P, PK_P, PK_D)$:

the proxy first checks the validity of the delegation certificates for $k = 1, \ldots, |\mathscr{C}|$ does: if $0 \leftarrow \mathsf{Vfy}^{**}(PK_{k-1}, 100 \parallel \mathscr{PK} \parallel \mathscr{FK} \parallel \mathscr{W}, \mathsf{cert}_k)$, it returns $\bot$ and rejects this delegation, where cert_k means the kth entry in the list $\mathscr{C}$. Otherwise, first generate a fresh proxy key pair $(PK'_P, SK'_P) \leftarrow \mathsf{Kg}^{**}(1^k)$ and run $\mathscr{FK} \leftarrow \mathsf{push}(\mathscr{FK}, PK'_P)$. Then compute $\mathsf{cert} \overset{\$}{\leftarrow} \mathsf{Sign}^{**}(SK_P, 100 \parallel \mathscr{PK} \parallel \mathscr{FK} \parallel \mathscr{W})$. Finally, run $\mathscr{PK} \leftarrow \mathsf{push}(\mathscr{PK}, PK_P)$, $\mathscr{W} \leftarrow \mathsf{push}(\mathscr{W}, W)$, $\mathscr{C} \leftarrow \mathsf{push}(\mathscr{C}, \mathsf{cert})$; set $PSK = (\mathscr{FK}, \mathsf{cert}, SK'_P)$ and output the delegation information $(\mathscr{PK}, \mathscr{W}, \mathscr{C}, PSK)$.

(iii) $\mathsf{PSign}^{**}(\mathscr{PK}, \mathscr{W}, \mathscr{C}, PSK, m)$:

$\Sigma \overset{\$}{\leftarrow} \mathsf{Sign}^{**}(SK'_P, 101 \parallel m)$, output the proxy signature $(\mathscr{PK}, \mathscr{W}, \mathscr{C}, P\Sigma := \Sigma)$.

(iv) $\mathsf{PVfy}^{**}(\mathscr{PK}, \mathscr{FK}, \mathscr{W}, \mathscr{C}, m, P\Sigma)$:

V first checks the validity of the delegation certificates, for $k = 1, \ldots, |\mathscr{C}|$ does $\mathsf{Vfy}^{**}(PK_{k-1}, 100 \parallel \mathscr{PK} \parallel \mathscr{FK} \parallel \mathscr{W}, \mathsf{cert}_k)$ or $\mathsf{Vfy}^{**}(PK_{k-1}, 100 \parallel \mathscr{PK} \parallel \mathscr{FK} \parallel \mathscr{W}, \mathsf{cert}_k)$ dependent on the current certificate generated by Del^{**} or PKg^{**}, respectively. If all the verifications pass then return $\mathsf{Vfy}^{**}(PK'_P, 101 \parallel m, P\Sigma)$.

B. Security

We now analyze the security of the scheme SIG^{**}. This proof is roughly analogous to the proof of scheme SIG^{*}. However, because the proxy key is independent of the long-term secret key, we have to permit more queries to the adversary, such as a redelegation of a user's proxy key. Here we adapt Schuldt et al.'s [15] security model, EU-CMA-PKE which is the strongest notion for the proxy signature schemes (cf. Section 4 of [15] for detailed description), to the leakage-resilient cryptography setting, EU-CMLA-PKE. In the presence of leakage, we should care about what secret can be taken as input to the leakage function: long-term secret key, proxy key, or both? Our answer is both.

The detailed analysis is as follows.

Theorem B.1. *The proxy signature scheme SIG^{**} is EU-CMLA-PKE secure based on the security of the leakage-resilient FKPR signature scheme SIG.*

We show that if there exists a EU-CMLA-PKE adversary $\mathscr{A}$ which can break the security of the scheme SIG^{**}, then it can be used to construct a challenger $\mathscr{B}$ to break the security of the FKPR scheme SIG.

(I) Initially, $\mathscr{B}$ will be given a challenging public key PK' and can adaptively make signing query ($\mathscr{SQ}$) and leakage query ($\mathscr{LQ}$) in the experiment $\mathbf{Exp}^{\text{eu-cmla}}_{\mathsf{SIG}^{*}, \mathscr{B}}$. $\mathscr{B}$ first chooses a random $c \leftarrow \{0, 1\}$. If $c = 0$, $\mathscr{B}$ sets $PK^{*} := PK'$ and $SK^{*} := \emptyset$. Otherwise, $\mathscr{B}$ generates a fresh key pair $(PK^{*}, SK^{*}) \leftarrow \mathsf{Kg}^{**}$ and chooses random $i^{*} \leftarrow \{1, \ldots, q_d\}$ (where q_d is the number that $\mathscr{A}$ queries to the delegation oracle; $\mathscr{B}$ will use

PK' instead of a fresh key in the i^{*}th delegation query by $\mathscr{A}$). For both cases, $\mathscr{B}$ sends PK^{*} to $\mathscr{A}$ as the challenging public key of the experiment $\mathbf{Exp}^{\text{eu-cmla-pke}}_{\mathsf{SIG}^{*}, \mathscr{A}}$. Then it plays the experiment with $\mathscr{A}$.

(II) $\mathscr{A}$ may adaptively ask $\mathscr{B}$ for the following. When the queries by $\mathscr{A}$ need signing invocation of SK' corresponding to PK', $\mathscr{B}$ queries its own singing oracle $\mathscr{SQ}$, and we omit this implicit description in the following proof. In addition, $\mathscr{B}$ will maintain a set of lists $\mathsf{PskList}(*, *)$ which contains all proxy keys generated by $\mathscr{B}$ for the delegation chain with the public keys $\mathscr{PK}$ and warrants $\mathscr{W}$.

(i) Delegation to SK^{*}: $(\mathscr{PK}, \mathscr{FK}, \mathscr{W}, \mathscr{C})$

if $c = 0$, or $c = 1$ and this is not the i^{*}th delegation query, then $\mathscr{B}$ first runs $(PK, SK) \leftarrow \mathsf{Kg}^{**}(1^k)$, $\mathscr{FK} \leftarrow \mathsf{push}(\mathscr{FK}, PK)$ and set $SK_{\text{prx}} = SK$. If $c = 1$ and this is the i^{*}th delegation query, $\mathscr{B}$ runs $\mathscr{FK} \leftarrow \mathsf{push}(\mathscr{FK}, PK^{*})$ and set $SK_{\text{prx}} = \emptyset$. Then $\mathscr{B}$ computes $\mathsf{cert} \leftarrow \mathsf{Sign}^{**}(SK_{\text{prx}}, 100 \parallel \mathscr{PK} \parallel \mathscr{FK} \parallel \mathscr{W})$. Finally, store $PSK = (\mathscr{FK}, \mathsf{cert}, SK_{\text{prx}})$ in $\mathsf{PskList}(\mathscr{PK}, \mathscr{W})$.

(ii) Delegation from SK^{*}: this query can be divided into the following three cases.

(a) Delegation of SK^{*}: (PK_p, W)

$\mathscr{B}$ sets $\mathscr{PK} = \{PK^{*}, PK_p\}$, $\mathscr{FK} = \emptyset$, and $\mathscr{W} = \{W\}$. Then compute $\mathsf{cert} \leftarrow \mathsf{Sign}^{**}(SK^{*}, 100 \parallel \mathscr{PK} \parallel \mathscr{FK} \parallel \mathscr{W})$ and set $\mathscr{C} = \{\mathsf{cert}\}$. Finally return the delegation information $(\mathscr{PK}, \mathscr{FK}, \mathscr{W}, \mathscr{C})$ to $\mathscr{A}$.

(b) Redelegation of PSK: $(\mathscr{PK}, \mathscr{W}, \mathscr{C}, j, PK_p, W)$

$\mathscr{B}$ retrieves the jth proxy key $\mathsf{PskList}(\mathscr{PK}, \mathscr{W})$ and parses it as $(\mathscr{FK}, \mathsf{cert}, SK_{\text{prx}})$. Then run $\mathscr{PK} \leftarrow \mathsf{push}(\mathscr{PK}, PK_p)$, $\mathscr{W} \leftarrow \mathsf{push}(\mathscr{W}, W)$, compute $\mathsf{cert} \leftarrow \mathsf{Sign}^{**}(SK_{\text{prx}}, 100 \parallel \mathscr{PK} \parallel \mathscr{FK} \parallel \mathscr{W})$, and set $\mathscr{C} \leftarrow \mathsf{push}(\mathscr{C}, \mathsf{cert})$. Finally return the delegation information $(\mathscr{PK}, \mathscr{FK}, \mathscr{W}, \mathscr{C})$ to $\mathscr{A}$.

(c) Self-delegation of SK^{*}: $(\mathscr{PK}, \mathscr{W}, \mathscr{C}, j, W)$

(1) if $\mathscr{PK}$ and $\mathscr{W}$ are empty (i.e., self-delegation of SK^{*}), $\mathscr{B}$ constructs $\mathscr{PK} = \{PK^{*}, PK^{*}\}$, $\mathscr{FK} = \emptyset$, and $\mathscr{W} = \{W\}$ and sets $SK_{\text{sel}} = SK^{*}$ and $\mathsf{cert}_{\text{sel}} = \emptyset$.

(2) If $\mathscr{PK}$ and $\mathscr{W}$ (i.e., delegation of PSK), $\mathscr{B}$ retrieves the jth proxy key in $\mathsf{PskList}(\mathscr{PK}, \mathscr{W})$ and parses it as $(\mathscr{FK}, \mathsf{cert}, SK_{\text{prx}})$. Then compute $\mathscr{PK} \leftarrow \mathsf{push}(\mathscr{PK}, PK^{*})$, $\mathscr{W} \leftarrow \mathsf{push}(\mathscr{W}, W)$, and set $SK_{\text{sel}} = SK_{\text{prx}}$ and $\mathsf{cert}_{\text{sel}} = \mathsf{cert}$.

$\mathscr{B}$ then computes $\mathsf{cert} \leftarrow \mathsf{Sign}^{**}(SK_{\text{sel}}, 100 \parallel \mathscr{PK} \parallel \mathscr{FK} \parallel \mathscr{W})$. If $c = 0$ or $c = 1$ and this not the i^{*}th delegation query, $\mathscr{B}$ first runs $(PK, SK) \leftarrow \mathsf{Kg}^{**}(1^k)$, and construct $\mathscr{FK} \leftarrow \mathsf{push}(\mathscr{FK}, PK)$. Otherwise, $\mathscr{B}$ constructs $\mathscr{FK} \leftarrow \mathsf{push}(\mathscr{FK}, PK^{*})$, and set $SK = \emptyset$. Finally, $\mathscr{B}$ computes $\mathsf{cert} \leftarrow \mathsf{Sign}^{**}(SK_{\text{sel}}, 100 \parallel$

$\mathscr{PK} \parallel \mathscr{FK} \parallel \mathscr{W}$) and $\mathscr{C} \leftarrow \text{push}(\mathscr{C}, \text{cert})$, and then store the proxy key $\text{PSK} = (\mathscr{FK}, \text{cert}, \text{SK})$ in $\text{PskList}(\mathscr{PK}, \mathscr{W})$ and send the transcript $(\mathscr{PK}, \mathscr{FK}, \mathscr{W}, \mathscr{C})$ to $\mathscr{A}$.

(iii) Ordinary signing queries of SK^*: m_i

$\mathscr{B}$ returns $\text{Sign}^{**}(\text{SK}^*, 111 \parallel m)$.

(iv) Proxy signing queries of SK^*: $(\mathscr{PK}, \mathscr{W}, \mathscr{C}, j, m_i)$

$\mathscr{B}$ retrieves the jth proxy key in $\text{PskList}(\mathscr{PK}, \mathscr{W})$ and parses it as $(\mathscr{FK}, \text{cert}, \text{SK}_{\text{prx}})$. Then compute $P\Sigma \leftarrow \text{PSign}^{**}(\text{SK}_{\text{prx}}, 101 \parallel m_i)$ and return $(\mathscr{PK}, \mathscr{W}, \mathscr{C}, (\mathscr{FK}, P\Sigma))$ to $\mathscr{A}$.

(v) Proxy key exposure queries: $(\mathscr{PK}, \mathscr{W}, j)$

$\mathscr{B}$ retrieves the jth proxy key in $\text{PskList}(\mathscr{PK}, \mathscr{W})$ and parses it as $(\mathscr{FK}, \text{cert}, \text{SK}_{\text{prx}})$. If $\text{SK}_{\text{prx}} = \emptyset$, $\mathscr{B}$ aborts. Otherwise, $\mathscr{B}$ returns $(\mathscr{FK}, \text{cert}, \text{SK}_{\text{prx}})$ to $\mathscr{A}$.

(vi) Leakage queries: f_i:

$\mathscr{A}$ makes query f_i for the leakage information about the secret key sk (randomness is also included here) after each delegation protocol, ordinary signing, or proxy signing query. If the used secret key is chosen by $\mathscr{B}$, then $\mathscr{B}$ returns $\Lambda_i = f_i(\text{sk})$. Otherwise, $\mathscr{B}$ makes the same query to its own leakage oracle $\mathscr{LQ}$, it will be returned as valid leakage information Λ_i or $\bot$ if f_i is illegal. Finally, $\mathscr{B}$ returns it to $\mathscr{A}$.

Remark. The secret state for $\mathscr{A}$ can be divided into two kinds, the first one is that chosen by $\mathscr{B}$ in the experiment, and the second one is that unknown to $\mathscr{B}$, that is, SK' and the randomness used in the singing oracle $\mathscr{SQ}$. For the first one, $\mathscr{B}$ can directly answer $\mathscr{A}$ by itself. For the second one, similar to the proof in Theorem 1, $\mathscr{B}$ can make the same query to its leakage oracle $\mathscr{LQ}$.

(III) Finally, according to the assumption, $\mathscr{A}$ outputs a forgery for the challenging public key PK^* (with respect to the scheme SIG^{**}). It must be one of the following cases:

(1) ordinary signature (m^*, Σ^*);

(2) proxy signature $(m^*, (\mathscr{PK}, \mathscr{W}, \mathscr{C}, (\mathscr{FK}, P\Sigma^*)))$, where the last key in $\mathscr{FK}$ was not generated by $\mathscr{B}$;

(3) proxy signature $(m^*, (\mathscr{PK}, \mathscr{W}, \mathscr{C}, (\mathscr{FK}, P\Sigma^*)))$, where the $(i^* - 1)$th key in $\mathscr{FK}$ was not generated by $\mathscr{B}$;

(4) proxy signature $(m^*, (\mathscr{PK}, \mathscr{W}, \mathscr{C}, (\mathscr{FK}, P\Sigma^*)))$, where the last key in $\mathscr{FK}$ was generated by $\mathscr{B}$;

(5) proxy signature $(m^*, (\mathscr{PK}, \mathscr{W}, \mathscr{C}, (\mathscr{FK}, P\Sigma^*)))$, where the $(i^* - 1)$th key in $\mathscr{FK}$ was generated by $\mathscr{B}$.

We now show how the challenger $\mathscr{B}$ translates $\mathscr{A}$'s forgery as a forgery with respect to the FKPR scheme SIG. If $\mathscr{B}$ has flipped $c = 0$ which means that $\text{PK}^* := \text{PK}'$, then the first three cases correspond to the forgeries where $\mathscr{A}$ has forged a signature under the secret key SK', and hence $\mathscr{B}$ can translate them to a forged signature corresponding to the scheme SIG which can be analogous to that in the proof of Theorem 1.

Otherwise, if $\mathscr{A}$ outputs a forgery that belongs to the last two cases, $\mathscr{B}$ will abort.

If $c = 0$ which means that $\mathscr{B}$ sets PK' as the i^*th fresh proxy public key: in this case, if $\mathscr{A}$ outputs a forgery that belongs to the first three cases, then $\mathscr{B}$ will abort. Otherwise, the last two cases indicate that $\mathscr{A}$ has forged a signature under one of the keys generated by $\mathscr{B}$ in a delegation, but for which $\mathscr{A}$ has not received the corresponding secret key. In those two cases, $P\Sigma^*$ will be a valid signature under a key PK generated by $\mathscr{B}$ in some delegation query; that is, PK will be the last key in the list $\mathscr{FK}$ for a proxy key $(\mathscr{FK}, \text{cert}, \text{SK}_{\text{prx}})$ from some proxy key list $\text{PskList}(*, *)$. Therefore, with probability $1/q_d$, $\mathscr{B}$ can choose the right i^* such that $\text{PK} = \text{PK}'$. In this case, $\mathscr{B}$ outputs $P\Sigma^*$ as a valid forgery of the key PK' for the underlying signature scheme SIG.

From the above analysis, we can see that the challenger $\mathscr{B}$'s forgery with a nonnegligible probability is contradictory to the security of the FKPR scheme SIG (cf. Theorem 1 of [37]) and thus proves the security of the LRPS scheme SIG^{**}.

Disclosure

An abstract of this paper has been presented in the proceedings of the 5th International Conference on Intelligent Networking and Collaborative Systems (INCoS), IEEE, pp, 495–502, 2013 [42].

Conflict of Interests

The authors declare that there is no conflict of interests regarding the publication of this paper.

Acknowledgments

This research is supported by the National Natural Science Foundation of China (Grant no. 60970139), the Strategic Priority Program of Chinese Academy of Sciences (Grant no. XDA06010702), and the IIEs Cryptography Research Project. The authors would like to thank anonymous reviewers for their helpful comments and suggestions.

References

[1] W. Farmer, J. Gutmann, and V. Swarup, "Security for mobile agents: authentication and state appraisal," in *Computer Security—ESORICS 96: 4th European Symposium on Research in Computer Security Rome, Italy, September 25-27, 1996 Proceedings*, vol. 1146 of *Lecture Notes in Computer Science*, pp. 118–130, Springer, Berlin, Germany, 1996.

[2] P. Kotzanikolaous, G. Katsirelos, and V. Chrissikopoulos, "Mobile agents for secure electronic transactions," in *Recent Advances in Signal Processing and Communications*, pp. 363–368, World Scientific and Engineering Society Press, 1999.

[3] B. Lee, H. Kim, and K. Kim, "Secure mobile agent using strong non-designated proxy signature," in *Information Security and Privacy: Proceedings of the 6th Australasian Conference (ACISP '01), Sydney, Australia, July 11-13, 2001*, vol. 2119 of *Lecture Notes in Computer Science*, pp. 474–486, Springer, Berlin, Germany, 2001.

[4] B. Lee, H. Kim, and K. Kim, "Strong proxy signature and its applications," in *Proceedings of the Symposium on Cryptography and Information Security (SCIS '01)*, pp. 603–608, 2001.

[5] M. Mambo, K. Usuda, and E. Okamoto, "Proxy signatures: delegation of the power to sign messages," *IEICE Transactions on Fundamentals of Electronics*, vol. 79, pp. 1338–1353, 1996.

[6] G. Allée, S. Pierre, R. H. Glitho, and A. El Rhazi, "An improved itinerary recording protocol for securing distributed architectures based on mobile agents," *Mobile Information Systems*, vol. 1, no. 2, pp. 129–147, 2005.

[7] R. Aversa, B. Di Martino, N. Mazzocca, and S. Venticinque, "A skeleton based programming paradigm for mobile multi-agents on distributed systems and its realization within the MAGDA mobile agents platform," *Mobile Information Systems*, vol. 4, no. 2, pp. 131–146, 2008.

[8] K. Goto, Y. Sasaki, T. Hara, and S. Nishio, "Data gathering using mobile agents for reducing traffic in dense mobile wireless sensor networks," *Mobile Information Systems*, vol. 9, no. 4, pp. 295–314, 2013.

[9] Y. Wang, D. S. Wong, and H. Wang, "Employ a mobile agent for making a payment," in *Mobile Information Systems*, vol. 4, pp. 51–68, IOS Press, 2008.

[10] S. Parvin, F. K. Hussain, and S. Ali, "A methodology to counter DoS attacks in mobile IP communication," *Mobile Information Systems*, vol. 8, no. 2, pp. 127–152, 2012.

[11] H. U. Park and I. Y. Lee, "A digital nominative proxy signature scheme for mobile communication," in *Information and Communications Security: Third International Conference, ICICS 2001 Xian, China, November 13–16, 2001 Proceedings*, vol. 2229 of *Lecture Notes in Computer Science*, pp. 451–455, Springer, Berlin, Germany, 2001.

[12] S. Kim, S. Park, and D. Won, "Proxy signatures, revisited," in *Proceedings of the 1st International Conference on Information and Communication Security (ICICS '97)*, vol. 1334 of *Lecture Notes in Computer Science*, pp. 223–232, Springer, 1997.

[13] A. Boldyreva, A. Palacio, and B. Warinschi, "Secure proxy signature schemes for delegation of signing rights," *Journal of Cryptology*, vol. 25, no. 1, pp. 57–115, 2012.

[14] T. Malkin, S. Obana, and M. Yung, "The hierarchy of key evolving signatures and a characterization of proxy signatures," in *Advances in Cryptology—EUROCRYPT 2004*, vol. 3027 of *Lecture Notes in Computer Science*, pp. 306–322, Springer, Berlin, Germany, 2004.

[15] J. C. N. Schuldt, K. Matsuura, and K. G. Paterson, "Proxy signature secure against key exposure," in *Public Key Cryptography—PKC 2008: 11th International Workshop on Practice and Theory in Public-Key Cryptography, Barcelona, Spain, March 9-12, 2008. Proceedings*, vol. 4939 of *Lecture Notes in Computer Science*, pp. 141–161, Springer, Berlin, Germany, 2008.

[16] H. Wang and J. Pieprzyk, "Efficient one-time proxy signatures," in *Advances in Cryptology—ASIACRYPT 2003*, vol. 2894 of *Lecture Notes in Computer Science*, pp. 507–522, Springer, Berlin, Germany, 2003.

[17] F. Zhang, R. Safavi-Naini, and C. Y. Lin, "New proxy signature, proxy blind signature and proxy ring signature schemes from bilinear pairings," Tech. Rep. 2003/104, Cryptology ePrint Archive, 2003, http://eprint.iacr.org/.

[18] S. Goldwasser and S. Micali, "Probabilistic encryption," *Journal of Computer and System Sciences*, vol. 28, no. 2, pp. 270–299, 1984.

[19] M. Mambo, K. Usuda, and E. Okamoto, "Proxy signatures for delegating signing operation," in *Proceedings of the 3rd ACM Conference on Computer and Communications Security (CCS '96)*, pp. 48–56, ACM, March 1996.

[20] J. Y. Lee, J. H. Cheon, and S. Kim, "An analysis of proxy signatures: is a secure channel necessary?" in *Proceedings of the Cryptographers' Track at the RSA Conference, San Francisco, Calif, USA, April 2003*, Lecture Notes in Computer Science, pp. 68–79, Springer, 2003.

[21] Y. Dodis, J. Katz, S. Xu, and M. Yung, "Strong key-insulated signature schemes," in *Public Key Cryptography—PKC 2003*, vol. 2567 of *Lecture Notes in Computer Science*, pp. 130–144, Springer, Berlin, Germany, 2002.

[22] D. Brumley and D. Boneh, "Remote timing attacks are practical," *Computer Networks*, vol. 48, no. 5, pp. 701–716, 2005.

[23] P. Kocher, J. Jaffe, and B. Jun, "Differential power analysis," in *Advances in Cryptology—CRYPTO'99*, vol. 1666 of *Lecture Notes in Computer Science*, pp. 388–397, Springer, Berlin, Germany, 1999.

[24] E. Biham, Y. Carmeli, and A. Shamir, "Bug attacks," in *Advances in Cryptology—CRYPTO 2008*, vol. 5157 of *Lecture Notes in Computer Science*, pp. 221–240, Springer, Berlin, Germany, 2008.

[25] D. Boneh, R. A. DeMillo, and R. J. Lipton, "On the importance of checking cryptographic protocols for faults," in *Advances in Cryptology—EUROCRYPT'97*, vol. 1233 of *Lecture Notes in Computer Science*, pp. 37–51, Springer, Berlin, Germany, 1997.

[26] S. Micali and L. Reyzin, "Physically observable cryptography," in *Theory of Cryptography: Proceedings of the 1st Theory of Cryptography Conference (TCC '04), Cambridge, MA, USA, February 19—21, 2004*, vol. 2951 of *Lecture Notes in Computer Science*, pp. 278–296, Springer, Berlin, Germany, 2004.

[27] Z. Brakerski, Y. T. Kalai, J. Katz, and V. Vaikuntanathan, "Overcoming the hole in the bucket: public-key cryptography resilient to continual memory leakage," in *Proceedings of the IEEE 51st Annual Symposium on Foundations of Computer Science (FOCS '10)*, pp. 501–510, October 2010.

[28] Y. Dodis, K. Haralambiev, A. Lopez-Alt, and D. Wichs, "Cryptography against continuous memory attacks," in *Proceedings of the 51st Annual IEEE Symposium on Foundations of Computer Science*, pp. 511–520, 2010.

[29] K. Pietrzak, "A leakage-resilient mode of operation," in *Advances in Cryptology—EUROCRYPT '09*, vol. 5479 of *Lecture Notes in Computer Science*, pp. 462–482, Springer, Berlin, Germany, 2009.

[30] S. Garg, A. Jain, and A. Sahai, "Leakage-resilient zero knowledge," in *Advances in Cryptology—CRYPTO 2011*, vol. 6841 of *Lecture Notes in Computer Science*, pp. 297–315, Springer, Berlin, Germany, 2011.

[31] E. Kiltz and K. Pietrzak, "Leakage resilient ElGamal encryption," in *Advances in Cryptology—ASIACRYPT '10*, vol. 6477 of *Lecture Notes in Computer Science*, pp. 595–612, Springer, Berlin, Germany, 2010.

[32] M. Naor and G. Segev, "Public-key cryptosystems resilient to key leakage," in *Advances in Cryptology—CRYPTO 2009*, vol. 5677 of *Lecture Notes in Computer Science*, pp. 18–35, Springer, Berlin, Germany, 2009.

[33] S. S. M. Chow, Y. Dodis, Y. Rouselakis, and B. Waters, "Practical leakage-resilient identity-based encryption from simple assumptions," in *Proceedings of the 17th ACM Conference on Computer and Communications Security (CCS '10)*, pp. 152–161, ACM, October 2010.

[34] T. H. Yuen, S. S. M. Chow, Y. Zhang, and S. M. Yiu, "Identity-based encryption resilient to continual auxiliary leakage," in

Advances in Cryptology—EUROCRYPT 2012, vol. 7237 of *Lecture Notes in Computer Science*, pp. 117–134, Springer, Berlin, Germany, 2012.

[35] J. Alwen, Y. Dodis, and D. Wichs, "Leakage-resilient public-key cryptography in the bounded-retrieval model," in *Advances in Cryptology—CRYPTO 2009*, vol. 5677 of *Lecture Notes in Computer Science*, pp. 36–54, Springer, 2009.

[36] E. Boyle, G. Segev, and D. Wichs, "Fully leakage-resilient signatures," in *Advances in Cryptology—EUROCRYPT 2011*, vol. 6632 of *Lecture Notes in Computer Science*, pp. 89–108, Springer, Berlin, Germany, 2011.

[37] S. Faust, E. Kiltz, K. Pietrzak, and G. N. Rothblum, "Leakage-resilient signatures," in *Theory of Cryptography: 7th Theory of Cryptography Conference, TCC 2010, Zurich, Switzerland, February 9-11, 2010. Proceedings*, vol. 5978 of *Lecture Notes in Computer Science*, pp. 343–360, Springer, Berlin, Germany, 2010.

[38] J. Katz and V. Vaikuntanathan, "Signature schemes with bounded leakage resilience," in *Advances in Cryptology—ASIACRYPT 2009*, vol. 5912 of *Lecture Notes in Computer Science*, pp. 703–720, Springer, Berlin, Germany, 2009.

[39] T. Malkin, I. Teranishi, Y. Vahlis, and M. Yung, "Signatures resilient to continual leakage on memory and computation," in *Proceedings of the 8th Theory of Cryptography Conference (TCC '11)*, vol. 6597 of *Lecture Notes in Computer Science*, pp. 89–106, Springer, Providence, RI, USA, 2011.

[40] F. Tang, H. Li, Q. Niu, and B. Liang, "Efficient leakage-resilient signature schemes in the generic bilinear group model," Cryptology ePrint Archive 2013/785, 2013, http://eprint.iacr.org/.

[41] D. Boneh, C. Gentry, B. Lynn, and H. Shacham, "Aggregate and verifiably encrypted signatures from bilinear maps," in *Advances in Cryptology—EUROCRYPT 2003*, vol. 2656 of *Lecture Notes in Computer Science*, pp. 416–432, Springer, Berlin, Germany, 2003.

[42] F. Tang, H. Li, Q. Niu, and B. Liang, "Leakage-resilient proxy signatures," in *Proceedings of the 5th IEEE International Conference on Intelligent Networking and Collaborative Systems (INCoS '13)*, pp. 495–502, Xi'an, China, September 2013.

15

AirPrint Forensics: Recovering the Contents and Metadata of Printed Documents from iOS Devices

Luis Gómez-Miralles and Joan Arnedo-Moreno

Internet Interdisciplinary Institute (IN3), Universitat Oberta de Catalunya, Roc Boronat Street 117, 7th Floor, 08018 Barcelona, Spain

Correspondence should be addressed to Luis Gómez-Miralles; pope@uoc.edu

Academic Editor: David Taniar

Since its presentation by Apple, both the iPhone and iPad devices have achieved great success and gained widespread popularity. This fact, added to the given idiosyncrasies of these new portable devices and the kind of data they may store, opens new opportunities in the field of computer forensics. In 2010, version 4 of the iOS operating system introduced AirPrint, a simple and driverless wireless printing functionality supported by hundreds of printer models from all major vendors. This paper describes the traces left in the iOS device when AirPrint is used and presents a method for recovering content and metadata of documents that have been printed.

1. Introduction

Information technologies have grown rapidly in the last decades, changing the way we live, work, and communicate. Portable devices such as smartphones and tablets have evolved from simple phones and agendas into literally full-fledged, always-online computers. In this scenario, where mobile devices become ubiquitous, privacy and cybersecurity become a great concern since such devices may contain huge amounts of sensible valuable data about us: contacts, calendar, e-mails, and photographs as well as a pile of logs: phone calls, chat, geographic positions, and so forth.

The practice of digital forensics has needed to adapt quickly to the emerging mobile technologies. We once had a homogeneous personal computer market, mainly dominated by a few different Windows versions, with minor representations of Mac OS or Unix-based systems. Now we find that the most personal devices, the ones that always accompany their users and are more prone to contain sensitive information, run software environments which simply did not exist a few years ago, namely, Android and iOS. Furthermore, because of the competitive nature of the market, with each new version of these systems, new functionalities are added in order to appeal to a greater set of users and thus become their device of choice. However, some of these new features may manage

personal user data and are worth analyzing from a forensic investigation standpoint.

This paper focuses on the AirPrint feature of iOS devices (iPhone, iPad, and iPod Touch), which allows them to print wirelessly to compatible printers [1]. In a previous paper [2] we observed that printing a document through AirPrint leaves a trace in the filesystem of the iOS device in the form of a temporary file containing the printed content and with a specific metadata that allows for the identification of this precise kind of files in the filesystem. This paper extends our previous research to analyze if these temporary files can be recovered even in modern iOS versions which use hardware-based data encryption. Considering the rise of mobile devices and applications in general [3, 4] and the hundreds of millions iOS devices in particular [5], any available process which allows to recover user data becomes especially relevant from both a computer forensics and a privacy concern standpoint.

The main contribution of this paper is the exposition of a method to recover from an iOS device the contents and metadata of documents printed through AirPrint, even in modern devices which feature hardware-based data encryption. Analyzing the behavior of AirPrint posed an interesting challenge since iOS is a closed operating system and lacks public documentation about many internal aspects. In addition, even when the mobile threat landscape has been covered by other authors [6], there seems to be no additional research on how AirPrint works behind the scenes and the forensic traces it may leave. Several authors [7, 8] have reviewed the existing (mostly commercial) forensic investigation tools for iOS devices; however, analysis of AirPrint activity does not seem to be covered by any of the existing software solutions.

This paper is structured as follows. First of all, in Section 2, AirPrint and its mode of operation both from a user's and technical standpoints are presented. The analysis of the traces left by AirPrint and the information they contain is shown in Section 3. In Section 4, basic experiments are performed to assess the recoverability of the AirPrint traces in devices where encryption has been purposely disabled. Following, in Section 5, the recoverability is evaluated for the modern devices and OS versions that feature data encryption. Finally, concluding the paper, Section 6 summarizes the paper contributions and outlines further work.

2. Description of AirPrint Network Printing

Briefly explained, AirPrint is an iOS feature that allows applications to send content to printers using the iOS device's wireless connection. Apple is directly quoted [1]: "*AirPrint automatically finds printers on local networks and can print text, photos and graphics to them wirelessly over Wi-Fi without the need to install drivers or download software.*"

Apple announced AirPrint in September 2010. Two months later, iOS 4.2 was released for the iPhone, iPad, and iPod Touch, being the first iOS version to offer this feature to users. Its standard functionality allows printing only to specific, AirPrint-enabled printers. Nevertheless, as of July 2013 there were more than seven hundred AirPrint-enabled printer models in the market from sixteen different vendors [9]. Apple does not support sharing a common printer via the computer it is connected to, even when it was possible with some Mac OS X 10.6.5 beta versions; however, it can be done by using software tweaks such as AirPrintActivator [10].

Long before the introduction of AirPrint, different solutions [11, 12] tried to fill in this gap. Usually, such solutions involved iOS applications capable of opening different file formats and sending them to a desktop computer, running a companion application, which would in turn send the document to the printer itself. Some printer vendors developed specific clients; however, none of these solutions were ever widely spread among users. Currently, with AirPrint working out-of-the-box and embedded into all applications, it is hard to believe that new users will consider using a specific, usually paid application to handle printing, except maybe in some very particular environments, such as cases where the use of these kind of applications was consolidated before AirPrint was launched or some advanced capabilities are required by power users.

From a user's standpoint, Figure 1 summarizes the printing process, showing the screens it is actually possible to interact with, as seen on an iPhone.

In the client side (iOS device), AirPrint-enabled applications contain a "Print" button that, when pressed, will present an extremely simple menu (Figure 1(a)), with only two or three available options:

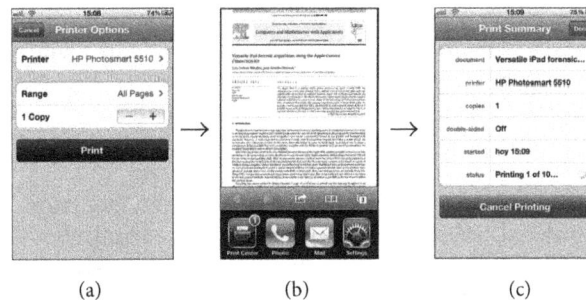

(a)   (b)   (c)

FIGURE 1: Step-by-step AirPrint options screen on an iPhone.

(1) *Printer:* this option opens a list of all AirPrint-enabled printers found in the local network, showing a "name" and "description" field for each one.

(2) *Range:* (optional) defaulting to "all pages", this option opens a selector which allows the user to choose a range of pages to be printed rather than all the document.

(3) *Copies:* this option specifies the number of copies to be printed.

(4) Depending on the printer features, additional parameters such as duplex printing can be controlled.

(5) *Print:* this button proceeds to send the job to the printer.

The user cannot specify any other kind of information usually available in printing menus, such as paper size or orientation and printing quality. Everything is automatically handled by AirPrint, using some default options. When the user chooses to "Print" the job, the device shows some brief messages ("*Contacting Printer*"; "*Preparing page (...) of (...)*"; "*Sending to Printer*"). However, depending on how long the print job is, these messages may be barely visible or last for several seconds.

After the job has been sent to the printer, the printing menu disappears and the application returns to its previous state. At this point, invoking the list of recently used applications, by double-clicking the device "Home" button, reveals a *Print Center* application (Figure 1(b)). Unless the user somehow knows this application has started running in background, it may be difficult for him to find it, since no active feedback is provided during the printing process, thus being invisible at casual glance.

Opening the *Print Center* application, the user can see the list of running and pending printing jobs, check their details, and cancel them (Figure 1(c)). When the last job finishes, that is, the moment the printer ejects the last page, the application closes and does not appear anymore in the list of recently used applications. As far as it is known, there is no way to open the *Print Center* as a standalone application. It is only executed while there are jobs being printed.

From a technical standpoint, the AirPrint service is known to use the standard IPP protocol at network level for printer management, and Bonjour/Zeroconf [13] for service discovery. A comprehensive description of the printing architecture and its underlaying API in iOS devices can be found in [14].

3. Forensic Traces Left by AirPrint

This section presents the preliminary information that must be considered before more in-depth forensic investigation may proceed. Mainly, it is important to assess whether any traces are left in the device after having printed a document using AirPrint, and if so, how they can be discovered, how they behave, and which useful information can be extracted from them. All this information was discovered through some basic experiments.

3.1. Preliminary Setup. For the experiments described in this section, the following equipment was used:

(i) iOS device #1: Apple iPhone 4, 16 GB (model `A1332`) running iOS 4.3.3 (`8J2`) and 5.0 (`9A334`), both jailbroken using the `redsn0w` software [15].

(ii) iOS device #2: Apple iPhone 3G, 8 GB (model `A1241`) running iOS 4.2.1 (`8C148`), jailbroken using `redsn0w` and enabling multitasking and AirPrint.

(iii) Desktop computer information: Apple MacBook Pro (model `MacBookPro5,1`) running Mac OS X 10.7.2 (`11C74`).

(iv) Printer: HP Photosmart 5510 B111a (model `CQ176B`) running firmware version `EPL2CN1122AR`.

(v) Wireless connectivity: all devices were connected to a wireless 802.11g network to perform the experiments.

(vi) Physical connectivity: all devices were using physical wires for AC power only.

3.2. The "Jailbreak" Process. Given that iOS enforces the device to run only code signed by Apple (downloaded from the App Store), during our experiments we used the "jailbreak" technique to bypass that restriction in order to have full access to the devices and be able to run shell commands on them. The jailbreak process is exempted from prosecution under the anticircumvention section of the U.S. Digital Millenium Copyright Act [16], and it has been very useful for forensic research in the past [17, 18].

Both iOS devices were jailbroken in order to gain full access and install an SSH server and basic UNIX tools.

3.3. Traces Found. Once we were able to execute code in the device, we invoked a series of commands before, during, and after printing, and we compared the results looking for remarkable differences. The most relevant commands used were:

(1) `find / -type (b,c,d,f,l,p,s)` for listing, respectively, all the block special devices, character special devices, directories, regular files, symbolic links, FIFOs, and sockets, in the filesystem.

(2) `netstat -an -f inet` for listing any current network connections. This would show active client-server activity as well as inactive servers awaiting for incoming petitions.

(3) `ps aux` for getting information about running processes.

By reviewing the lists of files and directories generated with the `find` commands explained above, it was observed that when a device prints via AirPrint for the first time the following folder is created:

`/var/mobile/Library/com.apple.printd/`

We observed that everytime a document is sent to a printer, a new file named `1.pdf` is created under this folder. With additional tests, it was observed that this PDF file exists in disk only while the document is being printed. The moment the printer ejects the last page and considers the job finished, the PDF file is deleted. This is also the moment at which the *Print Center* application disappears from the list of recently used applications.

By printing some documents and copying the resulting temporary PDF files to another location before their deletion and then examining them we observed that these files are regular PDF files with the same content that is being sent to the printer (no matter whether it was originally in PDF format or not). Hence, an obvious trace is being left in the filesystem, and it reflects exactly what was printed.

It must be noted that, in some of the preliminary tests, before the physical printer was available, we set up a virtual PDF printer in a Mac computer and shared, making it look like an AirPrint printer. It worked as expected and it was possible to print to it from an iPhone; however, the printing of the document (in this case, the generation of a file in the hard drive of the Mac) was much shorter than the actual printing of a page through a real printer with real ink and paper. We observed that this greatly reduces the chances of the temporary files being flushed to physical storage from the buffer cache in the iOS device before deletion and thus their chances of recoverability. Therefore, to obtain accurate results, the experiments need to be performed with a real printer.

3.4. Properties of the AirPrint Temporary Files. From the execution of the different tests we extracted the following conclusions. Unless otherwise specified, every finding applies to all available iOS versions with AirPrint support (versions 4.2 through 6.1 and possibly later versions as well).

(1) For every print job sent via AirPrint, a file with the name of `1.pdf` is created in the directory

`/var/mobile/Library/com.apple.printd/`

(2) This file is in PDF format, containing the document sent to the printer. This observed behavior is consistent across internal iOS applications (Mail, Safari) as well as third party ones (GoodReader, Papers, Keynote, ...). The only exception found is the iOS Photos app, which seems not to generate

any temporary files on disk when printing, thus not leaving these traces.

(3) The file 1.pdf is deleted as soon as the printing job finishes. The timing observed indicates that this happens not just after finishing the task of submitting the job to the physical printer, but after the document has been completely printed.

(4) When one job is being printed, subsequent jobs arriving to the queue generate files named 2.pdf, 3.pdf, and so forth. The behavior observed suggests that in iOS 4 the counter resets as soon as the queue is empty (if a new job arrives later it will be named 1.pdf again), whereas starting from iOS version 5 the counter seems to keep increasing (each new job gets a higher number even if the queue is empty) until the device reboots.

(5) When a job asks for more than one copy of the same document, the temporary PDF file contains only one copy of it. The information on the number of copies to be printed is being sent to the printer in a separate channel (standard PS commands or similar).

(6) When a page range is specified, the temporary PDF file contains only this page range. There is an exception when some applications print files that are themselves PDFs, which is studied later in Section 3.5.

3.5. PDF Metadata of the Temporary Files. Using a standard PDF reader, the metadata contents inside different temporary files generated by AirPrint were extracted and compared. The use of document metadata in forensic investigations has proven useful in different scenarios before [19–22]. A good analysis of the PDF format itself from a forensics point of view, considering its security and privacy aspects, can be found in [23].

Generally, all the temporary PDF files created by AirPrint can be identified as such by their metadata: they all show the same "PDF producer" entry ("*iPhone OS x.y.z Quartz PDFContext*," with *x.y.z* being the iOS version number), and the creation and modification dates both indicate the date and time when the document was sent to the printer (see Figure 2). Therefore, knowing what has been printed from a device looks as simple as recovering deleted PDF files from it and focusing on those with the strings "*iPhone OS x.y.z Quartz PDFContext.*" Moreover, the creation and modification dates contained inside those PDF files in the form of PDF metadata will indicate precisely the printing date and time.

Only one exception was found to this behavior. When printing PDF files, that is, when the content to be printed is itself in PDF format, some applications behave like described earlier, but others (including iOS built-in applications such as Safari or Mail) actually copy the original PDF file to the com.apple.printd directory as 1.pdf instead of generating a new PDF file with fresh metadata. In these cases, PDF metadata cannot be used to tell whether one given file is a trace from AirPrint printing: the only peculiar thing about that PDF file is the fact that it resides under such directory.

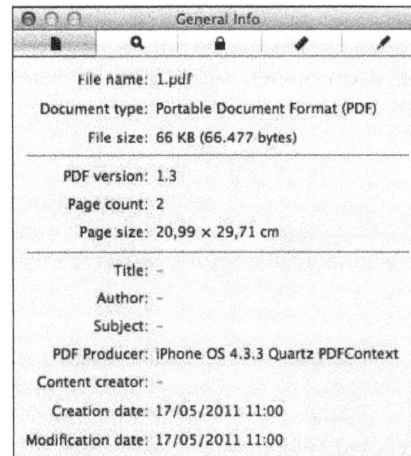

FIGURE 2: Metadata of one temporary PDF file generated by AirPrint, as shown by the OS X "Preview" tool.

Considering this, an eventual automated tool aimed at recovering the traces left by AirPrint should go further than just carving for PDF files: it should correctly interpret the internals of the HFSX filesystem and tell whether each recovered file existed within the com.apple.printd/ directory or somewhere else.

4. Recoverability of AirPrint Traces without iOS Data Protection

Given that the temporary PDF files generated by AirPrint only exist in the filesystem for a limited time and are deleted when the document has been printed, the possibility that, depending on disk scheduling and other factors, these files might never be actually flushed to the physical disk must be considered. In that case, they are unrecoverable by a subsequent forensic analysis. This section studies such possibility and assesses the probability that a given trace may be actually recovered at a later time.

4.1. Preliminary Setup. For the experiments described in this section, the following equipment was used:

(i) iOS device: Apple iPhone 3G, 8 GB (model A1241) running iOS 4.2.1 (8C148), jailbroken using redsn0w and enabling multitasking and AirPrint.

(ii) Desktop computer information: Apple MacBook Pro (model MacBookPro5,1) running Mac OS X 10.7.2 (11C74).

(iii) Printer: HP Photosmart 5510 B111a (model CQ176B) running firmware version EPL2CN1122AR.

(iv) Wireless connectivity: all devices were connected to a wireless 802.11g network to perform the experiments.

(v) Physical connectivity: all devices were using physical wires for AC power only.

By using older equipment (an iPhone 3G) in this set of experiments, we had a device without iOS "data protection" mechanisms (hardware-based encryption), which allowed us to analyze the behavior of AirPrint without having to avoid the added pitfall of encryption.

The AirPrint feature depends on the multitasking capabilities of the device, which are disabled by default in older models (such as an iPhone 3G). However, it is possible to enable those features during the jailbreak process using the `redsn0w` tool. Having AirPrint capabilities and an unencrypted filesystem made this device the perfect testbed for our experiments.

4.2. Experiments. There are several factors that increase the chances of the temporary PDF files being flushed to disk, making them potentially recoverable in the future. Some of these factors are listed as follows:

(i) Documents that take a long time to be printed because they have many pages or because they contain graphics. Note that, when sending a document to the printer, the user can control very few options: printer, page range, number of copies, and nothing else. There is neither an option for printing in "draft mode" nor one for using only the black cartridge, meaning that everything sent via AirPrint is printed in full color, good quality... and may require quite a lot of time to finish.

(ii) The quality of the wireless link between the iOS device and the printer. It can also affect the time needed to transmit and print the document.

(iii) Documents that are sent while there are other printing tasks running.

(iv) Periods of printing interruptions due to the need of human interaction, such as the printer running out of paper or out of ink.

In order to test the recoverability of AirPrint traces in the form of deleted temporary files, a series of experiments was run. The goals of these experiments were twofold. The first goal is to determine whether the temporary files generated by AirPrint are actually written to the physical disk at some point and thus may be recoverable after deletion. The second goal is to asses whether, even if the previous case is true, each of those temporary files would still be recoverable after generating more of them (i.e., what are the chances that the AirPrint temporary files overlap each other in disk, always overwriting the same space and thus making each other unrecoverable?).

Some steps in the testing process involved printing documents, while others were just aimed at simulating some casual user activity in the device (mail browsing, software update, and reboot). For those tests that involve printing, 10 documents of a given kind are printed in each test. Five of the tests involve printing, which means that, after completing all the tests, 50 documents had been printed.

After each of the tests, a dump of the iPhone filesystem was obtained. For those tests that involved printing, the dump process was not started until the printer had finished printing all of the submitted jobs sent.

The test was performed as follows:

(1) Various sets of 10 items each were printed using different applications (Safari, Photos, Mail, GoodReader, ...).

(2) A dump of the device storage was obtained after each set of 10 items.

(3) Additional batches of activity were performed in the device (download email, install a software update, and reboot the device) and additional storage dumps were obtained after performing each of these tests.

We transferred the disk image on-the-fly via Wi-Fi to the desktop computer using the method proposed by Zdziarski [18], based on the `dd` and `netcat` commands.

Each filesystem dump was analyzed in order to determine whether the temporary files generated by AirPrint at each stage were recoverable both immediately after their generation and at later stages.

With no data encryption in place, we were able to use the *file carving* [24] method for recovering deleted files. Carving is a data recovery method consisting of going through a raw data stream (a filesystem dump in this case) looking for possible "headers" and "footers" (beginnings and ends) of known, chosen file types, such as JPEG pictures, MPEG video files, and PDF documents. The more strict a file type specification is, the easier it is for the carving tool to identify and recover that file type. In addition, some tools perform sanity checks such as establishing a file size limit or checking each recovered item against its format specification to determine whether it may be corrupt.

One of the benefits of the carving technique is that it can be used, with more or less success, over any kind of data, be it a known or unknown filesystem, a portion of it, or something completely different, such as network captures or RAM memory dumps. Note, however, that many tools implement specific strategies for common filesystems in order to improve overall success rate and extract items only from the unallocated space, skipping the disk space used by normal existing files. This is something very useful when the user just wants to focus on recovering deleted files.

This technique was applied by means of the open source, widely used `photorec` tool [25], which has a long track of usefulness in this kind of scenarios [24, 26] even on mobile devices [27].

Every dump obtained during the tests was carved for PDF files using the default `photorec` parameters. Fine-tuning these parameters would have probably improved the recovery success rate; however, after the tests were completed, it was found unnecessary given that the results achieved were indeed positive.

4.3. Results. The results of trace extraction using the photorec tool were analyzed from two different standpoints:

(1) *Recoverability.* Tenths of the temporary PDF files generated by AirPrint were successfully recovered from each filesystem dump, which confirms that those files are indeed flushed to disk.

(2) *Persistence.* The artifacts created during our first tests were still recoverable after the last tests. This suggests that, probably due to iOS file allocation strategies, the temporary files generated by AirPrint are not likely to overwrite each other.

The results of these tests, which can be seen in more detail in [2], show that under iOS 4 and with no data encryption in place the temporary files generated by AirPrint are indeed written to disk and are potentially recoverable even after rebooting the device or turning it off.

Considering that the artifacts are stored in unallocated space, it is unavoidable that using the device for long periods of time reduces the chance of recovering such artifacts, as new data stored in the device may overwrite that disk space. However, the tests show that the probability does not decrease very quickly, and there is at least a good chance to recover most of the traces.

5. Impact of iOS Data Protection on the Recoverability of AirPrint Traces

In this section we describe a new set of experiments aimed at assessing whether the traces left by AirPrint in the device's filesystem can be recovered even when modern iOS data encryption mechanisms are in place.

5.1. Preliminary Setup. The experiments described in this section were carried out using the equipment described below:

(i) iOS device: Apple iPhone 3GS, 32 GB (model A1303) running iOS 6.1 (10B141), jailbroken using the evasi0n software [28].

(ii) Desktop computer: Apple iMac (model iMac13,2) running OS X 10.8.4 (12E55).

(iii) Printer: HP Photosmart 5510 B111a (model CQ176B) running firmware version: EPL2CN1122AR.

(iv) Wireless connectivity: all three devices were connected to a wireless 802.11g network to perform the experiments.

(v) Physical connectivity: the printer and desktop computer used physical wires for AC power. In addition, the iOS device was connected most of the time to the desktop computer using the standard Apple USB to 30-pin cable; this powered the device and served as the transmission channel when dumping the device internal storage to the desktop computer.

5.2. Mechanisms to Bypass iOS Data Protection. As we introduced in previous sections, all current iOS devices offer hardware-based encryption, backed with software support at the OS and application level. As noted by Casey et al. [29], "the increasing use of full disk encryption can significantly hamper digital investigations, potentially preventing access to all digital evidence in a case."

The data encryption mechanisms that Apple calls "data protection" were introduced in iOS version 4 (June 2010)

and stood publicly unbreakable for nearly one year until, in May 2011, a software firm announced a product capable of bypassing this encryption [30]. This product is restricted to "*established law enforcement, intelligence and forensic organizations as well as select government agencies.*"

At the same time, Bedrune and Sigwald published [31] details about iOS data protection shortly after they released the tools and source code capable of breaking this encryption [32]. Even if the device is locked with a passcode, the tools include a bruteforce script that runs in the iOS device itself and obtains the user-defined passcode, unless it is an alphanumeric code, something rarely seen in these devices, although supported. As of July 2013, their toolkit has been updated and works successfully with supported devices even under iOS version 6.1.

Nowadays most forensic tools support a similar functionality to one extent or another.

5.3. How iOS Data Protection Affects the Recovery of Deleted Files. One of the features of iOS encryption is that it relies on per-file encryption keys, which means that each file in the filesystem is encrypted using a different key. This, in turn, means that recovering a deleted file is more difficult than just retrieving the portion of disk space where the contents of this file reside: one must also recover the necessary filesystem metadata containing the encryption key for that given file.

For this reason, commercial tools, even when able to defeat encryption, do not recover deleted files so far. Instead, these tools usually opt for (a) examining the device backups stored in iTunes in a local computer rather than device itself or (b) just query the device for as much information as possible using standard APIs; for instance, get every sent and received SMS from the Messages application, list recent lookups from the Maps application, or query the Phone application for the recent calls log. Both these approaches, however, overlook any deleted data in the device, skipping a lot of information that could be relevant to the forensic investigator.

Given that the file contents are encrypted, the carving technique cannot be used to look for files (allocated or not). However, a smarter tool looking for deleted file/directory entries in the HFSX filesystem should succeed at recovering these files, even when their contents are encrypted.

Bedrune and Sigwald's ios_examiner tool [32] recently incorporated an undelete function which applies a novel technique [33] based on using the additional data stored in the filesystem's transaction journal in order to improve the recovery results. This tool is still very recent and under improvement, but it is expected that commercial forensic application developers may include it in later versions of their products or, at least, use similar techniques to allow forensic tools to analyze an encrypted filesystem.

5.4. Adapting the Iphone-Dataprotection *Toolkit.* As we started new experiments, we observed that it was certainly possible to recover deleted files from our test devices using the iphone-dataprotection toolkit. However, only certain file types were recovered (JPG pictures, SQLite databases, XML files, ...), whereas no PDF files were recovered at all.

```
magics = ["SQLite", "bplist", "<?xml", "\xFF\xD8\xFF", "\xCE\xFA\xED\xFE",
          "\x89PNG", "\x00\x00\x00\x1CftypM4A", "\x00\x00\x00\x14ftypqt", "\x25PDF-"]
```

ALGORITHM 1

```
knownExtensions = (".m4a", ".plist", ".sqlite", ".sqlitedb", ".jpeg", ".jpg",
                   ".png", ".db", ".json", ".xml", ".sql", ".pdf")
```

ALGORITHM 2

There are two modifications that must be applied to the software to have it recover PDF files.

The undelete algorithm used by the tool considers that a file is correctly recovered only if the initial bytes of the file match a given set of patterns. The stock list includes a limited set: SQLite databases, XML files, binary property lists, JPEG pictures, Mach-O executable binaries, PNG graphics, and M4A audio files. In order to have the tool recover deleted PDF files, the file signature of the PDF type must be added in hg/python_scripts/hfs/journal.py (lines 58-59) as shown in Algorithm 1.

In order to have these files stored in a separate directory, we declare .pdf as a known extension by modifying hg/python_scripts/nand/carver.py (line 119) as shown in Algorithm 2.

After performing this modification we observed that the ios_examiner tool successfully recovered (where technically possible) deleted PDF files. In fact it is even possible to acquire one or more dumps of the device's internal storage using the original (unmodified) tool, apply our described modifications afterwards, and then run the modified tool against the acquired images to recover any deleted PDF files they might contain, some of which can be traces left by the use of AirPrint, whereas others will be regular PDF files that have been deleted or reallocated in disk for whatever reason.

Given a set of recovered PDF files, we wrote a Perl script that outputs a CSV table indicating which of the files correspond indeed to contents printed through AirPrint, and if so, when were the documents printed and under which iOS version as shown in Algorithm 3.

5.5. Experiments. In this series of experiments we wanted to verify whether AirPrint traces were recoverable in scenarios where iOS data protection is enabled and analyze how the amount of free disk space affects the recoverability rate.

A detailed description of the whole testing process follows:

(1) Fill the device with some applications (GoodReader, plus Apple's Podcasts, iBooks, iTunes U, Find My Friends, and Find My iPhone) and multimedia content. Setup an iCloud account for activating the Find My iPhone service and start syncing email, contacts, calendars, reminders, Safari tabs, notes, photos, documents, and data.

(2) After a period of 24 hours for any massive syncing activity to take place, the device storage as reported in "Settings" is 4.1 GB available of a total of 28.3 GB (i.e., 15% free space).

(3) Reboot the device to start from a clean state.

(4) Use the GoodReader application to print a fixed set of 20 documents amounting to a total of 109 paper pages; send each document to the printer only when the previous one has been completely printed.

(5) Turn the device off and acquire forensic image #1.

(6) Boot the device. Remove all optional Apple applications as well as multimedia content (audio, video) and iCloud accounts added in step (1).

(7) After a period of 24 hours for any deletion activity to take place, the device storage as reported in "Settings" is 27.2 GB available of a total of 28.3 GB (i.e., 96% free space).

(8) Reboot the device to start from a clean state.

(9) Repeat step (4).

(10) Turn the device off and acquire forensic image #2.

(11) Recover AirPrint traces from both images and compare the results.

5.6. Results. Table 1 shows the results for each individual file. In each case, we were able to recover between 5% and 10% of the documents printed through AirPrint, extracting the following conclusions:

(i) It is possible to recover the full content of documents printed through AirPrint as well as relevant metadata such as print date and iOS version. In some cases the recovered PDF files may be corrupt but still contain details such as iOS version used and print date.

(ii) The low recoverability rate observed (5–10% in realistic scenarios) could be due to the disk scheduling algorithm used in the iOS operating system (in this particular version at least). This would also explain the fact that the success rate keeps constant regardless of the amount of free disk space.

(iii) At any particular iOS version (existing or future), a change in the disk scheduling subsystem could boost the success rate significantly. Additional work

```perl
#!/usr/bin/perl
print "Filename,iOS version,Print date\n";
while( $file = shift(@ARGV) ) {
  $ios = ";
  $date = ";
  $pdfinfo = `pdfinfo $file 2>&1`;
  @metadata = split( /\n/, $pdfinfo );
  if( @metadata[0] =~ m/iPhone OS.* Quartz PDFContext/ ) {
    ($ios = @metadata[0]) =~ s/.*iPhone OS ([0-9.]+) Quartz PDFContext/iOS \1/;
    ($date = @metadata[1]) =~ s/CreationDate: //;
    print "$file,$ios,$date\n";
  } else { print "$file,does not look like an AirPrint temporary file.\n"; }
}
```

ALGORITHM 3

TABLE 1: Recoverability of AirPrint temporary artifacts under iOS 6.

#	File	Size (bytes)	Size (pages)	Recovered?
1	000001.DOC	40.960	2	In image #1
2	000002.DOC	57.856	2	In image #2 and partially in #1
3	000003.DOC	55.808	2	No
4	000004.DOC	175.616	4	No
5	000005.DOC	180.736	5	No
6	000006.DOC	67.584	5	No
7	000007.DOC	179.200	30	No
8	000008.PPT	302.592	12	No
9	000009.PDF	39.586	4	No
10	000010.PDF	120.441	4	No
11	000011.PDF	31.367	1	No
12	000012.PDF	22.857	6	No
13	000013.PDF	38.638	2	No
14	000015.PDF	55.964	2	No
15	000016.PDF	150.586	4	No
16	000018.PDF	94.424	9	No
17	000019.PDF	124.152	3	No
18	000020.PDF	4.755	2	No
19	000021.PDF	4.521	2	No
20	000022.PDF	21.235	8	No

is needed to assess whether the results observed are kept consistent across different device models and iOS versions.

The traces of the printing activity from step (9) should be more easily recoverable given that in step (7) we tried to improve the recoverability rate by freeing most disk space (to reduce the probability that some new file, log entry, etc. could overwrite the AirPrint traces once they've been deleted) and by reducing most of the device's background activity (iCloud syncing, e-mail activity, . . .). Hence, the second image should contain a higher number of AirPrint traces than the first one.

In contrast to what could be reasonably expected, we observed that in each case we recovered only one or two AirPrint temporary files out of the 20 possible. It could be thought that only the latest jobs are being recovered; however

the traces we found corresponded to the first jobs sent to the printer rather than the last ones.

We performed additional experiments introducing circumstances such as print interruptions due to lack of paper and loss of network link between the device and the printer. In such circumstances, the temporary files remain much longer in the iOS device's filesystem and will persist for an undetermined amount of time (even some time after rebooting the device). Under these conditions we saw the success rate increase to 15%.

6. Conclusions and Future Work

This paper analyzes the forensic traces left by usage of the AirPrint functionality in iOS based devices. We have developed a method which leverages publicly available tools to recover from an iPhone or iPad the contents of documents that have been printed using the standard AirPrint feature. The recovery of these artifacts can be valuable from the point of view of a forensic investigation in scenarios such as information leak or distribution of inadequate content; however it could also pose a privacy risk to the user community.

The traces described could persist even after the original file has been deleted, or if the original file resides inside some "vault-type" application which protects its contents on disk with an additional layer of encryption.

With modern iOS 6 data encryption mechanisms in place, the described method still succeeded in recovering between 5 and 15% of the documents printed through AirPrint. We believe the success rate can depend on factors such as the disk scheduling strategy, and thus different iOS versions could throw different results.

Considering the use case of AirPrint in domestic scenarios, probably home users will not be particularly concerned about this finding, a possible exception to this being explicit graphic content. In this aspect it is interesting to note that Apple's stock Photos application specifically did not generate, in our experiments, the temporary files described in this paper, whereas other 3rd party applications offering added security to store this kind of information are likely to generate the standard AirPrint traces when printing documents. As a general solution, in order to limit the possibility of recovering data from the filesystem, some techniques aimed at performing a secure deletion could be adopted, such as the one presented in [34] for the Android OS.

Further work must be carried out to assess whether it is possible to capture the network traffic generated while printing using techniques similar to [35] and recover the contents of the documents being printed. It would also be interesting to extend the research presented in this paper across a wider range of devices, iOS versions, and 3rd-party applications and to examine if similar issues affect other printing solutions for iOS devices and other mobile devices.

Conflict of Interests

The authors declare that there is no conflict of interests regarding the publication of this paper.

Acknowledgment

This work was partly funded by the Spanish Government through projects TIN2011-27076-C03-02 "CO-PRIVACY" and SMARTGLACIS (TIN2014-57364-C2-2-R).

References

[1] Apple, *Apple's AirPrint Wireless Printing for iPad, iPhone & iPod touch Coming to Users in November, 2010*, Apple, 2010, http://www.apple.com/uk/pr/library/2010/09/15airprint.html.

[2] L. Gomez-Miralles and J. Arnedo-Moreno, "Analysis of the forensic traces left by airprint in apple iOS devices," in *Proceedings of the 27th International Conference on Advanced Information Networking and Applications Workshops (WAINA '13)*, pp. 703–708, IEEE, Barcelona, Spain, March 2013.

[3] O. Bohl, S. Manouchehri, and U. Winand, "Mobile information systems for the private everyday life," *Mobile Information Systems*, vol. 3, no. 3-4, pp. 135–152, 2007.

[4] S. Caballé, F. Xhafa, and L. Barolli, "Using mobile devices to support online collaborative learning," *Mobile Information Systems*, vol. 6, no. 1, pp. 27–47, 2010.

[5] T. Cook, Apple WorldWide Developers Conference keynote, 2013, http://www.apple.com/apple-events/june-2013/.

[6] A. Castiglione, R. de Prisco, and A. de Santis, "Do you trust your phone?" in *E-Commerce and Web Technologies*, vol. 5692 of *Lecture Notes in Computer Science*, pp. 50–61, Springer, Berlin, Germany, 2009.

[7] A. Hay, D. Krill, B. Kuhar, and G. Peterson, "Evaluating digital forensic options for the apple iPad," in *Advances in Digital Forensics VII*, vol. 361 of *IFIP Advances in Information and Communication Technology*, pp. 257–273, Springer, Boston, Mass, USA, 2011.

[8] A. Levinson, B. Stackpole, and D. Johnson, "Third party application forensics on Apple mobile devices," in *Proceedings of the 44th Hawaii International Conference on System Sciences (HICSS '11)*, pp. 1–9, January 2011.

[9] Apple, *iOS: AirPrint 101*, Apple, 2011, http://support.apple.com/kb/ht4356.

[10] Netputing, *AirPrint Activator*, 2011, http://netputing.com/airprintactivator.

[11] Avatron Software Inc, *Print Sharing*, 2011, http://avatron.com/apps/print-sharing/.

[12] EuroSmartz, PrintCentral for iPad, iPhone or iPod Touch, 2012, http://mobile.eurosmartz.com/products/printcentral.html.

[13] D. Steinberg and S. Cheshire, *Zero Configuration Networking: The Definitive Guide*, O'Reilly, 2005.

[14] Apple Inc, How Printing Works in iOS, 2011, https://developer.apple.com/library/ios/.

[15] iPhone Dev Team, redsn0w, 2011, http://blog.iphone-dev.org/post/5239805497/tic-tac-toe/.

[16] US Copyright Office, *Rulemaking on Exemptions from Prohibition on Circumvention of Technological Measures that Control Access to Copyrighted Works*, US Copyright Office, 2010, http://www.copyright.gov/1201/2010/.

[17] J. R. Rabaiotti and C. J. Hargreaves, "Using a software exploit to image RAM on an embedded system," *Digital Investigation*, vol. 6, no. 3-4, pp. 95–103, 2010.

[18] J. Zdziarski, *iPhone Forensics: Recovering Evidence, Personal Data, and Corporate Assets*, O'Reilly Media, 2008.

[19] F. Buchholz and E. Spafford, "On the role of file system metadata in digital forensics," *Digital Investigation*, vol. 1, no. 4, pp. 298–309, 2004.

[20] A. Castiglione, A. De Santis, and C. Soriente, "Taking advantages of a disadvantage: digital forensics and steganography using document metadata," *Journal of Systems and Software*, vol. 80, no. 5, pp. 750–764, 2007.

[21] A. J. Clark, "Document metadata, tracking and tracing," *Network Security*, vol. 2007, no. 7, pp. 4–7, 2007.

[22] M. S. Olivier, "On metadata context in database forensics," *Digital Investigation*, vol. 5, no. 3-4, pp. 115–123, 2009.

[23] A. Castiglione, A. De Santis, and C. Soriente, "Security and privacy issues in the portable document format," *Journal of Systems and Software*, vol. 83, no. 10, pp. 1813–1822, 2010.

[24] M. I. Cohen, "Advanced carving techniques," *Digital Investigation*, vol. 4, no. 3-4, pp. 119–128, 2007.

[25] CGSecurity, *PhotoRec, Digital Picture Recovery*, 2009, http://www.cgsecurity.org/wiki/PhotoRec.

[26] S. L. Garfinkel, "Forensic feature extraction and cross-drive analysis," *Digital Investigation*, vol. 3, pp. 71–81, 2006.

[27] I. Pooters, "Full user data acquisition from Symbian smart phones," *Digital Investigation*, vol. 6, no. 3-4, pp. 125–135, 2010.

[28] Y. D. Wang, N. Bassen et al., evasi0n, 2013, http://evasi0n.com/.

[29] E. Casey, G. Fellows, M. Geiger, and G. Stellatos, "The growing impact of full disk encryption on digital forensics," *Digital Investigation*, vol. 8, no. 2, pp. 129–134, 2011.

[30] ElcomSoft, ElcomSoft investigates iPhone hardware encryption, provides enhanced forensic access to protected, 2011, http://www.elcomsoft.com/PR/eppb110524en.pdf.

[31] J. B. Bedrune and J. Sigwald, *iPhone Data Protection in Depth*, HITB, Amsterdam, The Netherlands, 2011.

[32] J. B. Bedrune and J. Sigwald, iPhone data protection tools, 2011, http://code.google.com/p/iphone-dataprotection/.

[33] A. Burghardt and A. J. Feldman, "Using the HFS+ journal for deleted file recovery," *Digital Investigation*, vol. 5, supplement, pp. S76–S82, 2008.

[34] A. Castiglione, G. Cattaneo, G. De Maio, and A. De Santis, "Automatic, selective and secure deletion of digital evidence," in *Proceedings of the 6th International Conference on Broadband and Wireless Computing, Communication and Applications (BWCCA '11)*, pp. 392–398, October 2011.

[35] A. Castiglione, G. Cattaneo, G. de Maio, and A. de Santis, "Forensically-sound methods to collect live network evidence," in *Proceedings of the 27th IEEE International Conference on Advanced Information Networking and Applications (AINA '13)*, pp. 405–412, Barcelona, Spain, March 2013.

16

A Study on the Distributed Antenna Based Heterogeneous Cognitive Wireless Network Synchronous MAC Protocol

Lian-Fen Huang,[1] **Sha-Li Zhou,**[1] **Yi-Feng Zhao,**[1] **and Han-Chieh Chao**[2,3]

[1]*Department of Communication Engineering, Xiamen University, Xiamen, Fujian 361005, China*
[2]*Institute of Computer Science & Information Engineering and Department of Electronic Engineering, National Ilan University, I-Lan, Taiwan*
[3]*Department of Electrical Engineering, National Dong Hwa University, Hualien, Taiwan*

Correspondence should be addressed to Yi-Feng Zhao; zhaoyf@xmu.edu.cn

Academic Editor: Ilsun You

This paper introduces distributed antennas into a cognitive radio network and presents a heterogeneous network. The best contribution of this paper is that it designs a synchronous cognitive MAC protocol (DAHCWNS-MAC protocol: distributed antenna based heterogeneous cognitive wireless network synchronous MAC protocol). The novel protocol aims at combining the advantages of cognitive radio and distributed antennas to fully utilize the licensed spectrum, broaden the communication range, and improve throughput. This paper carries out the mathematical modeling and performance simulation to demonstrate its superiority in improving the network throughput at the cost of increasing antenna hardware costs.

1. Introduction

DAS (distributed antenna system) is used as an extension of the outdoor cellular mobile system in early stage, which is widely used for indoor or blind spot coverage. However the research of MIMO (multiple input multiple output) technology has become more and more sophisticated, which provides a broader space for the further development of DAS. This network structure can improve the wireless signal covering ability and system capacity and obtain high power efficiency. Because of these advantages, the DAS has been considered to be a key way of multiple antennas accessing in future mobile communication. Most current researches on the distributed antenna are focused on how distributed antennas can be used in the cellular mobile network physical layer to improve network capacity. Introducing DAS into WLAN (wireless local area network) can fully utilize its physical advantages to greatly improve the performance of WLAN, which is an attractive research area in the future. However there is scant literature on how distributed antenna can be used in WLANs. Reference [1] thinks that the fixed channel allocation method is not flexible and fair enough for multicell WLAN. New systems replace APs (access points) with distributed antennas using a control center. The control center dynamically adjusts the channel assignment scheme according to a judgment standard determined by the throughput, fairness, and other factors. Reference [2] designs the control center internal structure used in the distributed antenna WLAN system. The author thinks that each channel needs a corresponding processing unit. The more nodes the antenna services, the more channels it needs. The antenna therefore allocated more processing units.

This paper studies distributed antenna applications in the MAC (media access control) layer and designs a distributed antenna-based synchronous MAC protocol to sense the spectrum and transmit data.

Numerous literatures have studied spectrum sensing, producing some novel algorithms. References [3, 4] proposed a multitaper method (MTM) to detect the spectrum, with the advantages of low complexity and high detection accuracy. References [5, 6] proposed a detection method based on the signal covariance matrix. The ratio of two statistics is used to judge if the primary user appears. References [7, 8] studied the generalized likelihood ratio test method. To eliminate the influence of noise uncertainty, it first uses the maximum likelihood algorithm to estimate the unknown noise power and then completes the detection algorithm.

In addition to spectrum sensing we can further use distributed antennas to locate the primary user and adopt different access methods according to positioning information. Reference [9] studied the power allocation problem under overlay/underlay hybrid access mechanism. Reference [10] introduced the distributed antenna into the cognitive radio network. It utilizes the distributed antenna to sense the spectrum and locate the primary user. The author designed an asynchronous cognitive MAAC-MAC protocol (multi-antenna asynchronous cognitive MAC).

This paper focuses on how the distributed antenna can be used in data transmission and does not explore its usage in spectrum sensing and positioning. In future studies we can continue this research in different directions.

In order to design an appropriate network architecture for DAS, this paper studies the hybrid wireless networks. Many documents have studied hybrid wireless networks [11–14]. Adding AP/BSs (access point/base stations) into an ad hoc network can combine the network advantages with ad hoc infrastructure. Currently related literatures are focused on a cellular network and ad hoc network combination. In [15] the whole area is partitioned into many cells and all cells use the TDMA (time division multiple access) scheme. The author proposes two hybrid routing strategies that combine direct transmission (ad hoc mode) and forwarding data through base stations (infrastructure mode). Reference [16] proposes a protocol to judge whether data needs to be forwarded or not in the hybrid network.

This paper introduces the distributed antenna into the cognitive radio network and designs a heterogeneous network consisting of an ad hoc network and a sparse network of distributed antennas. The new network utilizes the distributed antennas to sense the spectrum and transmit data. We design a synchronous cognitive MAC protocol (DAHCWNS-MAC protocol) that can improve the sensing performance and also broaden the communication range to increase the throughput compared to the original single-hop network.

2. Heterogeneous Network Model

2.1. Communication Scenario. The distributed antenna layout is shown in Figure 1. Distributed antennas are uniformly placed throughout the region. Seven antennas are used to cover the entire communication scenario. The seven antennas are connected to a control center via an optical fiber. Distributed antennas forward data only with the other complex processing work completed by the control center.

2.2. Network Architecture. In an 802.11DCF network with infrastructure all data is forwarded by the access point (AP). After adding distributed antennas the original AP coverage is divided into seven smaller cells, each of which is covered by an antenna. Within the coverage of each antenna the distance between nodes can be regarded as one hop and

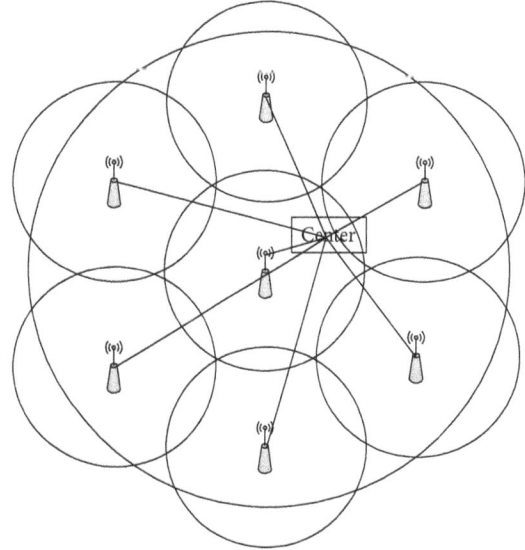

FIGURE 1: The layout of distributed antennas.

they can send data to each other directly. A node cannot communicate directly with its destination node when its destination node is not located in the same cell. In this case distributed antennas are used to forward data. We assume that nodes are uniformly placed in the region and each node accesses its nearest antenna. When a node and its destination node are located in the same antenna coverage they can send data directly. However, if a node and its destination node are not located in the same cell, they have to use distributed antennas to forward data. This is a heterogeneous network consisting of an ad hoc mode and infrastructure mode.

This heterogeneous network combines the advantages of two different networks. The network with infrastructure is easy to manage and does not have to exchange messages among nodes as in the ad hoc network. This reduces the extra overhead. It can also broaden the communication range through forwarding data by infrastructure. Under the ad hoc mode direct data transmission decreases the time consumed by forwarding, so the throughput can be improved.

The DAHCWNS-MAC protocol is designed for cognitive radio networks where nodes use idle licensed bands to communicate. The spectrum sensing is an important part of the protocol. What is different from the previous cognitive MAC protocols is that the DAHCWNS-MAC protocol uses distributed antennas instead of nodes to sense the spectrum. Results at all the antennas will be submitted to the control center for a final judgment result. The idle bands will be allocated to nodes by the control center. Nodes do not have to sense the spectrum by themselves as in the self-organized network, which reduces the exchanging sensing results overhead and decreases the node hardware requirements. Furthermore, the sensing performance is improved while utilizing the distributed antennas' macrodiversity.

We can utilize distributed antennas to locate primary users and adopt overlay/underlay hybrid access schemes through power control for more effective licensed band usage.

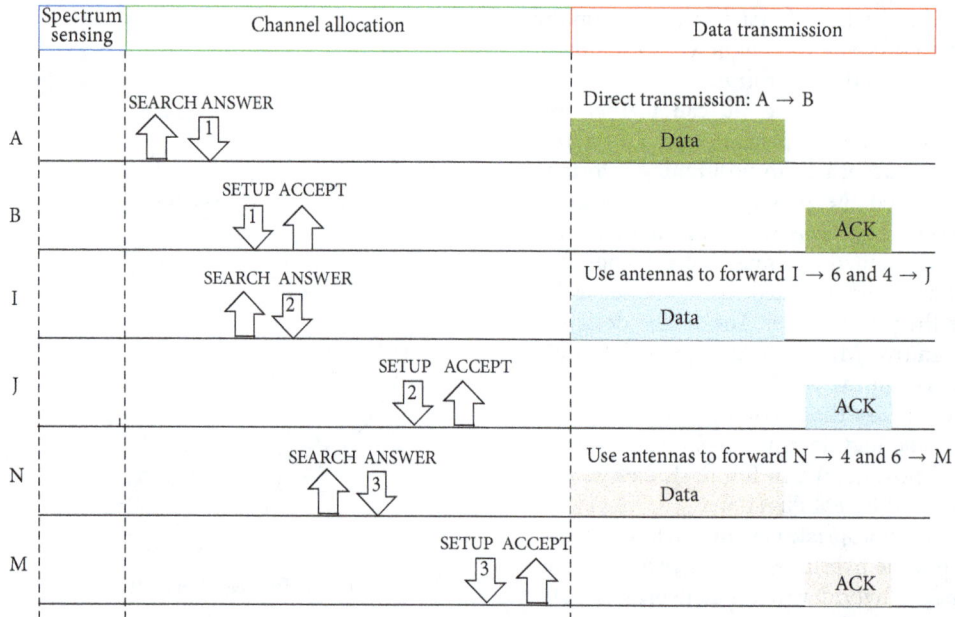

FIGURE 2: The time frame of DAHCWNS-MAC protocol.

Section 1 introduces some novel sensing algorithms and hybrid access schemes that are very useful in completing the DAHCWNS-MAC protocol in future research. Because this paper aims mainly at exploring how distributed antennas are used for data transmission, we no longer repeat the spectrum sensing and positioning parts of the protocol in the next article.

3. The Distributed Antenna Based Heterogeneous Cognitive Wireless Network Synchronous MAC Protocol (DAHCWNS-MAC Protocol)

In this section the DAHCWNS-MAC protocol will be described in detail.

3.1. Assumptions. Before presenting the specific protocol some necessary assumptions are summarized as follows.

(a) There are $N + 1$ licensed channels for use, all of which have the same bandwidth. Since no overlap occurs among channels, packets transmitted on different channels will not affect each other. The control center knows how many licensed channels can be used in advance.

(b) One of $N + 1$ licensed channels is used as the control channel. This can be the unlicensed band in practice and thus free from interference from the primary users.

(c) Each CU (cognitive user) is equipped with a single cognitive radio. This radio can either transmit or receive, but it cannot do both simultaneously.

(d) All antennas can forward data correctly. The control center has different processing units corresponding to different antennas, so it can parallel process data from different antennas. The control center also knows which nodes exist within the antenna coverage.

(e) All nodes are strictly synchronous. They always start and finish a beacon interval at the same time.

(f) Distributed antennas can sense the spectrum precisely to obtain all idle channels.

3.2. Protocol Design. The DAHCWNS-MAC protocol is a synchronous MAC protocol. The whole time can be divided into frames with fixed length. The time frame can be separated into three parts: sensing, channel allocation, and transmission phases. The DAHCWNS-MAC time frame is depicted in Figure 2.

(1) The first part of the DAHCWNS-MAC protocol is the sensing phase. Distributed antennas sense spectrum during this period. Each antenna detects N channels independently and submits its result to the control center. The control center will get final judgment result through data fusion and allocates idle channels to nodes to communicate in the next part of the protocol.

(2) The second part of the DAHCWNS-MAC protocol is the channel allocation phase. This phase is based on the CSMA/CA mechanism. All nodes contend with each other for the right to use channels. Since this network is a heterogeneous network it is necessary to determine whether nodes need distributed antennas to forward data. The specific process is as follows. Nodes that have data to transmit contend to send SEARCH frames to their nearest antenna on the control channel. If the competition succeeds, the corresponding antenna will submit the frame to the control center and then stores it. At the same time the control center

will reply with an ANSWER frame to the node which includes the number of allocated channels (randomly selected from the remaining channels until no idle channels remain). The above is the uplink part of this phase.

Upon receiving the SEARCH frame the control center will find the number of distributed antennas to which the destination node belongs and check if the sending and receiving nodes are located in the same antenna coverage area. If so, then in the next data transmission phase there is no need for distributed antennas to forward data. The sending and receiving nodes can communicate directly. If not, data packets are forwarded by distributed antennas. The control center records the result if a pair of nodes need data forwarding and in the next data transmission phase it can use the result to judge whether to forward data packets from this node. Next the control center checks if the antenna to which the receiving node belongs is idle. If the antenna is idle a SETUP frame will be sent to the receiving node immediately which contains the same channel number as the ANSWER frame. Otherwise, the SETUP frame will be sent until the antenna becomes idle. Upon the receiving node getting the SETUP frame it will reply with an ACCEPT frame, which means the handshake is completed. The above is the downlink part of the channel allocation phase.

Because the uplink and downlink capacity are unbalanced in networks with infrastructure, to ensure downlink transmission completion, the protocol does not use a competition scheme during the downlink, which means the nodes do not need to perform backoff. Specifically, there is no need for backoff before sending the SETUP frame.

The transceiver node pair completes any message exchange necessary for the data transmission phase through a four-way handshake. As shown in Figure 3 the four-way handshake includes two parts: the uplink part and downlink part.

The DAHCWNS-MAC protocol makes the collision domain narrow to the antenna coverage. Cells covered by different antennas are independent and do not affect each other. Each antenna can work in parallel with others, which means different antennas can be in different transmission/receiving stages at the same time.

All operations during this phase are carried out on the control channel. The control channel may also be used for data transmission.

(3) The third part of the DAHCWNS-MAC protocol is the data transmission phase. The transceiver node pair which completes the four-way handshake can send data on the channel allocated to them. If they need help from distributed antennas in forwarding data the corresponding antenna will forward their data automatically. Otherwise, they will communicate directly until this time frame ends. A transceiver node pair that does not complete the four-way handshake cannot enter the data transmission phase.

The time frame is divided into three parts. As illustrated there are three pairs of nodes communicating: A to B, I to J, and N to M. For example, A wants to communicate with B and it sends SEARCH frame to the center. Then the center replies with ANSWER frame which allocates channel 1 to A. At the same time the center sends SETUP frame to B to tell B that

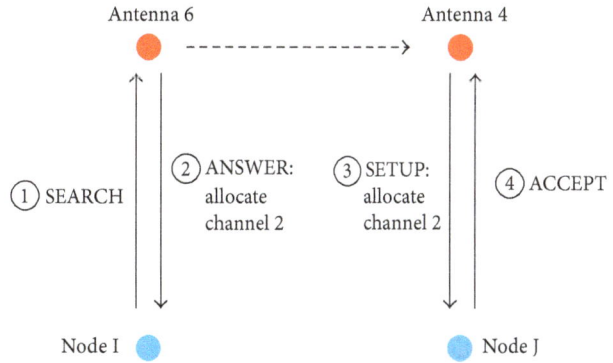

FIGURE 3: The four-way handshake.

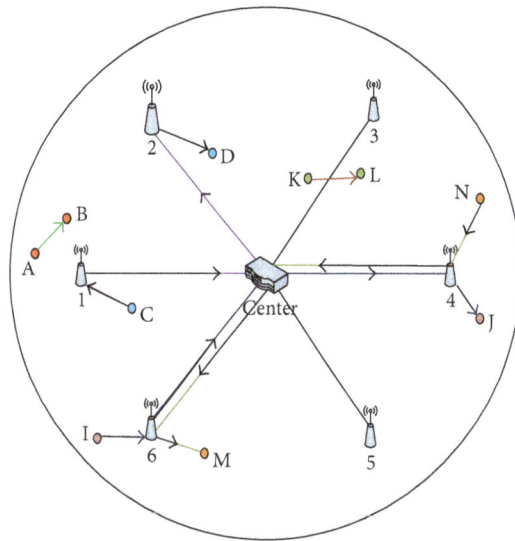

FIGURE 4: Transmission example.

the allocated channel is 1 and then B replies with ACCEPT frame to complete four-way handshake. Because nodes A and B are in one-hop distance, they can transmit directly in channel 1. I and J communicate on channel 2 through antenna forwarding. N and M communicate on channel 3 through antenna forwarding.

A specific transmission case is presented in Figure 4. For example, nodes A and B are located in antenna 1 coverage area. They can communicate directly because the control center allocated channel 1 to them. Nodes I and J are located in different antenna coverage areas, so they need antennas 4 and 6 to forward data with channel 2 assigned to them.

This figure uses the communication process between nodes I and J to explain how the four-way handshake works. The first step: node I sends SEARCH frame to antenna 6. The second step: the center replies with ANSWER frame which allocates channel 2 to I. The third step: at the same time antenna 4 sends SETUP frame to node J to tell J that the allocated channel is 2. The last step: node J replies with ACCEPT frame to complete four-way handshake.

3.3. The DAHCWNS-MAC Protocol Model. In this section a mathematical model for the DAHCWNS-MAC is presented to theoretically calculate its saturated throughput.

Since only transceiver nodes that complete a handshake can enter the data transmission phase, the key in calculating throughput is to calculate how many pairs of transceiver nodes complete the handshake during the channel allocation phase. The channel allocation phase is divided into uplink and downlink parts. How many times the downlink part is finished represents how many pairs of transceiver nodes complete the handshake. Therefore the objective is to calculate the downlink part throughput during the channel allocation phase.

Whether frames are sent in the downlink part depends on the uplink part. The SETUP frame is sent only when the SEARCH frame has been successfully received. Since seven distributed antennas operate independently, the uplink and downlink transmissions in different cells are independent of each other. Therefore, the average throughput for the seven cells should be the same. It is feasible to calculate the downlink throughput of one cell first and then multiply it by seven.

We assume that sending nodes always have data to transmit. Nodes contend to send frames in uplink and the transmission probability in each slot can be calculated using the two-dimension Markov chain model. Since there is no competition during downlink, the downlink transmission probability is equal to the probability of generating downlink data.

Several necessary variables are defined as follows. τ^{up} and τ^{down} represent the transmission probability in each slot of uplink and downlink, respectively. The average number of sending nodes in each cell is n.

In 802.11DCF network, the transmission probability in each slot can be calculated using the two-dimension Markov chain model according to [17]. The transmission probability of the uplink part should be the same. So

$$\tau^{\text{up}} = \frac{2\left(1 - 2p\right)}{\left(1 - 2p\right)\left(\text{CW} + 1\right) + p\text{CW}\left[1 - \left(2p\right)^m\right]}, \quad (1)$$

where CW is the size of the smallest contention window and m is the maximum backoff stage. The collision probability of uplink is

$$p = 1 - \left(1 - \tau^{\text{up}}\right)^{n-1}\left(1 - \tau^{\text{down}}\right). \quad (2)$$

There is no need in downlink for backoff before sending data. Frames are sent out as soon as the channel is idle. Therefore,

the transmission probability of downlink is equal to the probability of generating downlink data. The probability depends on the uplink transmission success. Only when the uplink transmission succeeds, which means the SEARCH frame has been successfully stored in center cache, will the corresponding exit antenna have data to send. So the downlink transmission probability is assumed to be equal to the probability of successful uplink transmission:

$$\tau^{\text{down}} = p_s^{\text{up}} p_{\text{tr}}^{\text{up}} = n\tau^{\text{up}}\left(1 - \tau^{\text{up}}\right)^{n-1}\left(1 - \tau^{\text{down}}\right). \quad (3)$$

Combining (1), (2), and (3), the τ^{up} and τ^{down} can be solved.

According to [18] the probability that at least one uplink frame is in transmission at some slot is

$$p_{\text{tr}}^{\text{up}} = 1 - \left(1 - \tau^{\text{up}}\right)^n. \quad (4)$$

Under the case that uplink frames exist which are in transmission, the probability that only one uplink frame is being transmitted is

$$p_s^{\text{up}} = \frac{n\tau^{\text{up}}\left(1 - \tau^{\text{up}}\right)^{n-1}\left(1 - \tau^{\text{down}}\right)}{1 - \left(1 - \tau^{\text{up}}\right)^n}. \quad (5)$$

Likewise, under the case that downlink frames exist which are in transmission, the probability that only one downlink frame is being transmitted is

$$p_s^{\text{down}} = \frac{\tau^{\text{down}}\left(1 - \tau^{\text{up}}\right)^n}{\tau^{\text{down}}}. \quad (6)$$

Four possible states exist in one slot as follows.

(1) The channel is idle and the probability is $\left(1 - p_{\text{tr}}^{\text{up}}\right)\left(1 - \tau^{\text{down}}\right)$. The length of a slot is σ.

(2) The uplink frame is transmitted successfully and the probability is $p_s^{\text{up}}\tau^{\text{down}}$. The average transmission time is T_s^{up}.

(3) The downlink frame is transmitted successfully and the probability is $p_s^{\text{down}}\tau^{\text{down}}$. The average transmission time is T_s^{down}.

(4) When a collision happens the probability is $p_{\text{tr}}^{\text{up}}\left(1 - p_s^{\text{up}}\right) + \tau^{\text{down}}\left(1 - p_s^{\text{down}}\right) - p_{\text{tr}}^{\text{up}}\tau^{\text{down}}$. The average time is T_c.

The downlink throughput is

$$s = \frac{\tau_{\text{down}}P_s^{\text{down}}E^{\text{down}}}{\left(1 - P_{\text{tr}}^{\text{up}}\right)\left(1 - \tau_{\text{down}}\right)\sigma + P_{\text{tr}}^{\text{up}}P_s^{\text{up}}T_s^{\text{up}} + \tau_{\text{down}}P_s^{\text{down}}T_s^{\text{down}} + \left(P_{\text{tr}}^{\text{up}}\left(1 - P_s^{\text{up}}\right) + \tau_{\text{down}}\left(1 - P_s^{\text{down}}\right) - P_{\text{tr}}^{\text{up}}\tau_{\text{down}}\right)T_c}. \quad (7)$$

The denominator of the above formula is the average length of one slot where the downlink transmission occupies a proportion:

$$r = \left(\tau_{\text{down}}P_s^{\text{down}}T_s^{\text{down}}\right)$$

$$\times \left(\left(1 - P_{\text{tr}}^{\text{up}}\right)\left(1 - \tau_{\text{down}}\right)\sigma + P_{\text{tr}}^{\text{up}}P_s^{\text{up}}T_s^{\text{up}}\right.$$

$$+ \tau_{\text{down}}P_s^{\text{down}}T_s^{\text{down}}$$

$$+ \left(P_{\text{tr}}^{\text{up}} \left(1 - P_s^{\text{up}} \right) + \tau_{\text{down}} \left(1 - P_s^{\text{down}} \right) \right.$$

$$\left. - P_{\text{tr}}^{\text{up}} \tau_{\text{down}} \right) T_c \right)^{-1} . \tag{8}$$

So the total number of transceiver nodes that can complete a handshake during the channel allocation phase is

$$\text{num} = 7 * \text{CA\_window\_length} \times \frac{r}{T_s^{\text{down}}}, \tag{9}$$

where CA_window_length is the length of the channel allocation phase.

The final saturation throughput is

$$s = \left(\frac{1}{7} \times \text{num} \times \frac{\text{data\_window\_length}}{T_{\text{data}}} \times l \times 8 + \frac{6}{7} \right.$$

$$\times \text{num} \times \frac{\text{data\_window\_length}}{2 \times T_{\text{data}}} \times l \times 8 \right) \tag{10}$$

$$\times \left(\text{interval\_length} \right)^{-1} ,$$

where data_window_length is the length of the data transmission phase, T_{data} is the transmission time of a data packet, l is the real payload in a data packet, and interval_length is the length of a time frame.

As nodes are randomly placed, the probability that sending and receiving nodes are located in the same cell is 1/7 at which time they can send data directly. However, the probability that sending and receiving nodes are located in a different cell is 6/7 at which time distributed antennas should help in forwarding data. The time required for forwarding data is doubled compared to that of sending directly.

4. Performance Simulation and Analysis

In this section we present the DAHCWNS-MAC protocol simulation results. We assume a time frame is fixed at 100 ms and the state of licensed channels changing at the beginning of each time frame and remaining unchanged until the time frame ends up. The related parameters are shown in Table 1. This paper uses C language and MATLAB 7.0 as simulation tools. The simulation results come from C language programming and the theoretic results come from MATLAB programming.

4.1. The Comparison of Mathematical Model and Simulation Throughput. The channel allocation phase length is fixed to 10 ms and the number of idle channels is unlimited (all the transceiver nodes pairs which complete four-way handshake can be assigned a channel). A comparison between the mathematical model and simulation results is shown in Figure 5. We can see that the outcome for the two is quite close. This demonstrates the accuracy of the mathematical model. As the number of sending nodes increases, the throughput first increases and then decreases. This is because when the number of nodes increases the collision probability increases as well. The increasing collision probability results in a

TABLE 1: The main simulation parameters.

Data channel rate	1 Mb/s
Control channel rate	1 Mb/s
Transmission delay	1 us
DIFS	50 us
SIFS	10 us
Slot time	50 us
Maximum backoff state (m)	4
Data packet length	512 bytes
Time frame length	100 ms
Simulation time	100 s

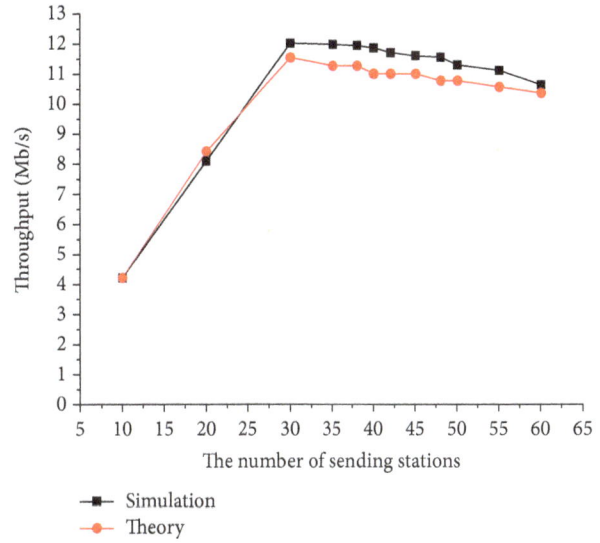

FIGURE 5: The saturation throughput of mathematical model and simulation.

decreasing number of nodes that can complete a handshake and enter the transmission phase. Therefore, the throughput eventually decreases.

4.2. The Relationship between Throughput and Channel Allocation Phase Length. Now let us analyze the channel allocation phase length effect on the saturation throughput. With 50 sending stations and 50 licensed channels, as shown in Figure 6, the saturation throughput increases as the time increases from 5 ms to 20 ms when the primary users' activity rate is 0. However, the saturation throughput decreases after 20 ms. This is because the number of nodes that complete a handshake is limited when the length is smaller than 20 ms. As the length becomes larger, the number of nodes that can complete a handshake to get a channel becomes larger, which results in an increase in the saturation throughput. However, the length of the time frame is fixed in the DAHCWNS-MAC protocol, which means the data transmission time will decrease when the channel allocation time increases. As the throughput reaches peak when the time is 20 ms because all transceiver nodes can get idle channels at this moment, increasing the channel allocation time results in the data

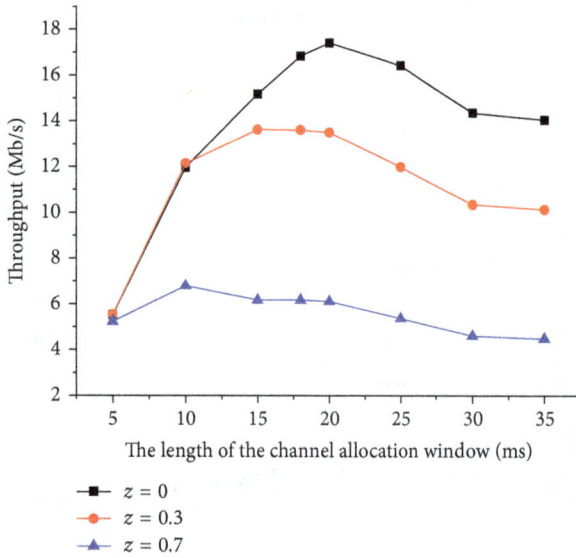

FIGURE 6: The relationship between throughput and channel allocation phase length.

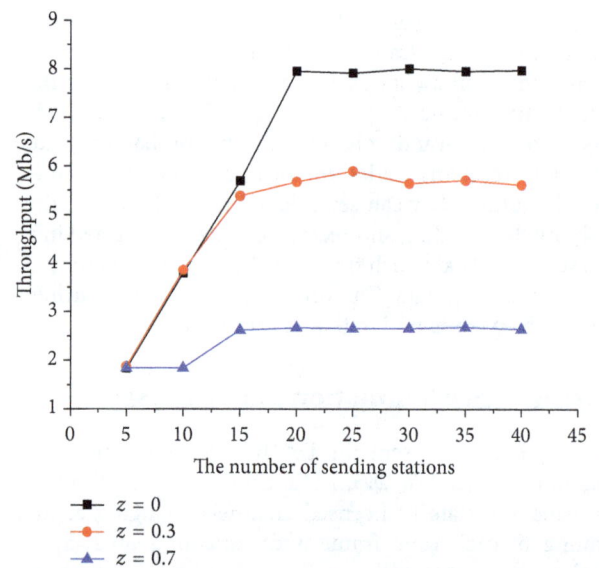

FIGURE 7: The relationship between throughput and number of licensed channels.

transmission time being reduced and the saturation throughput decreasing. When the primary users' activity rate gets higher, fewer idle channels become available. Idle channels are therefore assigned soon when the channel allocation time is small. For example, the throughput reaches the maximum when the channel allocation phase length is short.

4.3. The Relationship between Throughput and Number of the Licensed Channels.
The relationship between the number of licensed channels and throughput in the DAHCWNS-MAC protocol is shown in Figure 7. The number of sending stations is 20 and the channel allocation phase length is 15 ms. When the number of licensed channels increases, the saturation throughput increases under different primary users' activity rate, which means the proposed protocol can fully utilize licensed channels to communicate without interrupting primary users.

4.4. The Relationship between Throughput and Number of Sending Stations.
Figure 8 shows the impact of the number of sending stations on the DAHCWNS-MAC protocol's throughput. The number of licensed channels is 20 and the channel allocation phase length is 15 ms. As the number of sending stations increases, the throughput will increase and then remain stable when the primary users' activity rate is 0. This is because the number of node pairs that can complete a handshake is larger than 20 when the number of licensed channels is 20. The saturation throughput will reach the maximum when all channels are assigned. After this, because there are no more available channels, an increase in the number of sending stations will not result in an increase in the throughput. So the throughput remains stable.

4.5. Comparison to the C-MMAC Protocol.
Reference [19] proposes the C-MMAC (cognitive-multichannel MAC) protocol. This protocol is also a synchronous MAC protocol for

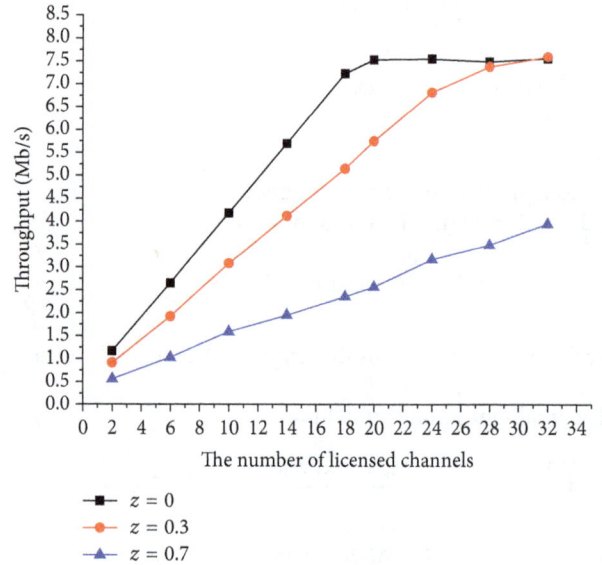

FIGURE 8: The relationship between throughput and number of sending stations.

cognitive radio network. The C-MMAC has some similarities with the DAHCWNS-MAC protocol: (1) The time frame of the C-MMAC protocol is divided into several parts (also include sensing phase, channel allocation phase, and data transmission phase) too and (2) all the nodes are strictly synchronous. The idea of designing DAHCWNS-MAC protocol is inspired by the C-MMAC protocol. However, the biggest difference between the two protocols is the network architecture. The C-MMAC protocol sets a single-hop network. The distance between two arbitrary nodes is within single-hop transmission range, and nodes can communicate directly. The network is self-organized, without infrastructure. Nodes have to sense the spectrum, exchange results, and negotiate channels by themselves. However, the protocol proposed in this paper is a hybrid network design. So this paper put these two protocols togethcr to compare.

The two protocols are compared next through simulation. Since C-MMAC protocol is designed for the single-hop scenario, the scheme used in C-MMAC protocol is assumed to be a one-cell DAHCWNS-MAC protocol network. We compare the performance under the single-hop setting first (in one cell). The performance under a multiple cells network will be compared later.

4.5.1. Comparison in Single Cell.
Nodes are placed randomly in one cell and communicate directly.

(1) 10 pairs of nodes are randomly placed in a cell, forming 10 communication streams. The number of licensed channels is 4. As shown in Figure 9 the throughput varies as the channel allocation phase length changes. When the primary user activity rate is 0 the C-MMAC protocol throughput is larger than that for the DAHCWNS-MAC at the beginning. This is because the C-MMAC protocol requires a three-way handshake (ATIM frame, ATIM-ACK frame, and ATIM-RES frame) to complete channel allocation while the DAHCWNS-MAC protocol requires a four-way handshake (SEARCH frame, ANSWER frame, SETUP frame, and ACCEPT frame). Under the same channel allocation window the C-MMAC protocol allows more nodes to complete a handshake to get available channels, which results in higher throughput. When the channel allocation window gets larger it allows most nodes to complete a handshake. The C-MMAC protocol also adopts a competition scheme during the data transmission phase, which results in a longer time to send data packets than the DAHCWNS-MAC protocol. The C-MMAC throughput is therefore smaller than that of the DAHCWNS-MAC protocol.

(2) 10 pairs of nodes are randomly placed in a cell, forming 10 communication streams. The channel allocation window length is fixed at 30 ms which is sufficient for 10 pairs of nodes to complete a handshake to get available channels. As shown in Figure 10 the throughput varies as the number of licensed channels changes. When the primary user activity rate is 0 all licensed channels are idle and the control channel can also be used to transmit data. The throughput of the DAHCWNS-MAC protocol reaches the maximum when the number of licensed channels is 9 and then remains stable. However, the nodes must sense the spectrum by themselves in the C-MMAC protocol. Not all channels are detected in order to save power. Each node randomly selects a channel to sense and exchange its results with other nodes to obtain the entire sensing results. Whether all channels can be sensed depends on the number of nodes. According to [19], the probability that all channels can be detected is 0.95 when there are 10 licensed channels and 50 nodes. When the number of licensed channels is equal to the number of nodes the probability falls to 0.036. This is why the C-MMAC throughput does not reach the maximum when the number of licensed channels increases to 9 (not all 9 channels have been detected). The maximal throughput does not come up until the number increases to 16 when the nodes can detect

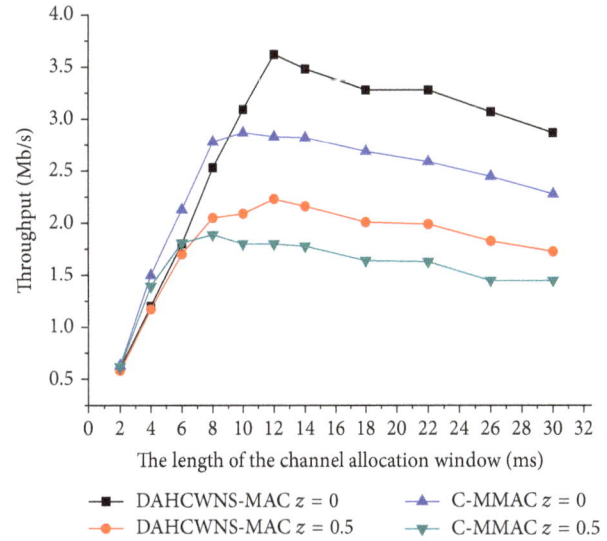

FIGURE 9: The comparison between the DAHCWNS-MAC protocol and the C-MMAC protocol as the channel allocation window length varies.

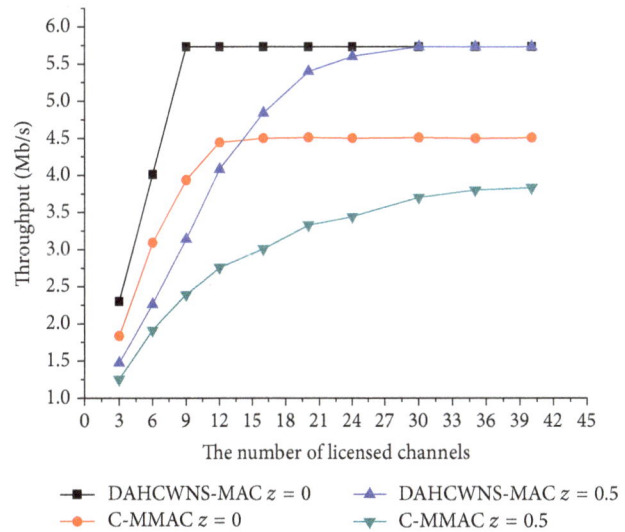

FIGURE 10: The comparison between the DAHCWNS-MAC protocol and the C-MMAC protocol as the number of licensed channels varies.

10 idle channels. The DAHCWNS-MAC protocol is free from this problem because distributed antennas undertake the spectrum sensing job and are able to detect all channels. The curve has the same trend when the primary user activity rate is 0.5.

4.5.2. Comparison in Multiple Cells.
The DAHCWNS-MAC protocol scheme is changed to seven cells while the C-MMAC protocol scheme remains the same.

(3) The number of nodes in the DAHCWNS-MAC protocol multiple cells scheme is more than seven times that of the C-MMAC single cell protocol. Assume that 35 pairs of nodes exist in the former scene and 5 pairs in the latter

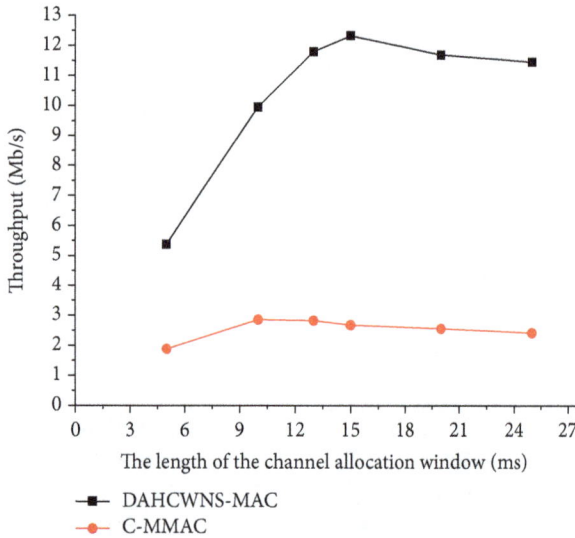

FIGURE 11: Comparison in multiple cells with different numbers of nodes.

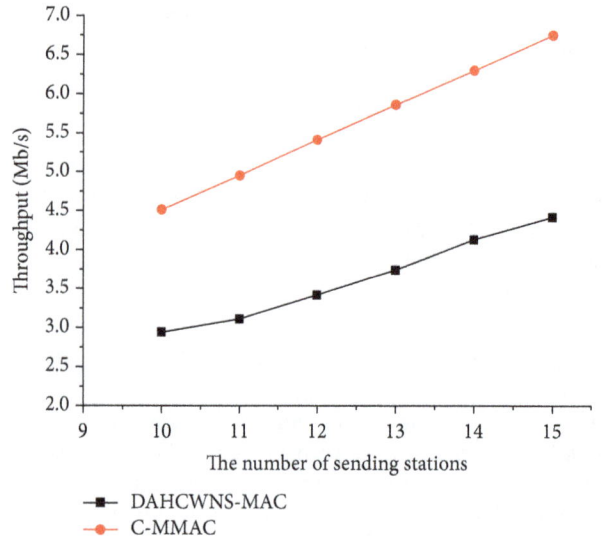

FIGURE 13: Comparison of multiple cells with the same number of nodes as the number of sending stations varies.

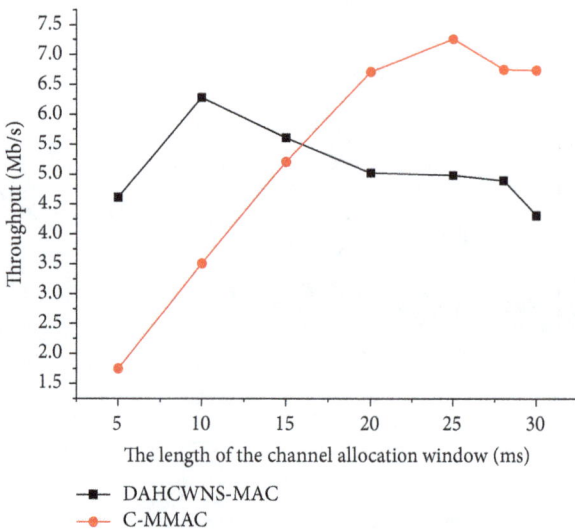

FIGURE 12: Comparison in multiple cells with the same number of nodes.

scheme. Ensure that nodes can be assigned channels as long as they complete a handshake. As shown in Figure 11, the DAHCWNS-MAC protocol throughput is not exactly seven times as large as that of the C-MMAC protocol. This is because the case nodes require distributed antennas to forward data which takes double the time of direct transmission.

(4) Assume 15 pairs of nodes are randomly placed in the two protocols, which means the number of nodes is the same in both schemes. Ensure that nodes can be assigned channels as long as they complete a handshake. As shown in Figure 12, the DAHCWNS-MAC protocol throughput is larger than that of the C-MMAC protocol when the channel allocation window length is relatively small. This is because operations in seven cells can be conducted in parallel, which leads to more nodes being able to complete a handshake. Since seven

cells can work independently, 15 pairs of nodes can complete a handshake within 10 ms and the throughput reaches peak at this moment. After this moment the increase in channel allocation window length only leads to a decrease in the DAHCWNS-MAC protocol throughput. As the window becomes larger the number of nodes that finish a handshake becomes larger in C-MMAC. Since nodes can communicate directly, not like the DAHCWNS-MAC where antennas are required to forward data, the throughput becomes gradually higher than that of the DAHCWNS-MAC and reaches the maximum at 25 ms.

(5) Assume the number of nodes in the two schemes is the same. The channel allocation window is fixed at 30 ms. The number of licensed channels is sufficient. As shown in Figure 13, because the case exists where nodes need antennas to forward data in DAHCWNS-MAC, the throughput is absolutely less than that of the C-MMAC protocol where nodes are put together in a single cell.

5. Conclusion

The paper introduced distributed antennas into the cognitive wireless network. The cognitive wireless network is designed to be a heterogeneous network consisting of an ad hoc network with a sparse network with infrastructure. A distributed antenna based synchronous MAC protocol (DAHCWNS-MAC protocol) is also presented for the proposed network. This protocol utilizes distributed antennas to sense the spectrum and transmit data, which can improve the sensing performance and increase network throughput. Every part of the protocol was described in detail and a mathematical model and performance simulation were presented. The proposed protocol combines the advantages of the ad hoc network and the network with infrastructure to fully utilize idle licensed channels to increase throughput. We compared the proposed protocol with the C-MMAC protocol to demonstrate that the

DAHCWNS-MAC protocol can broaden the communication range and increase the network throughput compared with the original single-hop network at the cost of increasing antenna hardware costs.

The introduction of distributed antennas into spectrum sensing can fully utilize the spatial resources at the expense of increasing the antenna hardware costs. It can also overcome hidden/exposed terminal problems to a certain extent and improve sensing performance.

In addition to the spectrum sensing, we can further let the distributed antennas be used to locate the primary user and adopt different access methods according to positioning information. For example, the hybrid underlay/overlay access scheme can be adopted through power control, which can greatly improve the channel utilization ratio and increase network throughput.

Furthermore, in the DAHCCWNS-MAC protocol the center allocates the same channel to communication pair which is not flexible enough. In future work we can take the channel state information into consideration and allocate different channels to sending node and receiving node, respectively, according to in-time channel state to further improve network performance.

This paper focused on how the distributed antenna can be used in data transmission and did not explore its usage in spectrum sensing and positioning. In future studies we can continue to study on these aspects to complete the DAHCWNS-MAC protocol.

Conflict of Interests

The authors declare that there is no conflict of interests regarding the publication of this paper.

Acknowledgments

The work presented in this paper was partially supported by 2011 National Natural Science Foundation of China (Grant no. 61172097), 2014 National Natural Science Foundation of China (Grant no. 61371081), and 2012 Natural Science Foundation of Fujian (Grant no. 2012J01424).

References

[1] P. Yue, X. Yi, and Z.-J. Liu, "A novel wireless network architecture and its radio frequency assignment mechanism for WLAN based on distributed antenna system using radio over free space optics," in *Proceedings of the International Conference on Information Science and Technology (ICIST '11)*, pp. 488–492, March 2011.

[2] Z. Xu, C. Zhou, and J. Wang, "A novel cell architecture based on distributed antennas for mobile WiMAX systems," in *Proceedings of the 4th IEEE International Conference on Circuits and Systems for Communications (ICCSC '08)*, pp. 172–176, May 2008.

[3] D. J. Thomson, "Spectrum estimation and harmonic analysis," *Proceedings of the IEEE*, vol. 70, no. 9, pp. 1055–1096, 1982.

[4] S. Haykin, D. J. Thomson, and J. H. Reed, "Spectrum sensing for cognitive radio," *Proceedings of the IEEE*, vol. 97, no. 5, pp. 849–877, 2009.

[5] Y. Zeng, C. L. Koh, and Y.-C. Liang, "Maximum eigenvalue detection: theory and application," in *Proceedings of the IEEE International Conference on Communications (ICC '08)*, pp. 4160–4164, Beijing, China, May 2008.

[6] Y. Zeng and Y.-C. Liang, "Eigenvalue-based spectrum sensing algorithms for cognitive radio," *IEEE Transactions on Communications*, vol. 57, no. 6, pp. 1784–1793, 2009.

[7] A. Taherpour, M. Nasiri-Kenari, and S. Gazor, "Multiple antenna spectrum sensing in cognitive radios," *IEEE Transactions on Wireless Communications*, vol. 9, no. 2, pp. 814–823, 2010.

[8] R. Zhang, T. J. Lim, Y. C. Liang, and Y. Zeng, "Multi-antenna based spectrum sensing for cognitive radios: a GLRT approach," *IEEE Transactions on Communications*, vol. 58, no. 1, pp. 84–88, 2010.

[9] H. Yao, *Research on the Spectrum Sensing and Resource Allocation in Cognitive Radio Networks*, Beijing University of Posts and Telecommunications, 2011.

[10] C. Zhao, L. Huang, Z.-L. Gao, S. Zhou, D. Guo, and H.-C. Chao, "performance analysis of the multiple antenna asynchronous cognitive MAC protocol in cognitive radio network for IT convergence," *Intelligent Automation and Soft Computing*, vol. 20, no. 1, pp. 61–75, 2014.

[11] E. Y. Kang, H. Park, and J. Chae, "A hybrid message delivery scheme for improving service discovery in mobile ad-hoc networks," *Journal of Internet Technology*, vol. 13, no. 6, pp. 879–890, 2012.

[12] Y.-X. Lai, C.-F. Lai, Y.-M. Huang, and H.-C. Chao, "Multi-appliance recognition system with hybrid SVM/GMM classifier in ubiquitous smart home," *Information Sciences*, vol. 230, pp. 39–55, 2013.

[13] L. Zhou, H.-C. Chao, and A. V. Vasilakos, "Joint forensics-scheduling strategy for delay-sensitive multimedia applications over heterogeneous networks," *IEEE Journal on Selected Areas in Communications*, vol. 29, no. 7, pp. 1358–1367, 2011.

[14] C.-W. Chiang, "Two novel genetic operators for task matching and scheduling in heterogeneous computing environments," *Journal of Internet Technology*, vol. 13, no. 5, pp. 773–784, 2012.

[15] B. Liu, Z. Liu, and D. Towsley, "On the capacity of hybrid wireless networks," in *Proceedings of the 22nd Annual Joint Conference on the IEEE Computer and Communications Societies*, vol. 2, pp. 1543–1552, April 2003.

[16] R. S. Chang, W. Y. Chen, and Y. F. Wen, "Hybrid wireless network protocols," *IEEE Transactions on Vehicular Technology*, vol. 52, no. 4, pp. 1099–1109, 2003.

[17] G. Bianchi, "Performance analysis of the IEEE 802.11 distributed coordination function," *IEEE Journal on Selected Areas in Communications*, vol. 18, no. 3, pp. 535–547, 2000.

[18] N. Liu, *Wireless Local Area Networks (WLAN)—Principle, Technique and Application*, Xidian University Press, Xi'an, China, 2007.

[19] M. Yu, The Study of Cognitive Multi-Channel MAC Protocol Based on Spectrum Sensing, 2009.

17

A Distributed TDMA Slot Scheduling Algorithm for Spatially Correlated Contention in WSNs

Ashutosh Bhatia and R. C. Hansdah

Department of Computer Science and Automation, Indian Institute of Science, Bangalore 560012, India

Correspondence should be addressed to Ashutosh Bhatia; ashutosh.b@csa.iisc.ernet.in

Academic Editor: David Taniar

In WSNs the communication traffic is often time and space correlated, where multiple nodes in a proximity start transmitting simultaneously. Such a situation is known as *spatially correlated contention*. The random access method to resolve such contention suffers from high collision rate, whereas the traditional distributed TDMA scheduling techniques primarily try to improve the network capacity by reducing the schedule length. Usually, the situation of *spatially correlated contention* persists only for a short duration, and therefore generating an optimal or suboptimal schedule is not very useful. Additionally, if an algorithm takes very long time to schedule, it will not only introduce additional delay in the data transfer but also consume more energy. In this paper, we present a distributed TDMA slot scheduling (DTSS) algorithm, which considerably reduces the time required to perform scheduling, while restricting the schedule length to the maximum degree of interference graph. The DTSS algorithm supports unicast, multicast, and broadcast scheduling, simultaneously without any modification in the protocol. We have analyzed the protocol for average case performance and also simulated it using Castalia simulator to evaluate its runtime performance. Both analytical and simulation results show that our protocol is able to considerably reduce the time required for scheduling.

1. Introduction

A wireless sensor network (WSN) is a collection of sensor nodes distributed over a geographical region to monitor events of interest in the region. To effectively exchange data among multiple sensor nodes, WSNs employ the medium access control (MAC) protocol to coordinate the transmission over the shared wireless radio channel. Many times in WSNs, communication traffic is space and time correlated; that is, all the nodes in the same proximity transmit at the same time. Such a situation is known as *spatially correlated contention*. There exist many applications and protocols in WSNs, where the situations of *spatially correlated contention* can occur. Some of them are as follows.

(i) *Event Detection*: whenever an event occurs, all the nodes that sense the event will start transmitting the details of the event to the base station. Typical examples of such situations are the detection of earthquake and wildfire in WSNs for disaster recovery, fall-and-posture detection in healthcare WSNs [1], and intrusion detection [2] in WSNs for military

applications, in which sensor nodes only have data to send when a specific event occurs. As multiple nodes that detect the event are quite possibly in close proximity of each other, they would share the same transmission medium. Eventually, if all the nodes report the event at the same time, the situation would lead to *spatially correlated contention*.

(ii) *Multicast Communication*: in WSNs, the applications should be configured and updated in the sensor nodes multiple times during the lifetime of the network. An update by transmitting the content to each individual sensor node separately would be very inefficient and would consume a lot of resources such as bandwidth and energy. In this situation, multicasting provides an efficient configuration and update of applications running over sensor nodes by reducing the number of transmitted packets. Another example of multicast communication in WSN is on-demand data collection, where the base station (sink node) sends a data query to a prespecified group of nodes asking them to send their sensory data. The WSN is usually multihop in nature, and therefore direct transmission of multicast messages from the sink is not possible. To achieve this, the sensor nodes also work as routers and forward the received multicast packet to their one-hop neighbors. This simultaneous forwarding of the same packet in a proximity by multiple routers leads to *spatially correlated contention*. A detailed discussion on multicast in WSN can be found in [3].

(iii) *Routing Protocols*: the on-demand routing protocols, for example, Ad Hoc On-Demand Distance Vector Routing (AODV) [4], try to find the appropriate path from source to destination only when the data transfer is required. This process is called route discovery and it is typically achieved by broadcasting a route request message in the network and consequently leads to the collision of request message due to its simultaneous forwarding by the neighboring nodes.

(iv) *Clock Synchronization*: clock synchronization protocols, for example [5], typically use message passing mechanism to share their local time information with other neighboring nodes. Since, initially, there is no coordination between the nodes, they may transmit the protocol message simultaneously with high probability, and therefore the transmitted messages might collide. This can considerably delay the process of synchronization.

(v) *Tree Based Convergecasting*: convergecast, that is, gathering of information towards a central node, is important communication paradigms across all application domains in WSN. This is mainly accomplished by constructing an efficient tree in terms of delay, energy, and bandwidth. The algorithm for construction of such a tree typically involves simultaneous transmission of protocol messages by the sensor nodes and hence causes *specially correlated contention*.

Thus the above discussion suggests that the MAC protocols for WSNs should effectively handle the correlated contention. MAC protocols for WSNs can be mainly classified into two major categories, namely, random access based and schedule access based. Random access methods do not use any topology or clock information and resolve contention among neighboring nodes for every data transmission. Thus, it is highly robust to any change in the network. But its performance under high contention suffers because of high overhead in resolving contention and collisions [6]. Contention causes message collisions, which are very likely to occur when traffic is *spatially correlated*. This, in turn, degrades the data transmission reliability and wastes the energy of sensor nodes.

A MAC protocol is contention-free if it does not allow any collisions. Assuming that the clocks of sensor nodes are synchronized, data transmissions by the nodes are scheduled in such a way that no two interfering nodes transmit at the same time. Early works [7–9] on scheduling are centralized in nature and normally need complete topology

information, and therefore, they are not scalable. To overcome the difficulty of obtaining global topology information in large size networks, many distributed slot assignment schemes [10–14] have been proposed. The primary objective of traditional distributed TDMA scheduling techniques is to improve the network capacity by reducing the schedule length. It is effective for the kind of applications where a fixed schedule can be used for a sufficiently longer time. All the scenarios for *correlated contention* discussed previously occur in form of sessions and the nodes require a time slot to transmit a sequence of data, only during these sessions. Moreover, the same schedule cannot be reused for multiple future sessions, because at that time the network topology might have changed, due to dynamic channel conditions and occasional sleeping of sensor nodes to conserve their energy. For example, in [15], a different set of nodes are selected as routers (to equally distribute the consumption of energy among sensor nodes) every time the algorithm for construction of data collection tree is executed, and therefore this changes the network topology. This suggests that the scheduling has to be performed for every instance of *correlated contention*, and therefore the effective benefit of reducing schedule length vanishes. If an algorithm takes too long to perform scheduling, as compared to the duration of *correlated contention*, it will not only degrade the QoS (e.g., delay in detection of event at the base station) but will also lead to poor bandwidth utilization and higher energy consumption. The preceding discussion emphasizes that in order to effectively handle the *correlated contention*, the TDMA scheduling algorithms should take very less time to perform scheduling.

In this paper, we propose a distributed TDMA slot scheduling (DTSS) algorithm for WSNs. The primary objective of DTSS algorithm is to reduce the time required to perform scheduling while restricting the schedule length to maximum degree of interference graph. The proposed algorithm is unified in the sense that the same algorithm can be used to schedule slots for different modes of communication, namely, unicast, multicast, and broadcast. In addition, the DTSS algorithm also supports heterogeneous mode of communication, where simultaneously a few nodes can take a slot for unicast, while other nodes can take it for multicast or broadcast purpose. In DTSS algorithm, a node is required to know only the IDs of its intended receivers, instead of all its two-hop neighbors. Also, in DTSS algorithm, the nodes in a neighborhood can take different slots simultaneously, if the resultant schedule is feasible. This is unlike the class of greedy algorithms where ordering between the nodes puts a constraint on the distributed algorithm to run sequentially and restricts the parallel implementation of the algorithm. The DTSS algorithm does not make use of any ordering among the nodes.

The rest of the paper is organized as follows. Section 2 discusses the related work. Section 3 gives the assumptions we make in the design of our algorithm, introduces some definitions, and explains the basic idea of our algorithm. In Section 4, we present a detailed description of the DTSS algorithm. Section 5 gives the proof of correctness of the DTSS algorithm. Section 6 presents the average case complexity analysis of the DTSS algorithm. Section 7 discusses the simulation results and performance comparison of DTSS algorithm with existing work. Section 8 concludes the paper with suggestions for future work.

2. Related Work

The broadcast scheduling problem to find optimal solution is NP-complete [16]. A different, but related, problem to TDMA node slot assignment is the problem of TDMA edge slot assignment, in which radio links (or edges) are assigned time slots, instead of nodes. Finding the minimum number of time slots for a conflict-free edge slot assignment is also an NP-complete problem [17]. In [18], another specific scheduling problem for wireless sensor network converge-cast transmission is considered in which the scheduling problem is to find a minimum length frame during which all nodes can send their packets to access point (AP), and it is shown to be NP-complete. Previous work [7–9, 19] on scheduling algorithms primarily focuses on decreasing the length of schedules. They are centralized in nature and normally need complete topology information and are, therefore, not scalable.

Cluster based TDMA protocols in [20, 21] prove to be having good scalability. The common feature of these protocols is to partition the network into some clusters, in which cluster heads are responsible for scheduling their members. However, cluster based TDMA protocols introduce intercluster transmission interference because clusters created by distributed clustering algorithms are often overlapped and several cluster heads may cover the same nodes.

Moscibroda and Wattenhofer [11] have proposed a distributed graph coloring scheme with a time complexity of $O(\rho \log n)$, where ρ is the maximum node degree and n is the number of nodes, in the network. The scheme performs distance-1 coloring such that adjacent nodes have different colors. Note that this does not prevent nodes within two hops of each other from being assigned the same color potentially causing hidden terminal collisions between such nodes. The NAMA protocol in [10] has proposed a distributed scheduling scheme based on hash function to determine the priority among contending neighbors. A major limitation of this hashing based technique is that even though a node gets a higher priority in one neighborhood, it may still have a lower priority in other neighborhoods. Thus the maximum slot number could be of $O(n)$. Secondly, since each node calculates the priority of all of its two-hop neighbors for every slot, it leads to $O(n^2)$ computational complexity, and hence, the scheme is not scalable for large network with resource constraint nodes. SEEDEX [22] uses a similar hashing scheme as NAMA based on a random seed exchanged in a two-hop neighborhood. In SEEDEX, at the beginning of each slot, if a node has a packet ready for transmission, it draws a "lottery" with probability p. If it wins, it becomes eligible to transmit. A node knows the seeds of the random number generators of its two-hop neighbors, and hence it also knows the number of nodes (including itself) n, within two hops which are also eligible to transmit. It then transmits with probability $1/n$. This technique is also called topology independent

scheduling. In this case, collisions may still occur if two nodes select the same slot and decide to transmit.

Another distributed TDMA scheduling scheme, called DRAND [12], proposes a distributed randomized time slot scheduling algorithm based on centralized scheduling scheme RAND [9]. DRAND is also used within a MAC protocol, called Zebra-MAC [23], to improve performance in sensor networks by combining the strength of scheduled access during high loads and random access during low loads. The runtime complexity of DRAND is $O(\delta)$, where δ is the maximum size of a two-hop neighborhood in a wireless network. The simulation results presented by the author show that the runtime actually becomes $O(\delta^2)$ due to unbounded message delays. FPRP [14], Five-Phase Reservation Protocol, is a distributed heuristic TDMA slot assignment algorithm. FPRP is designed for dynamic slot assignment, in which the real time is divided into a series of pairs of reservation and data transmission phases. For each time slot of the data transmission phase, FPRP runs a five-phase protocol for a number of times (cycles) to pick a winner of each slot. In another distributed slot scheduling algorithm, DD-TDMA [13], a node i decides slot j as its own slot if all the nodes with ID less than the ID of node i have already decided their slot, where j is the minimum available slot. The scheduled node i broadcasts its slot assignment to one-hop neighbors. Then the one-hop neighbors of node i broadcast this information to update two-hop neighbors. This process is repeated in every frame until all nodes are scheduled.

The protocol in [24] proposes a contention-free MAC for correlated contention, which does not assume global time reference. The protocol is based on local frame approach where each node divides time into equal sized frames. Each frame is further divided into equal sized time slots; a time slot corresponds to the time duration of sending one message. The basic idea is that each node selects a slot in its own frame such that selected slots of any two-hop neighbor nodes must not overlap. The protocol assumes that a node can detect a collision if two or more nodes (including itself) within its transmission radius attempt to transmit, which has its own practical limitations with wireless transceivers. These scheduling algorithms [12–14] commonly have the following issues.

(i) All algorithms use greedy approach for graph colouring which is inherently sequential in nature and put a constraint on distributed algorithm to run sequentially. This restricts the parallel implementation of the algorithm. Because of large runtime of these protocols, they are more suitable for wireless networks where interference relationship or network topology does not change for a long period of time.

(ii) They perform two-hop neighbor discovery, which adds considerable additional cost to runtime to perform scheduling. Additionally, the two-hop neighbors are calculated based on transmission range instead of interference range, which is normally higher than the transmission range.

(iii) They perform either broadcast (node) scheduling or unicast (link) scheduling but not both and also do not consider multicast scheduling separately.

Finally, a classification of different scheduling algorithms based on problem setting, problem goal, type of inputs, and solution techniques can be found in [25].

3. Our Approach to TDMA Scheduling in WSNs

In many applications such as weather monitoring, intrusion detection, sensor nodes are usually static. In this work also, we assume them to be static. Also, it is assumed that, for any task in an application, every node knows its receivers. Before a task begins its execution, the DTSS algorithm is executed to generate a TDMA schedule. After the task is finished, the TDMA schedule is discarded. We assume that each node in the WSN has a unique identifier. All the nodes in a WSN have some processing capability along with a radio to enable communication among them. Each node uses the same radio frequency. The communication capability is bidirectional and symmetric. The mode of communication between any two neighboring nodes is half-duplex; that is, only one node at a time can transmit. The transmission is omnidirectional; that is, each message sent by a node i is inherently received at all the nodes determined by its transmission range.

Timeline is divided into fixed size frames and each frame is further divided into fixed number of time slots, called schedule length. The nodes are assumed to be synchronized with respect to slot 0 and are aware of the slot size and the schedule length. The time of slot 0 is defined by the node which starts the scheduling process. To better understand the proposed algorithm, we introduce the following definitions.

Definition 1. The interference set N_i of a node i is defined as the set of nodes which are within the interference range of node i. That is, we say that a node $k \in N_i$ if it cannot successfully receive any message transmitted by any other node, at the same time when node i is also transmitting. Moreover, if only node i has transmitted in a slot, then a node in N_i may or may not receive the message successfully.

Note that the interference set N_i is different from the set of one-hop neighbors which depends upon the transmission range of node i. Usually, the interference range is higher than the transmission range.

Definition 2. The receiver set R_i of a node i is defined as the set of intended receivers of node i.

The size of the set R_i, $|R_i|$ depends upon the type of communication, namely, unicast, multicast, or broadcast transmission. Note that $R_i \subseteq N_i$. The DTSS algorithm assumes that only the subset R_i is known to the node i instead of all its two-hop neighbors.

Definition 3. The sender set S_i of a node i is defined as the set of nodes j such that $i \in R_j$.

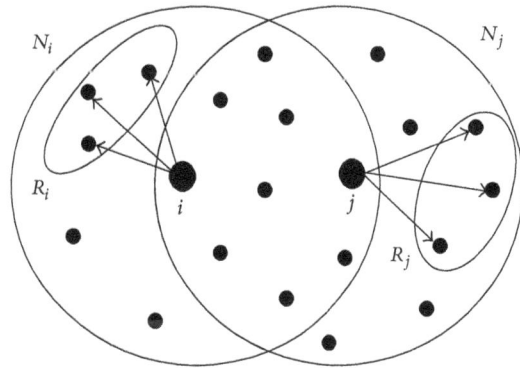

FIGURE 1: Example of nodes i and j that do not conflict even if they are in the interference range of each other.

A node need not know the set S_i before the start of the algorithm. It can be populated when the node receives protocol messages with destination ID as its own ID.

Definition 4. The interference graph $G = (V, E)$ of a WSN is defined as follows. V is the set of nodes in the WSN, and E is the set of edges, where edge $e = (i, j)$ exists if and only if $N_i \cap R_j \neq \phi \vee N_j \cap R_i \neq \phi$. The number of edges with which a node is connected to the other nodes is called the degree of the node.

Note that $i \in R_j$ or $j \in R_i$ is also possible. We say that node i and node j conflict and are adjacent to each other, if there exists an edge between them. An edge $e = (i, j)$ exists if and only if node i and node j cannot take the same slot. Two nodes cannot take the same slot, if the transmission of one node interferes at one of the receivers of the other node. The conflict between nodes depends not only upon their respective positions and transmission power but also on the type of communication, namely, unicast, multicast, or broadcast. Two nodes within the interference range of each other ($i \in N_j \vee j \in N_i$) can even take the same slot, if their transmissions do not interfere at each other's receivers (Figure 1). Therefore, our definition of interference graph is free from well known exposed-node problem. This fact is usually ignored by most of the existing algorithms. On the other hand, two nodes which are not in the interference range of each other ($i \notin N_j \wedge j \notin N_i$) cannot transmit simultaneously, if their transmissions interfere at each other's receivers. In this manner, our definition of interference graph is also free from the hidden-node problem.

A sensor node requires a slot to transmit data packets such that data can be received successfully at all of its receivers without any interference.

The following two types of conflict relations can exist, between a pair of nodes.

Strong-Conflict Relation. Two nodes i and j have strong-conflict relation if $N_i \cap R_j \neq \phi \wedge N_j \cap R_i \neq \phi$.

Weak-Conflict Relation. Two nodes i and j have weak-conflict relation if either $N_i \cap R_j \neq \phi$ or $N_j \cap R_i \neq \phi$, but not both.

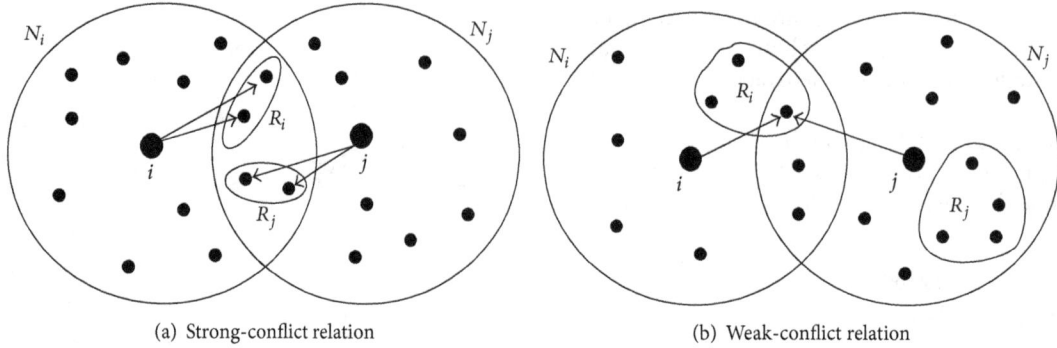

(a) Strong-conflict relation   (b) Weak-conflict relation

FIGURE 2: Conflict relationship between nodes.

Figures 2(a) and 2(b) depict the situation when node i and node j have strong-conflict and weak-conflict relations, respectively. In case of weak-conflict relation, if $N_j \cap R_i \neq \phi$ but $N_i \cap R_j = \phi$, we say that node j is stronger than node i and denote it by $j \rightarrow i$.

Definition 5. The interference degree Δ of a WSN is defined as the maximum of all degrees of nodes in the interference graph G of the WSN.

The DTSS algorithm runs in $O(\Delta)$ time. In case of broadcast transmission; Δ and δ are roughly the same, where δ is the size of two-hop neighborhood set. But for unicast or multicast transmission $\Delta \leq \delta$.

The TDMA slot scheduling problem can be formally defined as the problem of assignment of a time slot to each node, such that if any two nodes are in conflict (strong or weak), they do not take the same time slot. Such an assignment is called a feasible TDMA schedule. A feasible TDMA schedule which takes minimum number of slots is called an optimal TDMA schedule. Our goal in this paper is to develop an algorithm which can find a feasible but not necessarily optimal TDMA schedule and to minimize the time required to perform scheduling.

The basic idea of the DTSS algorithm is as follows. For each slot s in a frame, each node i checks whether it can take the slot s, by sending a request message to the first receiver in R_i with slot probability $P(s)$, which depends upon the remaining number of free slots in the frame (slots which are not taken by others) known at node i at the current time. If node i receives response message from the first receiver, then it blocks the slot and tries to get the responses from the remaining receivers in R_i using the same slot in subsequent frames. After receiving the response from all the receivers, it assigns the time slot s to itself; otherwise, it unblocks the slot, as soon as the response from one of the receivers is not received, and repeats the above process all over again. In case of unicast communication, node i can directly assign the slot to itself as soon as it receives the response message from its receiver j instead of blocking the slot. This is because receiving a response message from node j tells that no other node k adjacent to i in G has also blocked the same slot; otherwise, the REQ message transmissions of nodes i and k

would have collided at node j. An adjacent node of a node i in a graph is a node that is connected to i by an edge.

Once a slot is assigned to a node i, it continuously transmits at the same slot in subsequent frames. This would ensure that a conflicting node j in G cannot assign the same slot to itself, because of the collision between the transmission of node i and node j at one of the receivers in R_i. Furthermore, the nodes in R_i also propagate this information to next hop through their own transmissions. Note that the receivers of messages transmitted by the nodes in R_i are adjacent to i in G.

When a node j hears, from one of the receivers of node i, that slot s is blocked by node i, it leaves the slot temporarily and avoids further collisions to increase the chance of getting the slot by node i. Similarly, when node j hears, from one of the receivers of node i, that slot s is assigned to node i, it leaves the slot permanently and increases its slot probability for other free slots.

4. The DTSS Algorithm

In this section, we describe the proposed DTSS algorithm for TDMA slot scheduling problem as defined in Section 2. The number of slots, $\mathcal{N}$, in a frame is taken to be at least Δ. The slots are numbered from 1 to $\mathcal{N}$. Table 1 summarizes the set of data structures and variables maintained by a node i, to implement the algorithm. The DTSS algorithm uses two protocol messages, namely, request (REQ) and response (RES), for signaling purpose. The RES message is sent by a node, whenever its ID is the same as the destination ID in the received REQ message. The REQ/RES messages contain four fields, namely, source ID, destination ID, $L2$, and state. The value of field $L2$ in both REQ and RES messages is the copy of corresponding local variable. The value of field state in REQ message is the same as the value of the local variable nr while its value in RES message contains the value of field state as received in the corresponding REQ message. The field $L2$ in REQ/RES message is used to inform a node j about the slots which are already taken by other nodes conflicting with node j, whereas the field state helps the nodes to know the status of the node from where the REQ message has been transmitted. The higher level description of the DTSS algorithm is shown in the pseudocode given in Algorithm 1. The pseudocode

```
if i.slot = i.b_slot = null and s ∉ (L3 or L1) then
    With probability, P(s) do
        send REQ(i, rx_ID, L2, |Ri|)
        rx_ID = Ri → next
    End do
end if
if i.slot = s or i.b_slot = s then
    send REQ(i, rx_ID, L2, nr)
    rx_ID = Ri → next
end if
//perform channel listening
if i receives a REQ(j, dest_ID, L2, state) then
    if REQ.dest_ID = i then
        add REQ.j in Si,
        send RES(i, j, L2, REQ.state)
    end if
    if j ∈ Ri and REQ.state = 0 then
        slot s has been taken by node j, add s to L1
        if j ∈ Si add s to L2
        end if
    end if
    if j ∈ Si or j ∈ Ri and REQ.state ≠ 0 then
        slot s is blocked, add s to L3
    end if
end if
if i receives a RES(j, dest_ID, L2, state)
    if RES.dest_ID = i then
        if nr = |Ri| then i.b_slot = s
        end if
        nr = nr − 1
        if nr = 0 then i.slot = s
        end if
    else
        if RES.state = 0 then
            slot is taken by RES.dest_ID, add s to L1
            if RES.dest_ID ∈ Si add s to L2
            end if
        else
            slot s is blocked, add s to L3
        end if
    end if
end if
if i.slot = null and not received the RES for transmitted REQ message then
    i.b_slot = null, nr = |Ri|
end if
if s ∈ L3
    Remove s from L3 if blocked duration has expired
end if
```

ALGORITHM 1: DTSS algorithm.

describes the DTSS algorithm as executed on each node i at the current slot s.

Each node i, contending for a time slot, passes through several states. Figure 3 shows the finite state transition diagram for a node i. Initially, node i enters *contention state* (CS), where it sends a REQ message in the current time slot s, with probability $P(s)$. We call this probability as slot probability and it is equal to $1/(\mathcal{N} - |L1|)$. On receipt of a REQ message at a node j from node i, it sends a RES message immediately in the current slot s and also adds the node i to S_j if its ID is the same as the destination ID in the received REQ message. The duration of slot is kept sufficiently large to carry out the transmission of a pair of REQ and RES messages. The destination ID j of REQ message transmitted by node i in CS state can be any node from the set R_i. If a node i receives a RES message at time slot s in response to the REQ message sent by it and $|R_i| > 1$, then it blocks the time slot s and enters the *verification state* (VS). However, if $|R_i| = 1$, it assigns

TABLE 1: The set of data structures and variables maintained by a node i.

Notation	Description
R_i	A list of receivers of node i, maintained as circular linked list.
S_i	A list of transmitters of node i, constructed dynamically on receipt of REQ messages.
L1	A list of slots which are already taken by the nodes adjacent to i in G.
L2	A list of slots which are already taken by the nodes in S_i. Note that $L2 \subseteq L1$.
L3	A list of slots which are currently blocked by the nodes adjacent to i in G.
$\mathcal{N}$	Number of slots in a frame.
i.slot	Slot assigned to node i.
i.b_slot	Slot blocked by node i.
rx_ID	Current receiver node in R_i.
nr	Remaining number of receivers from where responses are required to be obtained by node i.
$P(s)$	Probability of transmitting a request message in slot s by node i, while contending for the slot.

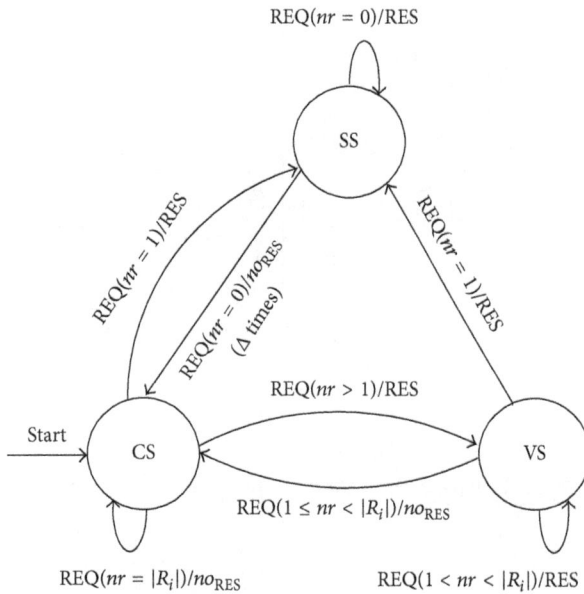

FIGURE 3: State transition diagram of a node i.

the slot to itself and enters the scheduled state (SS) directly. In VS state, it sends REQ messages one by one to the remaining nodes in R_i at the same time slot s in subsequent frames, by setting the pointer rx_ID to the next node in the list R_i. Furthermore, it does not transmit in slots other than s, while it is in VS state. If the node i successfully receives the RES messages from all of its receivers in R_i, it assigns the slot to itself and enters the *scheduled state* (SS); otherwise, it goes back to the CS state and starts the process all over again. In SS state, the node i always transmits REQ message in slot s so that no other node can take the same slot and also it does not transmit at slots other than s. The destination of REQ message

in SS state is selected in a round robin fashion among the nodes in R_i. Moreover, it does not progress to the next receiver until it receives a RES message from the current receiver. If a node i does not receive RES message consecutively Δ times from the same receiver node, then it goes back to the CS state.

If a node i in CS state receives a REQ message from node j at slot s with state > 0 and $j \in S_i$ or $j \in R_i$, then it adds the slot s in the list L3. If a node i in CS state receives a REQ message from node j at slot s with state $= 0$ and $j \in R_i$ it adds the slot s in the list L1 and also updates the slot probability of other slots not in L1 as $1/(\mathcal{N} - |L1|)$. Additionally, if $j \in S_i$, then it also adds the slot s in the list L2.

If a node i in CS state receives a RES message in response to REQ message from itself and $nr = |R_i|$, then it blocks the slot. Then it decreases nr by 1, and if nr becomes zero, it assigns the slot to itself. If a node i in CS state receives a RES message from node j in response to REQ message from node k at slot s with state > 0, then it adds the slot s in the list L3. A node does not transmit its own REQ messages in the slots belonging to L3 for the number of subsequent frames specified in the state field of the received RES message. This allows node k in VS state to successfully transmit its remaining REQ messages and subsequently move to SS state. If a node i in CS state receives a RES message from node j in response to REQ message from node k at slot s with state $= 0$, it adds the slot s in the list L1 and also updates the slot probability of other slots not in L1 as $1/(\mathcal{N} - |L1|)$. Node i permanently leaves the slots in L1 and does not transmit any further REQ messages in these slots. Additionally, if $k \in S_i$, then it adds the slot s in the list L2.

It could be possible that a node j does not receive the transmission of a REQ message from node i or RES message in response to the REQ message from node i in slot s because REQ/RES messages could get lost. In this situation, node j would not come to know that the slot s is either blocked or taken by node i until the transmissions of REQ/RES message at the same slot in the next frame. To avoid this delay, the nodes in R_i convey the same information through the field L2 of their own REQ messages transmitted in slots other than s. In this case, while a node is trying to take a slot, it is helping others to know the slots which are already taken by other conflicting nodes.

Finally, a node j, with $j \to i$, can enter into state SS, while node i is already in state SS. This is because the transmission of REQ messages from node i in slot s cannot interfere at any of the nodes in R_j, and therefore, node j will receive the RES messages from all of its receivers and move to state SS. On the other hand, the REQ messages sent by node i in SS state will collide at one or more receivers in R_i due to transmission from node j and therefore node i will not receive the corresponding RES messages. The above situation is shown in Figure 2(b). However, this can only happen if node j is not aware that the slot is already taken by node i. To avoid this, if the RES is not received by node i for consecutive Δ times, it leaves the slot by adding it to the list L1 and comes back to the CS state. If node j is not in SS state, it cannot collide with the transmission of node i in the same slot consecutively Δ times. This ensures that node i will only leave the slot s, if $j \to i$ and j is in SS state.

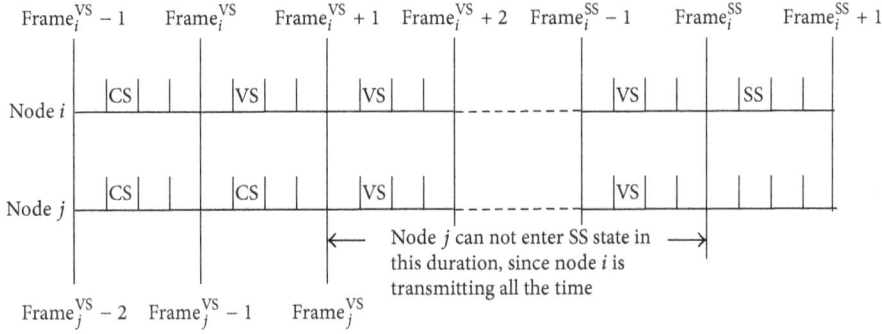

FIGURE 4: Node j cannot enter SS state since node i is continuously transmitting REQ messages from frame index $frame_i^{VS}$.

5. Correctness of the DTSS Algorithm

In this section, we prove that the schedule created by the DTSS algorithm is a feasible TDMA schedule. In a feasible schedule, two conflicting nodes will not transmit in the same time slot. That is, two conflicting nodes will not be assigned the same time slot by the DTSS algorithm. This happens because after the execution of the DTSS algorithm is completed, only one node (among the conflicting nodes) will remain in the SS state for a particular time slot. In the following, we prove this fact as Theorems 6 and 7 for strong-conflict and weak-conflict relationship, respectively.

Theorem 6. *If two nodes i and j have strong-conflict relationship, then they cannot be in SS state for the same time slot, at any time during the execution of DTSS algorithm.*

Proof. Let $frame_i^{VS}$ and $frame_i^{SS}$ be the frame indexes when node i enters VS and SS state, respectively, for a time slot k. It is possible that $frame_i^{VS} = frame_i^{SS}$ if $|R_i| = 1$. Furthermore, let $frame_j^{VS}$ and $frame_j^{SS}$ be the corresponding frame indexes for node j for the same time slot k. We can assume that once a node enters VS state from CS state, it remains in VS state until it enters SS state. If it is not so, then it goes back to CS state, and the argument can be repeated. It is to be noted that both cannot enter SS state at the same frame index without at least one of them going into VS state first. Now the following three cases arise.

Case 1 ($frame_i^{VS} < frame_j^{VS}$). In this case, only node i can enter SS state provided it has got response from all nodes $k \in N_j \cap R_i$ prior to $frame_j^{VS}$. There is no way node j can enter SS state since node i will be continuously transmitting REQ messages from frame index $frame_i^{VS}$, and node j cannot get response from any node in $N_i \cap R_j$ (Figure 4). Hence, in this case only node i will be able to enter SS state.

Case 2 ($frame_i^{VS} = frame_j^{VS}$). In this case, both node i and node j will be transmitting REQ messages from frame index $frame_i^{VS}$ onwards. Therefore, node i will not be able to get response from any node in $N_j \cap R_i$. Similarly, node j will not get response from any node in $N_i \cap R_j$. So, the node which does not receive the response first will go back to CS state. As

a result neither node i nor node j will be able to enter SS state as Case 2.

Case 3 ($frame_i^{VS} > frame_j^{VS}$). This case is similar to Case 1 except that now node j can enter SS state provided it satisfies the corresponding condition.

From the above argument, it is clear that only one of nodes i or j will be in SS state for the same time slot during the execution of the DTSS algorithm. Hence, the theorem is proved. □

In case of weak-conflict relationship, it could be possible that while a node i is already in SS state, another node j (stronger than node i) can enter SS state for the same time slot, during the execution of DTSS algorithm. Therefore, Theorem 6 does not sufficiently prove the correctness of DTSS algorithm when weak-conflict relationship also exists between the nodes.

Theorem 7. *If two nodes i and j have weak-conflict relationship, then eventually only one of nodes i and j will remain in SS state for the same time slot after the execution of the DTSS algorithm is completed.*

Proof. Let $frame_i^{VS}$, $frame_i^{SS}$, $frame_j^{VS}$, and $frame_j^{SS}$ be frames indexes of nodes i and j as in Theorem 6. Also, without loss of generality assume that $N_i \cap R_j \neq \phi$ and $N_j \cap R_i = \phi$. That is, node i is stronger than node j. As in Theorem 6, we also assume that, after entering VS state, both remain there until they enter SS state. It is to be noted that both cannot enter SS state at the same frame index directly from CS state, without at least one of them going into VS state first. In this case also, the following three cases arise.

Case 1 ($frame_i^{VS} < frame_j^{VS}$). This case is similar to Case 1 of Theorem 6, and only node i will be able to enter SS state, and node j will not be able to enter SS state.

Case 2 ($frame_i^{VS} = frame_j^{VS}$). In this case also, only node i will be able to enter SS state, and node j will not be able to enter SS state.

Case 3 ($frame_i^{VS} > frame_j^{VS}$). In this case, node i can enter SS state anyway. However node j can also enter SS state provided it has got response from all the nodes in $N_i \cap R_j$ before

TABLE 2: The set of notations used in Section 6.

Notation	Description
X_i	The time slot at which ith node enters SS state.
Y_i	Number of time slots between times X_i and X_{i-1}
P_{succ}	The probability that only one node transmits a REQ message in a slot after and the message is received successfully.
q_i	The probability of a node i entering SS state, in a round (frame).
q	The probability of a node with ID 1 entering SS state, in a round (the subscript 1 omitted from sake of clarity).
q_{min}	Minimum value of q.
$q(k)$	The probability of node with ID 1 entering SS state at slot k in a round.
q_{sum}	$\sum_{k=1}^{\mathcal{N}} q(k)$.
β_i	The number of 1's in row i of matrix B.
α_j	The number of 1's in column j of matrix B, excluding first row.
$\pi_{i,j}$	The transition probability from state i to state j of DTMC, presented in Section 6.
τ_i	The number of rounds required to reach state "n" in DTMC, starting from state "i."

frame index $frame_i^{VS}$. Let us assume that node j satisfies this condition. Now nodes i and j can enter SS state in any order. Assume that both nodes i and j are in SS state; then node j would not be able to get response continuously Δ times from a node in $N_i \cap R_j$. As a result, node j would move back to CS state, and it would not be able to enter SS state again. $\quad\square$

6. Complexity Analysis of DTSS Algorithm

In this section, we evaluate the expected runtime of DTSS algorithm, that is, the time when all nodes in the network reach SS state. Table 2 summarizes the set of notations used in this section. First, we consider the situation, when all nodes in the network interfere with each other's transmission; that is, the interference graph G is complete. This situation mainly occurs in single-hop WSNs. In this case, only a single transmission of REQ message in a slot can be successful, and therefore, nodes can enter SS state one at a time in each slot, as shown in Figure 5. Furthermore, we assume that $|R_i| = 1$, for each node i. In this case nodes directly enter SS state without entering the VS state. The analysis can be further extended for the case when $|R_i| > 1$. Initially, every node transmits REQ message with probability $1/\mathcal{N}$ in every slot.

Let the time slot at which ith node enters SS state be X_i. Note that X_i is a random variable. Clearly, X_n is the time slot when last node enters SS state, which is exactly the desired runtime of DTSS algorithm. Let $Y_i = X_i - X_{i-1}$. In this case,

$$X_i = X_{i-1} + Y_i,$$

$$EX_i = EX_{i-1} + EY_i = \sum_{j=1}^{i} EY_j. \tag{1}$$

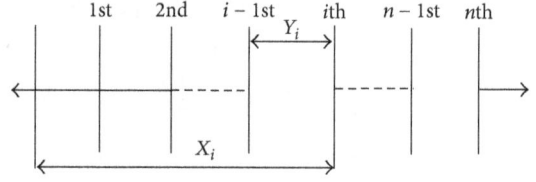

FIGURE 5: Sequence of slot assignment in a single-hop WSN. The nodes cannot enter state simultaneously in a slot.

Theorem 8. EX_n is $O(n)$ for single-hop WSNs.

Proof. At time slot X_{i-1}, exactly $i - 1$ nodes are in SS state and for the remaining $n - i + 1$ nodes which are not in SS state, set their slot probability to $1/(n - i + 1)$, for unoccupied slots. Let P_{succ} be the probability that only one node transmits a REQ message in a slot after time slot X_{i-1} and the message is received successfully at the intended receiver. Note that the REQ message could be lost not only due to collisions but also because of channel impairment. Therefore, a successful packet transmission also depends upon packer error rate (PER). Y_i is a geometric random variable with success probability P_{succ} and expectation $1/P_{succ}$. The upper bound of EY_n (runtime) can be calculated as follows:

$$P_{succ} = \binom{n - i + 1}{1} * \frac{1}{n - i + 1}$$

$$* \left(\frac{n - i}{n - i + 1} \right)^{n-i} * (1 - \text{PER})$$

$$= \left(\frac{n - i}{n - i + 1} \right)^{n-i} * (1 - \text{PER}),$$

$$EY_i = \frac{1}{P_{succ}} = \left(\frac{n - i + 1}{n - i} \right)^{n-i} * \frac{1}{(1 - \text{PER})}$$

$$= \left(1 + \frac{1}{n - i} \right)^{n-i} * \frac{1}{(1 - \text{PER})}, \tag{2}$$

$$EY_i \leq \frac{e}{(1 - \text{PER})},$$

$$\because \left\{ \left(1 + \frac{1}{n - i} \right)^{n-i} \right\},$$

is monotonically increasing and converges to e

$$EX_n = \sum_{j=1}^{n} EY_j \leq \frac{1}{1 - \text{PER}} \sum_{j=1}^{n} e = \frac{ne}{1 - \text{PER}} = O(n).$$

$\quad\square$

Now, we will consider a more generalized situation when not all nodes in the network interfere with each other's transmission; that is, the interference graph G is not necessarily complete. Again we assume that $|R_i| = 1$. The above situation mainly occurs in multihop WSNs. Further, we assume that the graph G is regular with degree $\mathcal{N} - 1$. Note that $\mathcal{N}$ is always taken to be greater than Δ, the maximum degree of interference graph. Therefore, assuming the graph to be regular with degree $\mathcal{N}-1$ will give the worst case analysis; that is, the expected runtime of DTSS algorithm for a nonregular interference graph with $\Delta = \mathcal{N} - 1$ will always be less than or equal to the expected runtime for a regular interference graph of degree $\mathcal{N} - 1$.

We will first find out for an arbitrary node i what the minimum value is that q_i can take, irrespective of the slot probabilities of other nodes in the network. This will help us to set an upper bound on the expected time, required by any node, to reach SS state.

Let us rearrange the IDs of the nodes in the following manner.

(i) The ID of node i is changed to 1.

(ii) The IDs of nodes adjacent to node i in G would be changed from 2 to $\mathcal{N}$. The ordering among these nodes could be arbitrary.

(iii) The IDs of all other nodes would become $\mathcal{N} + 1$ to n. The ordering among these nodes could be arbitrary.

Note that, the above rearrangement will not change the probability of node i entering SS state in a round. Our task is to find out q_1 instead of q_i. Further, we can also omit the subscript 1 from q_1 for sake of clarity.

Let the probability of node with ID 1 (after rearrangement) entering SS state at slot k in a round be $q(k)$. The value of $q(k)$ depends upon the transmission probability of node 1 in slot k and the transmission probabilities of its adjacent nodes in the same slot. Consider

$$q(k) = p_1(k) * \prod_{j=2}^{\mathcal{N}} \left(1 - p_j(k)\right). \tag{3}$$

The probability that node 1 can enter SS state in a round is equal to the probability that it can enter SS state in at least one slot of the round. Therefore, q can be written in terms of $q(k)$ as

$$q = 1 - \prod_{k=1}^{\mathcal{N}} \left(1 - q(k)\right). \tag{4}$$

In order to find the minimum value of q, we define another term q_{sum} as a function of $q(k)$'s as follows:

$$q_{\text{sum}} = \sum_{k=1}^{\mathcal{N}} q(k). \tag{5}$$

We know that, for a constant sum, the product can be maximized when the sum is partitioned equally [26]. Therefore, for a constant value of q_{sum}, q can achieve its minimum value q_{min}, if $q(l) = q(m), 1 \leq l, m \leq \mathcal{N}$. Obviously, q_{min} is a function of q_{sum}.

Theorem 9. *Let* $q_{\text{min}} = \mathcal{F}(q_{\text{sum}})$. *Then* $\mathcal{F}$ *is a monotonically increasing function.*

Proof. Let x and y be the two values of q_{sum}, such that $x > y$. We know that q_{min} is achieved when $q(l) = q(m), 1 \leq l, m \leq \mathcal{N}$. Let $c1$ and $c2$ be the corresponding values of $q(k)$, for all

k with respect to x and y. In this case, the value of $\mathcal{F}(x)$ and $\mathcal{F}(y)$ would be $1 - (1 - c1)^{\mathcal{N}}$ and $1 - (1 - c2)^{\mathcal{N}}$, respectively (4). Therefore,

$$x > y \implies \mathcal{N}c1 > \mathcal{N}c2 \implies c1 > c2$$

$$\implies (1 - c1) < (1 - c2) \implies (1 - c1)^{\mathcal{N}} < (1 - c2)^{\mathcal{N}}$$

$$\implies 1 - (1 - c1)^{\mathcal{N}} > 1 - (1 - c2)^{\mathcal{N}}$$

$$\implies \mathcal{F}(x) > \mathcal{F}(y).$$

$$\tag{6}$$

$\square$

It is clear from Theorem 9 that, to find q_{min}, we need to first minimize the q_{sum}. Let us define a binary square matrix, B, of size $\mathcal{N}$, in the following manner:

$$b_{i,j} = \begin{cases} 1, & \text{if } p_i(j) > 0 \\ 0, & \text{otherwise.} \end{cases} \tag{7}$$

The matrices P and B show an example of probability matrix and its corresponding binary transformation for $\mathcal{N} = 3$. Consider

$$P = \begin{pmatrix} \frac{1}{3} & \frac{1}{3} & \frac{1}{3} \\ \frac{1}{2} & \frac{1}{2} & 0 \\ \frac{1}{2} & 0 & \frac{1}{2} \end{pmatrix}, \qquad B = \begin{pmatrix} 1 & 1 & 1 \\ 1 & 1 & 0 \\ 1 & 0 & 1 \end{pmatrix}. \tag{8}$$

Let $\beta_i = \sum_{j=1}^{\mathcal{N}} b_{i,j}$ (number of 1's in row i of matrix B) and $\alpha_j = \sum_{i=2}^{\mathcal{N}} b_{i,j}$ (number of 1's in column j of matrix B, excluding first row). The $q(k)$ can be rewritten in terms of $b_{j,k}$ and β_j, $1 \leq j \leq \mathcal{N}$, as follows:

$$q(k) = \frac{b_{1,k}}{\beta_1} \prod_{j=2}^{\mathcal{N}} \left(1 - \frac{b_{j,k}}{\beta_j}\right). \tag{9}$$

Let B_{min} be the matrix for which the value of q_{sum} is minimum. To find out the properties of B_{min}, we start with the hypothesis that q would be minimum, if none of the nodes adjacent to node 1 is in SS state. This implies that node 1 is still transmitting in all the slots with probability $1/\mathcal{N}$; that is, $b_{1,k} = 1$, for all k, $1 \leq k \leq \mathcal{N}$. Now, we will present two lemmas based on the above hypothesis; this hypothesis will be used to find out the properties of B_{min} in Theorem 12, where we also explain the need for it.

Lemma 10. *For a given instance of matrix B, let $b_{1,k} = 1$, for all k, $1 \leq k \leq \mathcal{N}$, and for a slot j, $q(j) \leq q(k)$, for all $k \neq j$. Then, for any row i, q_{sum} reduces or remains the same, if $b_{i,j}$ is changed from 1 to 0.*

Proof. Let $q_{\text{sum}}^{\text{old}}$ and $q_{\text{sum}}^{\text{new}}$ be the respective sums before and after the conversion of $b_{i,j} = 1$ to 0. We need to show that $q_{\text{sum}}^{\text{old}} \geq q_{\text{sum}}^{\text{new}}$. Similarly, $q^{\text{old}}(j)$ and $q^{\text{new}}(j)$ can be defined.

Since $b_{1,k} = 1$, for all k, $1 \leq k \leq \mathcal{N}$, the $q^{\text{old}}(j)$ can be written as

$$q^{\text{old}}(j) = \frac{1}{\mathcal{N}} \prod_{k=2}^{\mathcal{N}} \left(1 - \frac{b_{k,j}}{\beta_k}\right)$$

$$= \frac{1}{\mathcal{N}} \left(\prod_{k=2, k \neq i}^{\mathcal{N}} \left(1 - \frac{b_{k,j}}{\beta_k}\right)\right) * \left(1 - \frac{1}{\beta_i}\right) \qquad (10)$$

and since $b_{i,j}$ becomes 0, after the conversion, $q^{\text{new}}(j)$ would be

$$q^{\text{new}}(j) = \frac{1}{\mathcal{N}} \left(\prod_{k=2, k \neq i}^{\mathcal{N}} \left(1 - \frac{b_{k,j}}{\beta_k}\right)\right). \qquad (11)$$

Therefore, from (10) and (11), we get

$$q^{\text{new}}(j) - q^{\text{old}}(j) = \frac{q^{\text{old}}(j)}{\beta_i - 1}. \qquad (12)$$

Similarly, for all other slots $k \neq j$ and $b_{i,k} = 1$,

$$q^{\text{new}}(k) - q^{\text{old}}(k) = -\frac{q^{\text{old}}(k)}{(\beta_i - 1)^2}. \qquad (13)$$

To show that $q_{\text{sum}}^{\text{old}} \geq q_{\text{sum}}^{\text{new}}$, we calculate $q_{\text{sum}}^{\text{new}} - q_{\text{sum}}^{\text{old}}$ as follows:

$$q_{\text{sum}}^{\text{new}} - q_{\text{sum}}^{\text{old}}$$

$$= \sum_{k=1}^{\mathcal{N}} q^{\text{new}}(k) - \sum_{k=1}^{\mathcal{N}} q^{\text{old}}(k)$$

$$= \sum_{k=1}^{\mathcal{N}} \left(q^{\text{new}}(k) - q^{\text{old}}(k)\right)$$

$$= \left(\sum_{k \neq j, b_{i,k}=1} q^{\text{new}}(k) - q^{\text{old}}(k)\right) + \left(q^{\text{new}}(j) - q^{\text{old}}(j)\right)$$

$$= \left(\sum_{k \neq j, b_{i,k}=1} -\frac{q^{\text{old}}(k)}{(\beta_i - 1)^2}\right) + \frac{q^{\text{old}}(j)}{\beta_i - 1}$$

$$= \left(\sum_{k \neq j, b_{i,k}=1} \frac{-q^{\text{old}}(j) + \left(q^{\text{old}}(j) - q^{\text{old}}(k)\right)}{(\beta_i - 1)^2}\right) + \frac{q^{\text{old}}(j)}{\beta_i - 1}$$

$$= \left(\sum_{k \neq j, b_{i,k}=1} \frac{-q^{\text{old}}(j)}{(\beta_i - 1)^2}\right)$$

$$+ \left(\sum_{k \neq j, b_{i,k}=1} \frac{\left(q^{\text{old}}(j) - q^{\text{old}}(k)\right)}{(\beta_i - 1)^2}\right) + \frac{q^{\text{old}}(j)}{\beta_i - 1}. \qquad (14)$$

Since the number of 1's in row i is β_i, the number of terms in the first summation of above equation would be exactly $\beta_i - 1$. Therefore,

$$q_{\text{sum}}^{\text{new}} - q_{\text{sum}}^{\text{old}} = \sum_{k \neq j, b_{i,k}=1} \frac{q^{\text{old}}(j) - q^{\text{old}}(k)}{(\beta_i - 1)^2} \leq 0,$$

$$\qquad (15)$$

$$\because q(j) \leq q(k), \quad \forall k \neq j.$$

□

Lemma 11. *For a given instance of matrix B, let $b_{1,k} = 1$, for all k, $1 \leq k \leq \mathcal{N}$, and $\beta_i = 2$, for all i, $2 \leq i \leq \mathcal{N}$. Then the following holds. For any two columns j and k and, for any row i, such that $\alpha_j > \alpha_k$, $b_{i,j} = 1$ and $b_{i,k} = 0$, q_{sum} either reduces or remains the same if the values of $b_{i,j}$ and $b_{i,k}$ are interchanged.*

Proof. Consider $q_{\text{sum}}^{\text{old}}$, $q_{\text{sum}}^{\text{new}}$, $q^{\text{old}}(j)$, and $q^{\text{new}}(j)$ as defined in Lemma 10. We need to show that $q_{\text{sum}}^{\text{old}} \geq q_{\text{sum}}^{\text{new}}$. Here, $q^{\text{old}}(j) = (1/\mathcal{N}) * 1/2^{\alpha_j}$, $q^{\text{old}}(k) = (1/\mathcal{N}) * 1/2^{\alpha_k}$, $q^{\text{new}}(j) = (1/\mathcal{N}) * 1/2^{\alpha_j - 1}$, and $q^{\text{new}}(k) = (1/\mathcal{N}) * 1/2^{\alpha_k + 1}$. Therefore,

$$q_{\text{sum}}^{\text{new}} - q_{\text{sum}}^{\text{old}} = \left(q^{\text{new}}(j) + q^{\text{new}}(k)\right) - \left(q^{\text{old}}(j) + q^{\text{old}}(k)\right)$$

$$= \left(q^{\text{new}}(j) - q^{\text{old}}(j)\right) + \left(q^{\text{new}}(k) + q^{\text{old}}(k)\right)$$

$$= \left(\frac{1}{2^{\alpha_j - 1}} - \frac{1}{2^{\alpha_j}}\right) + \left(\frac{1}{2^{\alpha_k + 1}} - \frac{1}{2^{\alpha_k}}\right)$$

$$= \frac{1}{2^{\alpha_j}} - \frac{1}{2^{\alpha_k + 1}} \leq 0 \quad \because \alpha_j > \alpha_k.$$

$$\qquad (16)$$

□

Now we will try to prove that $B_{\min}$ should satisfy a few constraints, in terms of α_i and β_i, $1 \leq i \leq \mathcal{N}$, with the help of Lemmas 10 and 11.

Theorem 12. *$B_{\min}$ has the following properties:*

   (1) *$\beta_i = 2$, for all i, $2 \leq i \leq \mathcal{N}$;*

   (2) *for exactly two columns $j1$ and $j2$, $\alpha_{j1} = \alpha_{j2} = 1$ and for all other columns $k \neq j1, j2$, $\alpha_k = 2$.*

Proof. We prove both the properties for two different cases: $\beta_1 = \mathcal{N}$ and $\beta_1 \neq \mathcal{N}$.

Case 1 ($\beta_1 = \mathcal{N}$). The property (1) can be proved by contradiction. First, we show that $\beta_i \geq 2$, $2 \leq i \leq \mathcal{N}$. If $\beta_i = 1$ with $b_{i,j} = 1$, for some row i, then node i is in SS state. Therefore, node 1 should have stopped transmitting in slot j, that is, $b_{1,k} = 0$, which contradicts our assumption that $\beta_1 = \mathcal{N}$. Now, we show that $\beta_i \leq 2$, $2 \leq i \leq \mathcal{N}$. Let $\exists i : \beta_i > 2$ and $\mathcal{A}$ the set of column indexes k for which $b_{i,k} = 1$; then $\exists j \in \mathcal{A}$, such that $q(j) \leq q(k)$, for all $k \in A$, $j \neq k$. Therefore, by the virtue of Lemma 10, q_{sum} reduces or remains the same, if $b_{i,k}$ is changed from 1 to 0. The same process can be repeated till $\beta_i = 2$.

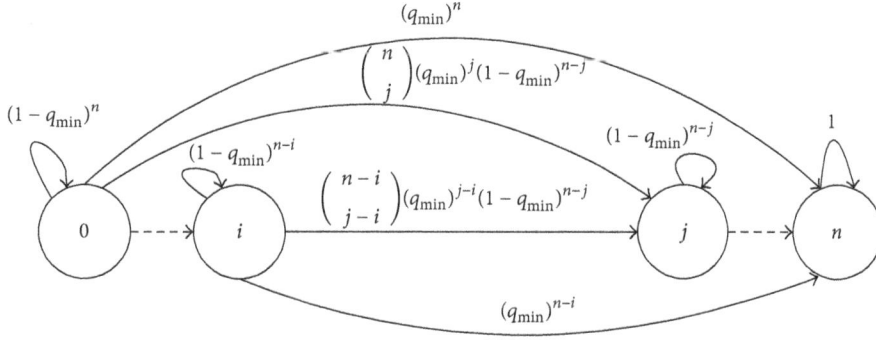

FIGURE 6: Discrete time Markov chain (DTMC) with number of nodes in SS state as random variable, assuming the probability of entering SS state, as q_{min}, for each node in the network.

The property (2) can also be proved by contradiction. We know that $\beta_1 = \mathcal{N}$ and $\beta_i = 2, 2 \leq i \leq \mathcal{N}$. Therefore, $\sum_{j=1}^{\mathcal{N}} \alpha_j = 2(\mathcal{N} - 1)$. First, we show that $\alpha_j \leq 2, 1 \leq j \leq \mathcal{N}$. For a column j, $\alpha_j > 2 \Rightarrow \exists k : \alpha_k < 2$; otherwise $\sum_{j=1}^{\mathcal{N}} \alpha_j$ would become less than $2(\mathcal{N} - 1)$. In this case, for any row i, such that $b_{\{i,j\}} = 1$ and $b_{\{i,k\}} = 0$, the value of $b_{\{i,j\}}$ and $b_{\{i,k\}}$ can be interchanged by virtue of Lemma 11. This proves that α_j could be either 0, 1 or 2, $1 \leq j \leq \mathcal{N}$. Since, any column can have at most two 1's, this implies that at most one column of type $\alpha_i = 0$ can exist and that also can be increased to 1 by virtue of Lemma 11. Furthermore, the number of columns of type $\alpha_i = 1$ cannot be one, since $2(\mathcal{N} - 1)$ is even. Finally, we can say that number of columns of type $\alpha_i = 1$ is exactly 2; otherwise, the total sum will be less than $2(\mathcal{N} - 1)$.

Case 2 ($\beta_1 \neq \mathcal{N}$). Let $q_{sum}^{case 1}$ and $q_{sum}^{case 2}$ be the corresponding summation for Cases 1 and 2, respectively. The value of $q_{sum}^{case 1}$ would be $(\mathcal{N} + 2)/4\mathcal{N}$. We will prove that $q_{sum}^{case 1} < q_{sum}^{case 2}$ by showing that any perturbation in the matrix corresponding to Case 1 will increase the value of q_{sum}. We have already proved, in Case 1, that any modification in any of the rows from 2 to row $\mathcal{N}$ and leaving row 1 unchanged will increase q_{sum}. Now, let us change a single entry $b_{1,k} = 1$ to 0; that is, node 1 has decided not to transmit in slot k. This only happens when at least one adjacent node i in G has gone to SS state for slot k, which implies that $b_{i,k} = 1$ and $b_{i,j} = 0$, for all $j \neq k$. Let us interchange the row i with row $\mathcal{N}$ and column k with column $\mathcal{N}$. In this case, $b_{1,\mathcal{N}} = 0, b_{1,j} = 1$, for all $j \neq \mathcal{N}$, and $b_{\mathcal{N},\mathcal{N}} = 1$ and $b_{\mathcal{N},j} = 0$, for all $j \neq \mathcal{N}$. Consider the submatrix of size $\mathcal{N} - 1$ times $\mathcal{N} - 1$. The minimum value of q_{sum} which can be achieved by this submatrix would be $(\mathcal{N} + 1)/4(\mathcal{N} - 1)$. Moreover, $q(\mathcal{N}) = 0$, because $b_{1,\mathcal{N}} = 0$. Therefore, $q_{sum}^{case 2} = (\mathcal{N} + 1)/4(\mathcal{N} - 1) > (\mathcal{N} + 2)/4\mathcal{N} = q_{sum}^{case 1}$. □

From Theorem 9, we know that the q_{min} can be achieved when q_{sum} is minimum and B_{min} should satisfy the properties as given in Theorem 12. Therefore,

$$q_{min} = 1 - \left(\frac{4 * \mathcal{N} - 1}{4 * \mathcal{N}}\right)^{(\mathcal{N}-2)} * \left(\frac{2 * \mathcal{N} - 1}{2 * \mathcal{N}}\right)^2. \quad (17)$$

The following matrix shows one of such B_{min} matrix for $\mathcal{N} = 4$:

$$B_{min} = \begin{pmatrix} 1 & 1 & 1 & 1 \\ 0 & 0 & 1 & 1 \\ 0 & 1 & 1 & 0 \\ 1 & 1 & 0 & 0 \end{pmatrix}. \quad (18)$$

To calculate the expected runtime of DTSS algorithm, we model the behavior of the system using a discrete time Markov chain (DTMC), with the number of nodes in SS state, X_t, at the beginning of round t, as a random variable. The transition probabilities, $\pi_{i,j}$, are defined as follows:

$$\pi_{i,j} = \begin{cases} \binom{n-i}{j-i} (q_{min})^{j-i} (1 - q_{min})^{n-j}, & j \geq i \\ 1, & i = j = n \\ 0, & \text{otherwise.} \end{cases} \quad (19)$$

In this DTMC (see Figure 6), all states are transient except state "n" which is an absorbing state. The probability of leaving a transient state "i" is always greater than 0; that is, $1 - \sum_{j>i} \pi_{i,j} > 0$. A transient state cannot be visited again, once it is left. This shows that the DTSS algorithm converges in a finite time. Let τ_i be the number of rounds required to reach state "n" starting from state "i." Our goal is to find $E[\tau_0]$, which can be calculated using the following recurrence relation:

$$E[\tau_i] = \begin{cases} 1 + \sum_{j=0}^{n} \pi_{i,j} E[\tau_j], & 1 \leq i \leq n-1 \\ 0, & i = n. \end{cases} \quad (20)$$

Note that the above DTMC is the approximation of actual stochastic process, where the transition probabilities not only depend upon the number of nodes in SS state, but also depend on exact nodes belonging to SS state.

We show that the value of $E[\tau_i]$ is greater than actual expected time required to reach state "n" starting from state "i" in DTMC, by proving that the transition probability of

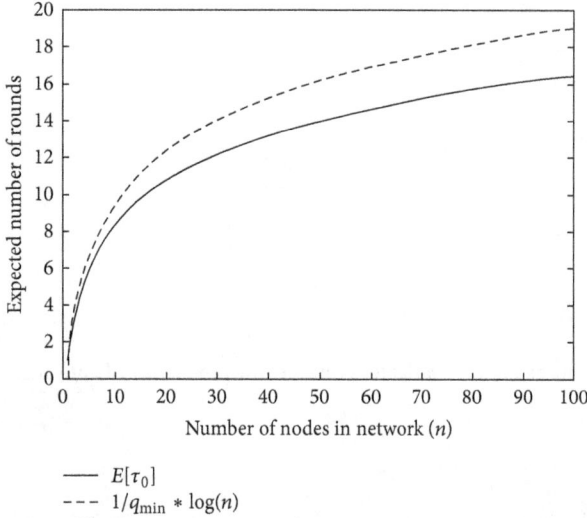

FIGURE 7: Runtime performance of DTSS algorithm in terms of number of rounds (frames) with respect to number of nodes in the network and its comparison with the function, $\log n/q_{min}$.

moving from i nodes in SS state to j nodes in SS state, in an actual stochastic process, is always greater than $\pi_{i,j}$. We know that the probability of each node moving from CS state to SS state in a round is always greater than q_{min} (17) and therefore the probability that, out of $n - i$ nodes in CS state, exactly $j - i$ nodes enter into the SS state is greater than $\binom{n-i}{j-i}(q_{min})^{j-i}(1 - q_{min})^{n-j} = \pi_{i,j}$.

Figure 7 shows the graph for $E[\tau_0]$ along with function $\log n/q_{min}$. The graph shows that $E[\tau_0]$ is upper bounded by $\log n/q_{min}$, and therefore, for a fixed frame size, $E[\tau_n]$ is $O(\log n)$. We know from (17) that q_{min} depends only upon $\mathcal{N}$, which is a measure of two-hop network density (δ).

Another method to analyze the expected runtime of DTSS algorithm is to calculate the expectation of maximum of all X_i's, where X_i is the time taken by node i to reach SS state. Consider

$$E[X_{max}] = E[\max(X_1, X_2, X_3, \ldots, X_n)]. \quad (21)$$

The X_is can be assumed as i.i.d (independent and identically distributed) geometric random variable with parameter q_{min}. In this case the $E[X_i]$ would be higher than the actual expected time to enter SS state, by node i. The value of $E[X_{max}]$ can be calculated as

$$E[X_{max}] = \sum_{k \geq 0} P([X_{max} > k])$$

$$= \sum_{k \geq 0} (1 - P(X_{max} \leq k))$$

$$= \sum_{k \geq 0} (1 - P(X_i \leq k)^n) \quad (22)$$

$$= \sum_{k \geq 0} \left(1 - \left(1 - (1 - \overline{q_{min}})^k\right)^n\right),$$

where $\overline{q_{min}} = 1 - q_{min}$. By considering the above infinite sum as right and left hand Riemann sum approximations [29] of the corresponding integral, we obtain

$$\int_0^\infty \left(1 - \left(1 - \overline{q_{min}}^k\right)^n\right) \leq E[X_{max}]$$

$$\leq 1 + \int_0^\infty \left(1 - \left(1 - \overline{q_{min}}^k\right)^n\right). \quad (23)$$

With the change of variable $u = 1 - \overline{q_{min}}^k$, we have

$$E[X_{max}] \leq 1 + \frac{1}{\log \overline{q_{min}}} \int_0^1 \frac{1 - u^n}{1 - u} du$$

$$= 1 + \frac{1}{\log \overline{q_{min}}} \int_0^1 \left(1 + u + \cdots + u^{n-1}\right) du$$

$$= 1 + \frac{1}{\log \overline{q_{min}}} \left(1 + \frac{1}{2} + \cdots + \frac{1}{n}\right) \quad (24)$$

$$\approx 1 + \frac{\log n}{\log \overline{q_{min}}}.$$

From (24), we can conclude that $E[X_{max}]$ is the $O(\log n)$, for a fixed neighborhood density, δ. We know from (17) that q_{min} depends only upon $\mathcal{N}$, which is a measure of δ. A more rigorous analysis on expectation of the maximum of IID geometric random variables can be found in [30].

7. Simulation Results

We have used Castalia simulator [27] to study the performance of DTSS algorithm. A multihop network, based on TelosB node hardware platform that uses CC2420 transceiver [28] for communication, is used in the simulation. The transceivers run at 250 kbps data rate and dbm transmission power which approximately gives 40 m of transmission range in the absence of interference. All nodes are distributed randomly within 250 m × 250 m area. Note that, at 250 kbps, it takes about 0.5 ms to transmit a packet of size 128 bits (80 bits for the MAC header, and 48 bits for $L2$ and state payload). Hence, we set TDMA time slots to a period of 1 ms, which is sufficiently long for the transmission of REQ/RES messages. The performance of protocol has been averaged over 100 simulation runs. The neighborhood size of the network is changed by varying the number of nodes from 50 to 300. This setup produces topologies with different neighborhood density, δ, values varying between 5 and 50.

Figure 8 shows the average number of slots taken by all the nodes to decide their slot in case of broadcast scheduling for frame sizes δ and 1.3δ, respectively. The error bars denote 95% confidence intervals. Figure 8 shows that runtime increases linearly with neighborhood density, δ. Given slot size as 1 ms, the total runtime for very high density network with $\delta = \mathcal{N} = 50$ is approximately 7 s. Furthermore, if we take more slots per frame, then runtime decreases and also confidence interval improves.

Figure 9 shows the average of the number of slots taken by all the nodes to decide their slot in case of broadcast

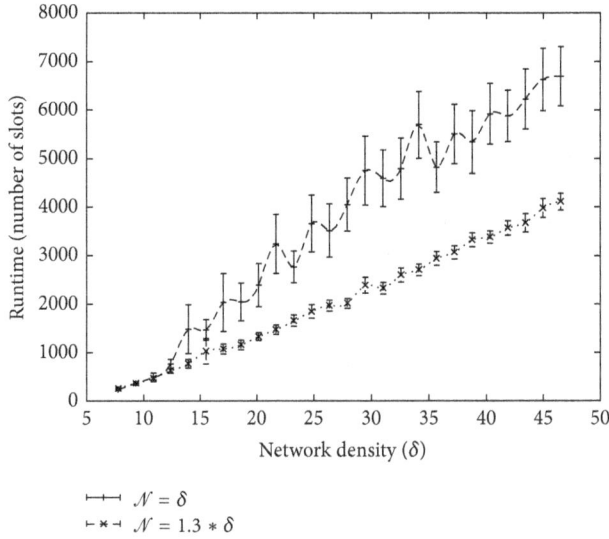

FIGURE 8: Runtime performance of DTSS algorithm with respect to network network density, δ, and the effect of taking the number of slots more than δ.

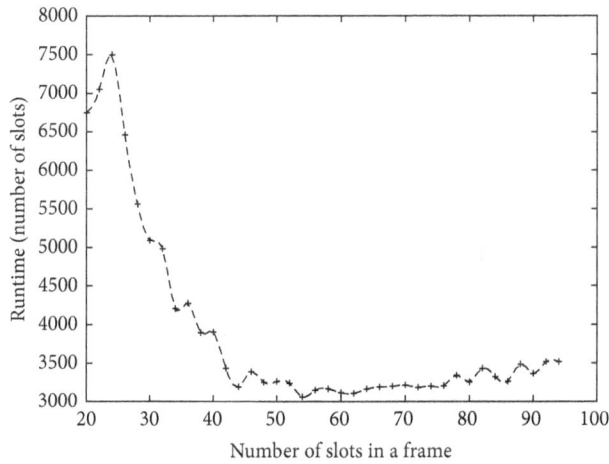

FIGURE 9: Runtime performance of DTSS algorithm with respect to number of slots in frame, $\mathcal{N}$.

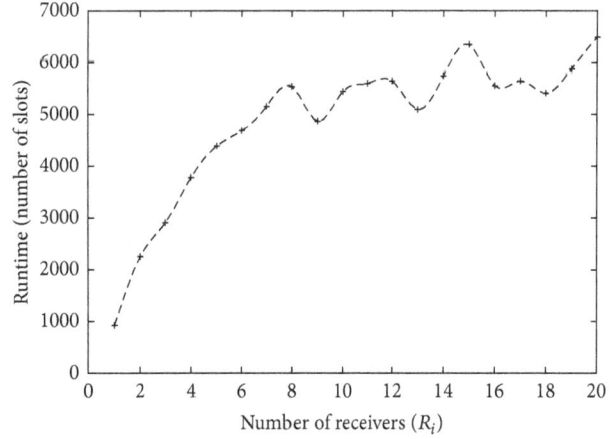

FIGURE 10: Runtime statistics of DTSS algorithm to show the performance with respect to unicast ($R_i = 1$), multicast, and broadcast ($R_i > 1$) mode of transmissions.

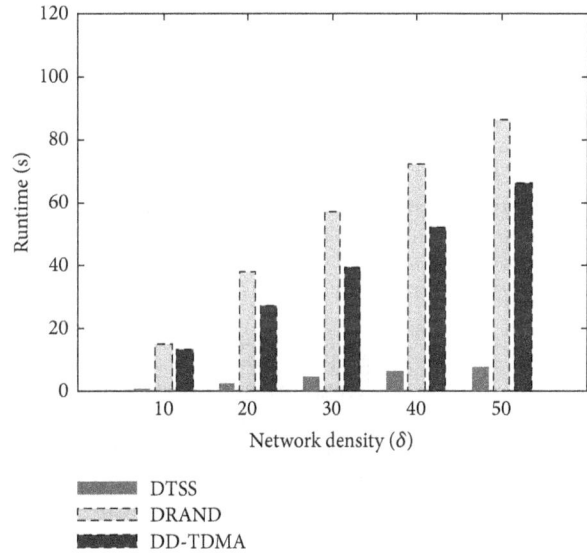

FIGURE 11: The runtime performance comparison of DTSS algorithm against DRAND and DD-TDMA, with respect to network density, δ.

scheduling for varying $\mathcal{N}$ values starting from δ. Figure 9 shows that the runtime reduces rapidly with small increase in $\mathcal{N}$ and further increase in Δ does not have much impact on runtime. This fact can be utilized as a tradeoff between runtime and frame length.

Figure 10 shows the average of the number of slots taken by all the nodes to decide their slot for varying the number of receivers (unicast to broadcast) with $\delta = \mathcal{N} = 40$. Figure 10 suggests that unicast or link scheduling can be performed in less than one second for a network with fairly high network density.

We now compare DTSS with DRAND [12] and DD-TDMA [13]. Figure 11 shows the performance results of DTSS along with DRAND and DD-TDMA with respect to runtime of each algorithm. The comparison is based on broadcast transmission because both DRAND and DD-TDMA only implement this mode of transmission. The primary reason

of getting less runtime is because the DTSS generates a feasible schedule when the number of available slots is already fixed, whereas other algorithms try to generate a suboptimal schedule by using greedy approach, which is inherently sequential. In case of unicast and multicast scheduling, the DTSS even takes lesser time to compute the schedule as compared to broadcast transmission. The number of slots taken by DTSS is always δ as shown in Figure 12, whereas the number of time slots taken by DRAND and DD-TDMA can be less than δ.

8. Conclusions and Future Work

For many applications in WSNs, efficiently handling the spatially correlated contention is an important requirement. The DTSS takes very less time to perform the scheduling

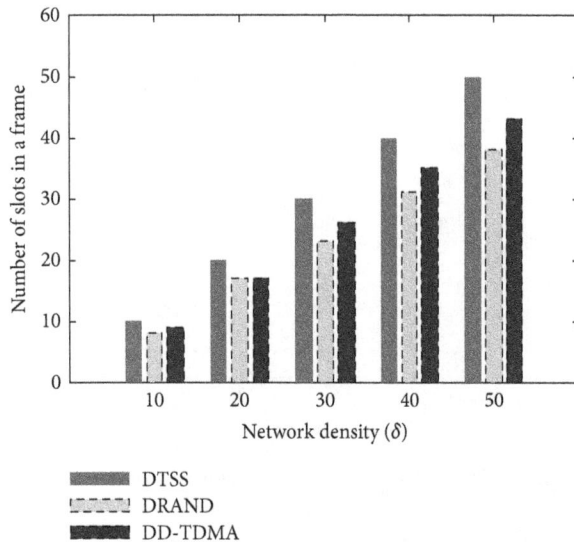

FIGURE 12: A comparison on frame size of DTSS algorithm against DRAND and DD-TDMA, with respect to network density, δ.

as compared to other existing distributed scheduling algorithms. We have shown that the runtime of DTSS algorithm is $O(n)$ and $O(\log(n))$ for single-hop and multihop WSNs, respectively, and therefore it is scalable for WSNs with large number of nodes. The interference model used by DTSS is more realistic than conventional protocol interference model. Additionally, the DTSS has a unique feature of unified scheduling in which simultaneously a few nodes can take a slot for unicast, while other nodes can take it for multicast or broadcast purpose. Although the number of slots taken by DTSS is bounded by Δ, further efforts can be applied to reduce the number of slots. In future, we plan to work on the variation of DTSS algorithm, for the situation, when nodes are not assumed to be synchronized before performing the slot scheduling.

Conflict of Interests

The authors declare that there is no conflict of interests regarding the publication of this paper.

References

[1] H. Alemdar and C. Ersoy, "Wireless sensor networks for healthcare: a survey," *Computer Networks*, vol. 54, no. 15, pp. 2688–2710, 2010.

[2] M. A. Maarof, M. A. Rassam, and A. Zainal.

[3] J. Silva, T. Camilo, A. Rodrigues, M. Silva, F. Gaudêncio, and F. Boavida, "Multicast in wireless sensor networks the next step," in *Proceedings of the 2nd International Symposium on Wireless Pervasive Computing (ISWPC '07)*, pp. 185–190, February 2007.

[4] C. E. Perkins and E. M. Royer, "Ad-hoc on-demand distance vector routing," in *Proceedings of the 2nd IEEE Workshop on Mobile Computing Systems and Applications (WMCSA '99)*, pp. 90–100, New Orleans, La, USA, February 1999.

[5] A. R. Swain and R. C. Hansdah, "A weighted average-based external clock synchronisation protocol for wireless sensor

networks," *International Journal of Sensor Networks*, vol. 12, no. 2, pp. 89–105, 2012.

[6] B. Hull, K. Jamieson, and H. Balakrishnan, "Mitigating congestion in wireless sensor networks," in *Proceedings of the 2nd International Conference on Embedded Networked Sensor Systems (SenSys '04)*, pp. 134–147, ACM, Baltimore, Md, USA, November 2004.

[7] G. Chakraborty, "Genetic algorithm to solve optimum TDMA transmission schedule in broadcast packet radio networks," *IEEE Transactions on Communications*, vol. 52, no. 5, pp. 765–777, 2004.

[8] C. Y. Ngo and V. O. K. Li, "Centralized broadcast scheduling in packet radio networks via genetic-fix algorithms," *IEEE Transactions on Communications*, vol. 51, no. 9, pp. 1439–1441, 2003.

[9] S. Ramanathan and E. L. Lloyd, "Scheduling algorithms for multihop radio networks," *IEEE/ACM Transactions on Networking*, vol. 1, no. 2, pp. 166–177, 1993.

[10] L. Bao and J. J. Garcia-Luna-Aceves, "A new approach to channel access scheduling for Ad Hoc networks," in *Proceedings of the 7th Annual International Conference on Mobile Computing and Networking (MobiCom '01)*, pp. 210–221, ACM, 2001.

[11] T. Moscibroda and R. Wattenhofer, "Coloring unstructured radio networks," in *Proceedings of the 17th Annual ACM Symposium on Parallelism in Algorithms and Architectures (SPAA '05)*, pp. 39–48, ACM, July 2005.

[12] I. Rhee, A. Warrier, J. Min, and L. Xu, "DRAND: distributed randomized TDMA scheduling for wireless ad-hoc networks," in *Proceedings of the 7th ACM International Symposium on Mobile Ad Hoc Networking and Computing (MobiHoc '06)*, pp. 190–201, ACM, May 2006.

[13] Y. Wang and I. Henning, "A deterministic distributed TDMA scheduling algorithm for wireless sensor networks," in *Proceedings of the International Conference on Wireless Communications, Networking and Mobile Computing (WiCOM '07)*, pp. 2759–2762, Shanghai, China, September 2007.

[14] C. Zhu and M. S. Corson, "A five-phase reservation protocol (FPRP) for mobile ad hoc networks," in *Proceedings of the IEEE 17th Annual Joint Conference of the IEEE Computer and Communications Societies (INFOCOM '98)*, vol. 1, pp. 322–331, IEEE, San Francisco, Calif, USA, March–April 1998.

[15] A. R. Swain, R. C. Hansdah, and V. K. Chouhan, "An energy aware routing protocol with sleep scheduling for wireless sensor networks," in *Proceedings of the 24th IEEE International Conference on Advanced Information Networking and Applications (AINA '10)*, pp. 933–940, IEEE, Perth, Australia, April 2010.

[16] E. Arikan, "Some complexity results about packet radio networks," *IEEE Transactions on Information Theory*, vol. 30, no. 4, pp. 681–685, 1984.

[17] S. Ramanathan, "A unified framework and algorithm for (T/F/C)DMA channel assignment in wireless networks," *Proceedings the 16th IEEE Annual Joint Conference of the IEEE Computer and Communications Societies (INFOCOM '97)*, vol. 2, pp. 900–907, 1997.

[18] S. C. Ergen and P. Varaiya, "TDMA scheduling algorithms for wireless sensor networks," *Wireless Networks*, vol. 16, no. 4, pp. 985–997, 2010.

[19] M. R. Palattella, N. Accettura, M. Dohler, L. A. Grieco, and G. Boggia, "Traffic aware scheduling algorithm for reliable low-power multi-hop IEEE 802.15.4e networks," in *Proceedings of the IEEE 23rd International Symposium on Personal, Indoor*

and Mobile Radio Communications (PIMRC '12), pp. 327–332, September 2012.

[20] S. Waharte and R. Boutaba, "Performance comparison of distributed frequency assignment algorithms for wireless sensor networks," in *Network Control and Engineering for QoS, Security and Mobility, III*, vol. 165, pp. 151–162, Springer, New York, NY, USA, 2005.

[21] O. Younis and S. Fahmy, "HEED: a hybrid, energy-efficient, distributed clustering approach for ad hoc sensor networks," *IEEE Transactions on Mobile Computing*, vol. 3, no. 4, pp. 366–379, 2004.

[22] R. Rozovsky and P. R. Kumar, "SEEDEX: a MAC protocol for ad hoc networks," in *Proceedings of the ACM International Symposium on Mobile Ad Hoc Networking and Computing (MobiHoc '01)*, pp. 67–75, October 2001.

[23] I. Rhee, A. Warrier, M. Aia, J. Min, and M. L. Sichitiu, "Z-MAC: a hybrid MAC for wireless sensor networks," *IEEE/ACM Transactions on Networking*, vol. 16, no. 3, pp. 511–524, 2008.

[24] C. Busch, M. Magdon-Ismail, F. Sivrikaya, and B. Yener, "Contention-free MAC protocols for wireless sensor networks," in *Distributed Computing: Proceedings of the 18th International Conference, DISC 2004, Amsterdam, The Netherlands, October 4–7, 2004*, vol. 3274 of *Lecture Notes in Computer Science*, pp. 245–259, Springer, Berlin, Germany, 2004.

[25] V. Gabale, B. Raman, P. Dutta, and S. Kalyanraman, "A classification framework for scheduling algorithms in wireless mesh networks," *IEEE Communications Surveys and Tutorials*, vol. 15, no. 1, pp. 199–222, 2013.

[26] I. M. Niven, *Maxima and Minima without Calculus*, Mathematical Association of America, Washington, DC, USA, 1981.

[27] Castalia: A simulator for Wireless Sensor Networks, https://castalia.forge.nicta.com.au/index.php/en/documentation.html.

[28] CC2420 Data Sheet, http://www.stanford.edu/class/cs244e/papers/cc2420.pdf.

[29] Riemann sum, http://en.wikipedia.org/wiki/Riemannsum.

[30] B. Eisenberg, "On the expectation of the maximum of IID geometric random variables," *Statistics & Probability Letters*, vol. 78, no. 2, pp. 135–143, 2008.

18

Efficient Service Selection in Mobile Information Systems

Shangguang Wang,[1] **Lei Sun,**[1] **Qibo Sun,**[1] **Xuyan Li,**[2] **and Fangchun Yang**[1]

[1]*State Key Laboratory of Networking and Switching Technology, Beijing University of Posts and Telecommunications, Haidian, Beijing 100876, China*
[2]*Basic Research Service Ministry of Science and Technology the People's Republic of China, Haidian, Beijing 100862, China*

Correspondence should be addressed to Shangguang Wang; sgwang@bupt.edu.cn

Academic Editor: David Taniar

With the rapid development of mobile wireless networks such as 4G and LET, ever more mobile services and applications are emerging in mobile networks. Faced with massive mobile services, a top priority of mobile information systems is how to find the best services and compose them into new value-added services (e.g., location-based services). Hence, service selection is one of the most fundamental operations in mobile information systems. Traditional implementation of service selection suffers from the problems of a huge number of services and reliability. We present an efficient approach to service selection based on computing QoS uncertainty that achieves the best solution in two senses: (1) the time cost for finding the best services is short and (2) the reliability of the selected services is high. We have implemented our approach in experiments with real-world and synthetic datasets. Our results show that our approach improves on the other approaches tested.

1. Introduction

Global mobile devices and connections grew to 7 billion in 2013, up from 6.5 billion in 2012; smartphones accounted for 77% of that growth, with 406 million net additions in 2013. The increasing number of mobile devices that are accessing mobile networks worldwide drives global mobile traffic growth. Cisco notes that globally smart traffic is going to grow from 88% of the total global mobile traffic to 96% by 2018. This is significantly higher than the ratio of smart devices and connections (54% by 2018) because on average a smart device generates much higher traffic than a nonsmart device.

The explosion of mobile traffic leads to a sharp drop in the quality of mobile service. To provide effective quality of service (QoS), service providers are fueling the growth of 4G deployments and adoption. Hence, service providers around the world are busy rolling out 4G networks to help them meet the growing end-user demand for better QoS such as more bandwidth and faster connectivity on the move. However, although 4G or LTE networks are faster, higher bandwidth, and more intelligent, when mobile device capabilities are combined with these networks, wide adoption of advanced mobile applications will increase mobile traffic even more. For example, in 2013 a 4G connection generated 14.5 times more traffic on average than a non-4G connection. Hence, to meet the fierce challenge from the explosion of mobile traffic, service providers also need to optimize bandwidth management and network monetization.

Although many notable network optimization schemes [1–6] have been proposed, most of them focus on network aspects and fail to consider services and applications of mobile information systems. Deploying next-generation mobile networks such as 4G or LTE requires greater service portability and interoperability. With the proliferation of mobile and portable devices, there is an imminent need for networks to allow all of these devices to be connected transparently, with the network providing high-performance computing and delivering enhanced real-time video and multimedia from mobile information systems. This openness will broaden the range of applications and services that can be shared and create highly enhanced mobile traffic. Hence, if an approach exists that can reduce the number of services and applications used, this will offload mobile traffic effectively from mobile information systems. We find that service composition technology and its application to mobile information systems may be a good alternative scheme.

Service composition involves the development of customized services, often by discovering, integrating, and executing existing services in a service-oriented architecture (SOA) [7]. A single service is usually not able to satisfy a user's service requirement. However, combining different existing services can provide a complex functional service for users; this is called *service composition*. In this way, services already existing in mobile information systems are orchestrated into new services that can reduce mobile traffic by shortening the number of interactions between services and users.

In the SOA paradigm of mobile information systems, these composite services are specified as abstract processes composed of a set of abstract services (called *service classes*). When a composite service starts to run, for each service class a concrete service (called a *service candidate*) is selected and invoked. Hence, service composition technology ensures loose coupling and design flexibility for many applications in mobile information systems [8].

As is well known, there are many mobile services in mobile information systems. Although their function attributes are the same or similar, their QoS values are different. For example, while Line, Kakao Talk, WeChat, and Viber can each be used as a mobile messaging application or service, the QoS (e.g., response time and throughput) of each is different. Hence, QoS plays an important role in determining which mobile service should be selected (called *service selection*) in mobile information systems [9]. Traditional service selection approaches (e.g., UDDI, Bluetooth) focus only on searching for services with specific functional attributes and do not provide any guarantee of QoS; this is unsuitable for a high-reliability (99.9999%) guarantee in mobile information systems.

Although many notable approaches to service selection including MIP [10], Heuristic [11], Hybrid [12], and LOEM [13] have been proposed in web service systems and have been shown to perform well in their respective contexts, these approaches should not be used in mobile information systems. The main reason is that they ignore the violent QoS fluctuations of mobile services (e.g., the response times of services change over time), so they cannot provide reliable mobile services for users in mobile information systems due to dynamic mobile service environments. Generally, service candidates participating in service selection are widely distributed in mobile information systems. These services originate from different service providers and run on different operations platforms. Any slight change in location, network environment, service requirement time, or other aspects will affect the QoS consistency of these service candidates (called *QoS uncertainty*). Hence, it is worth noting that a mobile service with consistently good QoS is typically more reliable than a mobile service with a large variance in its QoS [14]. Therefore, QoS uncertainty should be considered as an important criterion for reliable service selection in mobile information systems. Additionally, the obvious way to obtain the best service selection result is to enumerate all possible service compositions, which will cause poor real-time performance. Obviously, this is not acceptable for mobile information systems. Hence, to avoid unendurable computation delays, reducing the search space of service candidates with reliability guarantees is an important principle for service selection in mobile information systems.

Our proposed approach differs from existing approaches. Its main idea is to enhance reliability while reducing the solution space in the service selection process and to find a near-optimal but more reliable service composition in mobile information systems with lower computation time. That is, this novel service selection approach proposes to find the best reliable composite service faster. We first adopt variance theory to compute the QoS uncertainties, and then we prune less-reliable mobile services and select highly reliable ones from mobile information systems. Finally, Mixed Integer Programming is adopted to obtain the best composite service. We have experimented with our approach with 5825 real-world services and 10,000 synthetic mobile services. Our results show that our approach outperforms other approaches.

This paper is organized as follows. In Section 2, we introduce the concepts of service selection and give an example to illustrate it in mobile information systems. Section 3 describes our approach in detail, including computing QoS uncertainty, filtering uncertain services, and the service selection algorithm. The experimental results in Section 4 demonstrate the benefits of our approach relative to other approaches. Finally, Section 5 concludes the paper.

2. Service Selection Framework

To clarify the process of service selection, some preliminary background is introduced in this section. Some basic concepts are listed below.

(1) $S = \{s_1, s_2, \ldots, s_n\}$ ($n = 1, 2, 3, \ldots$) denotes a composition service that is to achieve a particular function that can satisfy a user's requirements. It is constructed by combining a plurality of service candidates that are selected from each service class. A service class $s_i \in S$ ($i = 1, 2, 3, \ldots$) ($1 \leq i \leq n$) often contains a number of service candidates $s_i = \{s_{i1}, s_{i2}, \ldots, s_{il}\}$, where l ($l = 1, 2, 3, \ldots$) is the number of service candidates with the same function but different values of their QoS attributes.

(2) $QS = \{q_1(S), q_2(S), \ldots, q_r(S)\}$ denotes the attribute vector of a composite service, where the value of $q_r(S)$ is aggregation of r ($r = 1, 2, 3, \ldots$) attribute values from all selected service candidates using QoS aggregation functions.

(3) $Qs_{ij} = \{q_1(s_{ij}), q_2(s_{ij}), \ldots, q_r(s_{ij})\}$ denotes the values of the service s_{ij}. Considering the actual features of mobile information systems, in this paper we consider only a sequential composition model. Table 1 lists the QoS aggregation functions of the sequential composition model.

(4) QoS utility functions for a mobile service $s_{ij} \in s_i$ and composite service S are defined as follows:

$$U\left(s_{ij}\right) = \sum_{k=1}^{r} \frac{Q_{i,k}^{\max} - q_k\left(s_{ij}\right)}{Q_{i,k}^{\max} - Q_{i,k}^{\min}} \cdot \omega_k,$$

TABLE 1: QoS aggregation functions.

QoS attributes	QoS aggregated functions
Response time Price	$q(S) = \sum_{i=1}^{n} q(s_i)$
Reputation	$q(S) = \dfrac{1}{n}\sum_{i=1}^{n} q(s_i)$
Reliability Availability	$q(S) = \prod_{i=1}^{n} q(s_i)$

$$U(S) = \sum_{k=1}^{r} \frac{Q_k^{\max} - q_k(S)}{Q_k^{\max} - Q_k^{\min}} \cdot \omega_k$$

(1)

with

$$Q_k^{\max} = \sum_{i=1}^{r} Q_{i,k}^{\max} \left(Q_{i,k}^{\max} = \max_{\forall s_{ij} \in s_i} q_k(s_{ij}) \right),$$

$$Q_k^{\min} = \sum_{i=1}^{r} Q_{i,k}^{\min} \left(Q_{i,k}^{\min} = \min_{\forall s_{ij} \in s_i} q_k(s_{ij}) \right),$$

(2)

where ω_k represents the user's preferences; it is the weight of each QoS attribute and satisfies $\sum_{k=1}^{r} \omega_k = 1$ ($0 < \omega_k < 1$); $Q_{i,k}^{\max}$ ($0 < k < r$) is the maximum value of the kth attribute in all service candidates of the service class s_i. Similarly, $Q_{i,k}^{\min}$ is the minimum value in the service class s_i; $Q_k^{\max}$ is the sum of every $Q_{i,k}^{\max}$ in the composition service S and $Q_k^{\min}$ is the sum of each $Q_{i,k}^{\min}$.

Based on the QoS utility function, we can calculate the global QoS attribute values of each candidate service by mapping the vector of QoS values Qs_{ij} into a single global QoS aggregated value Us_{ij}. We can then sort and rank all service candidates of each service class.

(5) $C = \{C_1, C_2, \ldots, C_m\}$ denotes the global QoS constraints in the service selection process, where m ($0 < m < r$) represents the number of QoS constraints.

We will present an example [13] from a mobile information system to motivate our work. As illustrated in Figure 1, a smartphone requests the latest video news from a service provider in which three tasks (service classes) are invoked: (1) a transcoding service that is used to transform various video formats into those supported by the smartphone, (2) a compression service to adapt the video content to the wireless link, and (3) a payment service used to pay the bill. The smartphone request is usually associated with a set of global QoS constraints C, for example, response time ≤ 1 second. Hence, the running service selection process can be divided into three stages as shown in Figure 1. First, a set of concrete mobile services, that is, $s_i = \{s_{i1}, s_{i2}, \ldots, s_{il}\}$, is obtained that ensures that the service candidates can meet the functional requirements of task $S = \{s_1, s_2, \ldots, s_n\}$ ($n = 3$) (transcoding, compression, and payment). Then, using the global QoS constraint set C, the composite solutions that satisfy the global QoS requirements are recorded. Finally, depending on the preferences of the smartphone end-user, an optimal mobile service selection that achieves the largest utility value $U(S)$ is found.

Suppose 100 service candidates are available for each task: transcoding, compression, and payment; namely, $|s_i| = 100$ ($i = 3$)s. Then, 10^6 combinations are possible for the service selection problem. Determining the optimal one requires enumerating all possible combinations (10^6) of service candidates, which can be expensive in terms of computation time. Additionally, assuring the reliability of the selected mobile services is also difficult because of the complexity.

3. Our Proposed Approach

We propose an efficient service selection approach (called *VMP*) with QoS uncertainty computing. VMP contains three phases. The first phase is computing QoS uncertainty, in which we adopt variance theory to transform the QoS values into a qualitative concept; it represents the degree of uncertainty of the QoS of a mobile service. The second phase is filtering uncertain services; we prune the uncertain service candidates using the qualitative concept, attempting to ensure the reliability of the service composition while reducing the search space of service selection. The final phase is service selection; we use a Mixed Integer Programming algorithm to select the most reliable composition service with shorter computation time.

3.1. Computing QoS Uncertainty. We first normalize the quantitative QoS values into the domain $[0, 1]$, which is convenient for data processing and uniform QoS attribute values. We then employ variance theory to compute the QoS uncertainty by transforming QoS quantitative values into two QoS qualitative concepts. Using the two qualitative concepts, a mobile service with consistently good QoS can be distinguished from other mobile services. For illustration, we first describe the following relevant concepts.

3.1.1. Data Normalizing. The normalization process limits the values to within a certain range (e.g., [0-1]); this is convenient for the QoS utility function. Data normalizing means that the original QoS values will be scaled proportionally. There are many ways to normalize, such as linear conversion, logarithmic conversion, and cotangent conversion. We adopt linear conversion. The specific formula is

$$y = \frac{(x - \text{Minvalue})}{(\text{Maxvalue} - \text{Minvalue})},$$

(3)

where x and y are the corresponding values before and after QoS data conversion, respectively; Maxvalue and Minvalue are the maximum and minimum of the original data, respectively.

3.1.2. Variance. Variance measures the deviation between the random variables and their mathematical expectation value. The variance can fully reflect the stability of the random variable. The larger the variance, the more dispersed the random variable's values relative to the expectation, and the greater the degree of disorder the sample data has. In this

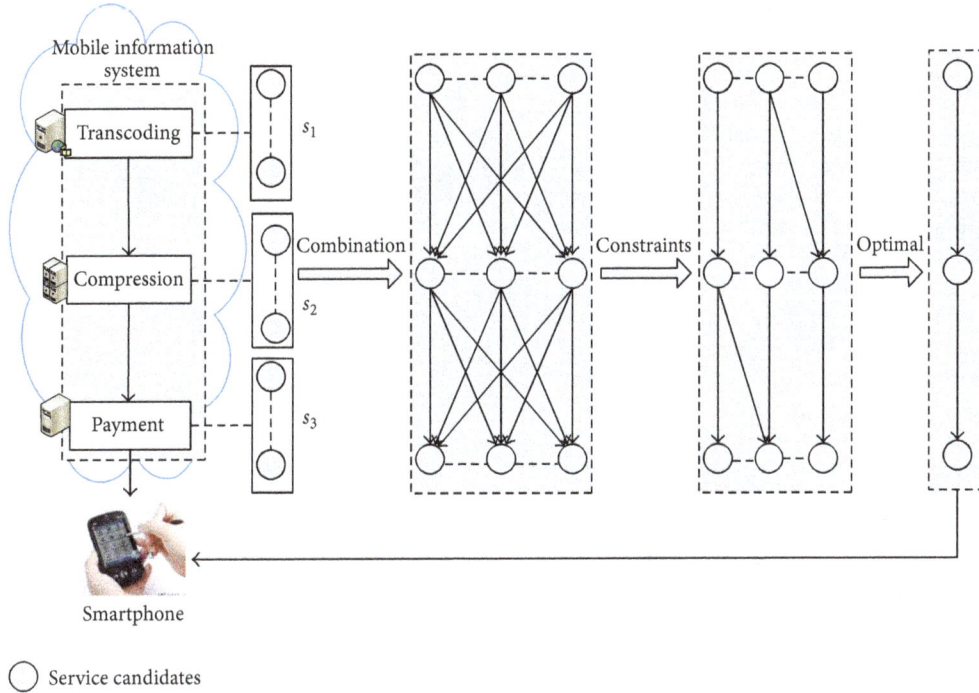

FIGURE 1: An example of service selection in a mobile information system.

paper we consider a real-world QoS historical value for a mobile service as a discrete random variable and employ the concept of variance to filter the mobile services by uncertainty. The following gives its definition.

Definition 1 (variance). Let X be the random variable. $\{X_1, X_2, \ldots, X_n\}$ is the range of X. Then the variance V can be obtained by the following:

$$V = \frac{1}{n-1} \sum_{i=1}^{n} \left(x_i - \overline{X} \right)^2, \tag{4}$$

with

$$\overline{X} = \frac{1}{n} \sum_{i=1}^{n} x_i, \tag{5}$$

where $\overline{X}$ represents the average of X.

Variance theory has been applied to many fields, such as financial markets and investment risk. It has achieved many positive results that provide a basis and reference for our approach. Using variance for abandoning uncertain mobile services will help us select the least uncertain mobile services.

3.2. Uncertain Services Filtering. Through the above QoS uncertainty computation, we can use the variance to filter the uncertain services. For mobile service filtering, we first describe QoS uncertainty through an example [14] as follows. We consider two services, W and T, which provide a similar mobile service. The performances of W and T are recorded in a series of transaction logs that capture the actual QoS values

TABLE 2: A set of service transactions [14].

| | Mobile service: W | | Mobile service: T |
ID	Response time (ms)	ID	Response time (ms)
w_1	24	t_1	13
w_2	27	t_2	34
w_3	26	t_3	36
w_4	33	t_4	45
w_5	35	t_5	44
w_6	25	t_6	35
w_7	31	t_7	15
w_8	26	t_8	12
w_9	29	t_9	33
w_{10}	37	t_{10}	18
w	29.3	t	28.5

delivered by each provider in practice. Because the dynamic environment causes some uncertainty of performance, there will be fluctuation in QoS among different transactions. For ease of illustration, although the actual number of transactions should be much larger, we only consider 10 transactions in each service. These historical transactions are represented in our example as $(w_1, \ldots, w_{10})$ and $(t_1, \ldots, t_{10})$ in Table 2, which lists the values of response time for these transactions along with their aggregated QoS value (w and t). The aggregated QoS values are obtained by averaging all of their corresponding transactions.

From Table 2, we find that the aggregate QoS value of service W is larger than that of service T; that is, $w > t$. In this case, most traditional service selection approaches usually select service T as a service component in a composition service. However, by analyzing the transactions provided by these two services more deeply we have discovered two facts that may be ignored in some existing approaches. (1) Service T's aggregate response time is slightly smaller than that of service W. However, the response times of six of service W's transactions $(w_2, w_3, w_4, w_5, w_6, w_9)$ are smaller than those of the corresponding three of service T's transactions $(t_2, t_3, t_4, t_5, t_6, t_9)$; that is, $w_2 < t_2, w_3 < t_3, \ldots, w_9 < t_9$. This means that, for most transactions, service W's response time was smaller than service T's. (2) Service T's response time is more volatile than that of service W; that is, the QoS value of service T has a large variance, while that of service W is more consistent and therefore more reliable.

In the service selection process, if we select service T as a service component, the actual response time of service T may deviate from t, which will lead to poor quality of the composite service or failure of service selection. Hence, selecting service W as a service component may be better because service W is more stable than service T. Thus, computing the QoS uncertainty, that is, how to distinguish one mobile service with stable QoS from many mobile services with high QoS uncertainty is an important aspect of the service selection process.

Consequently, we employ variance theory in this phase to compute the QoS uncertainty. Through QoS uncertainty, we transform QoS quantitative values into a QoS qualitative concept. Exploiting the qualitative concept, we can then distinguish a mobile service with a stable QoS from other mobile services. We first give the definition of coefficient of variation as follows.

We apply the variance definition above to Table 2. These response time values are quantitative QoS values expressed by ten real-world QoS historical values, that is, $(w_1, \ldots, w_{10})$ or $(t_1, \ldots, t_{10})$. The QoS uncertainty of each service can be expressed by its eigenvector, $EI = \{\overline{X}, V\}$. The resulting eigenvectors of services W and T are $EI_W = \{29.30, 191.70\}$ and $EI_T = \{28.5, 1331.50\}$. Because $191.70 \ll 1331.50$ (i.e., $V_W \ll V_T$), the QoS uncertainty of service W is smaller than that of service T. This means that the QoS of service W is more consistent than that of T. Thus, the service W should be selected as a service component rather than service T, which is different from the traditional approaches.

In this way, variance can help reduce the service selection search space and improve the reliability of service composition.

3.3. Service Selection. After computing QoS uncertainty and filtering uncertain services, some unreliable and redundant service candidates are pruned, and service candidates with consistently good QoS are found in each service class. The service selection solution must then find the most reliable service of each class within global QoS constraints. Recently, Mixed Integer Programming has been used by several researchers [10, 15, 16] to solve the service composition problem with good results. Here, a Mixed Integer Programming (MIP) model is used to solve the optimization problem of service selection based on the filtered services. We wish to optimize the following objective function; that is, we wish to

find the set of service candidates that makes the function a maximum:

$$
\text{Max } F(S)
$$

$$
= \sum_{k=1}^{r} \frac{QV_k^{\max} - \sum_{i=1}^{n} \sum_{j=1}^{l} x_{ij} \cdot q_k(s_{ij}) \cdot v_k(s_{ij})}{(QV_k^{\max} - QV_k^{\min})} \cdot \omega_k \tag{6}
$$

with

$$
QV_k^{\max} = \sum_{i=1}^{r} Q_{i,k}^{\max} \cdot V_{i,k}^{\max},
$$

$$
QV_k^{\min} = \sum_{i=1}^{r} Q_{i,k}^{\min} \cdot V_{i,k}^{\min},
$$

$$
\text{Subject to } \begin{cases} \sum_{i=1}^{n} \sum_{j=1}^{l} q_k(s_{ij}) \cdot x_{ij} \leq C_m, & 1 \leq m \leq r, \\ \sum_{j=1}^{l} x_{ij} = 1, & 1 \leq i \leq n, \ x_{ij} \in \{0, 1\}, \end{cases} \tag{7}
$$

where x_{ij} is a binary decision variable for representing whether a service candidate is selected. A candidate s_{ij} is selected if its corresponding variable x is set to 1 and not selected if 0; $q_k(s_{ij})$ represents the kth attribute value in service s_{ij} and $Q_k^{\max}$ and $Q_k^{\min}$ can be calculated by (2); $V_{i,k}^{\min}$ is the minimum variance value in the service class s_i and $V_k^{\max}$ is the sum of all $V_{i,k}^{\max}$ in the composition service S. Similarly, $V_k^{\min}$ is the sum of all $V_{i,k}^{\min}$ in the composition service S. r is the number of QoS attributes. m is the number of QoS constraints. n is the number of service classes. C_m represents the mth global QoS constraint with respect to the rth QoS attribute.

By solving the model using MIP solver methods, a list of reliable services can be obtained in each class for service brokers providing a composition service for users.

4. Experiments

We conducted experiments and compared our proposed approach with Global [10] and VGMIP in terms of computation time, reliability, and performance. Global is a state-of-the-art efficient service selection approach, the standard global optimization approach with all service candidates; VGMIP considers only half of the service candidates in the global optimization, those whose variances are relatively small and can find a near-optimal composite service with short computation time. Moreover, we also studied the parameters of our approach.

4.1. Experiment Setup. As is well known, there are few mobile service datasets, but to evaluate our approach, we needed two types of datasets. The first we chose is a real-world web service QoS dataset from [17–19] named WSDream. It contains nearly 2 million real-world QoS web service invocation records; each record contains two QoS attributes (i.e., *response time* and *throughput*; because we consider only the sequential composition model, we took *throughput* as the *price* attribute in our experiment) that were collected by

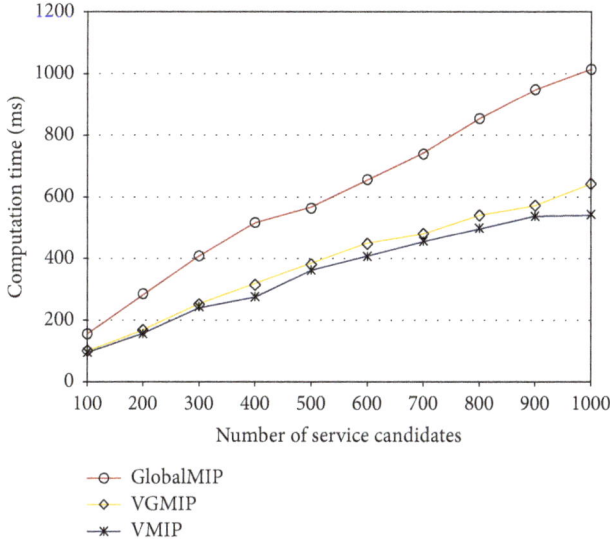

FIGURE 2: Computing time in WSDream dataset.

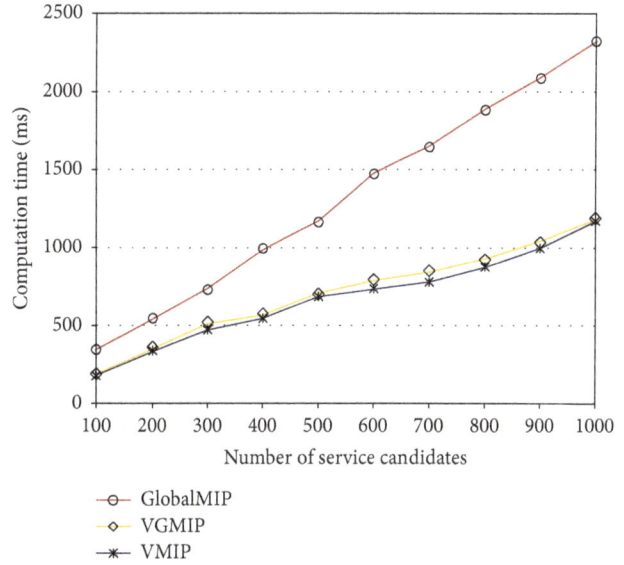

FIGURE 3: Computing time in Random dataset.

339 service users over 5825 web services. It also contains information about these 5825 web services and 339 users. To ensure the experiments results were not based on WSDream, we evaluated our approach with a synthetic dataset (named Random dataset) that contained 10,000 mobile services with two QoS attributes [20].

We conducted several QoS-aware service selection experiments. Each experiment consisted of a composition request with n service classes, l service candidates per class, and m global QoS constraints. We varied these parameters, and each unique combination of them represented one experiment. The number r of QoS attributes was set to two, as was m, the number of QoS constraints. The number of service candidates per service class varied in the range 100 to 1000; the weight for each of the two attributes was 0.5, and the CV was set to 50. The number of service classes, n, was fixed at five for the WSDream dataset and ten for the Random dataset. The number of historical transactions was set at 250 for the WSDream dataset and 500 for the Random dataset.

All of the experiments were performed on the same computer with an Intel(R) Xeon(R) 2.6 GHz processor, 32.0 GB of RAM, Windows Server 2008R2, and Matlab R2013a. We compare our approach (VMP) with the two notable approaches mentioned above.

4.2. Comparison of Computation Time Results.
In this experiment, we compared the computation time of our VMP approach with those of the other approaches.

As shown in Figures 2 and 3, we found that the computation time consumed by VMP was always lower than that of the other two approaches with increasing numbers of service candidates. This means that VMP can significantly reduce the time cost of service selection because its service search space is the smallest of the approaches.

4.3. Comparison of Reliability Results.
In this experiment, we designed the following reliability fitness function to measure service selection reliability.

Definition 2 (reliability fitness function). Reliability is the degree of deviation of the variance of the selected mobile services relative to the range of possible variance; that is,

$$\text{Reliability} = \frac{V_{\max} - \sum_{i=1}^{n} \sigma}{V_{\max} - V_{\min}} \times 100\%, \qquad (8)$$

where $V_{\max}$ is the maximum aggregated standard deviation of all selected services, $V_{\min}$ is the minimum aggregated standard deviation of all selected services, and σ is the standard deviation of the selected service in each service class.

By Definition 2, we compared VMP with Global and VGMIP on the reliability of service selection.

As shown in Figures 4 and 5, we found that no matter how many service candidates were there, the reliability of VMP was always higher than Global and VGMIP. These experimental results indicate that VMP can effectively avoid QoS uncertainty from service selection because the reliability of such services is very high. Thus, by using *variance* to monitor a service's historical QoS transactions, VMP effectively identifies which services have large variances in QoS and can then prune them.

4.4. Comparison of Performance Results.
In this experiment, the performance can be calculated by the following formula:

$$\text{performance} = F(S) \times 100\%. \qquad (9)$$

As shown in Figures 6 and 7, the performance of our approach was higher than 98% on average, which effectively demonstrates that VMP obtains a near-optimal service composition. Although the optimality of VMP was not 100%, the gap (2%) is small; users will not notice an issue in dynamic network environments. Users additionally will perceive little if any difference in the performance of the three approaches. However, the time cost of our approach is the lowest and

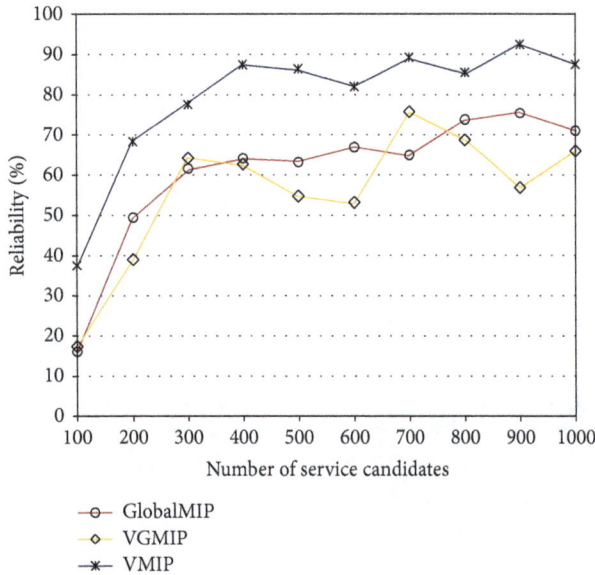

FIGURE 4: Reliability in WSDream dataset.

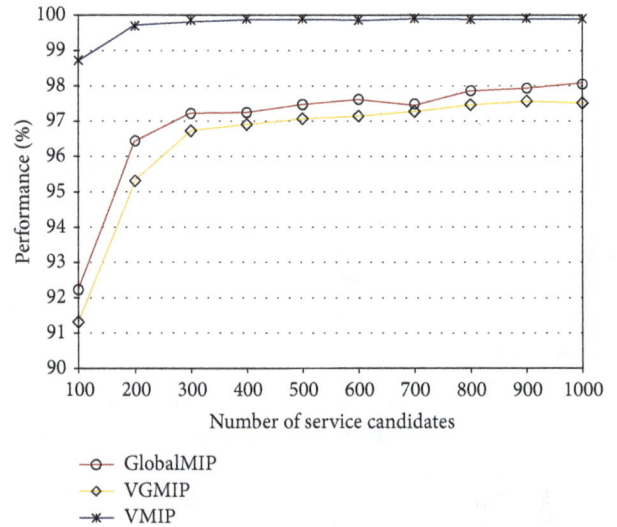

FIGURE 6: Performance in WSDream dataset.

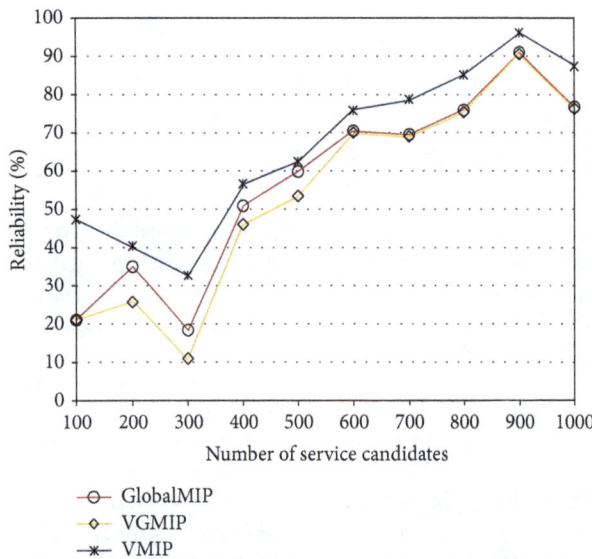

FIGURE 5: Reliability in Random dataset.

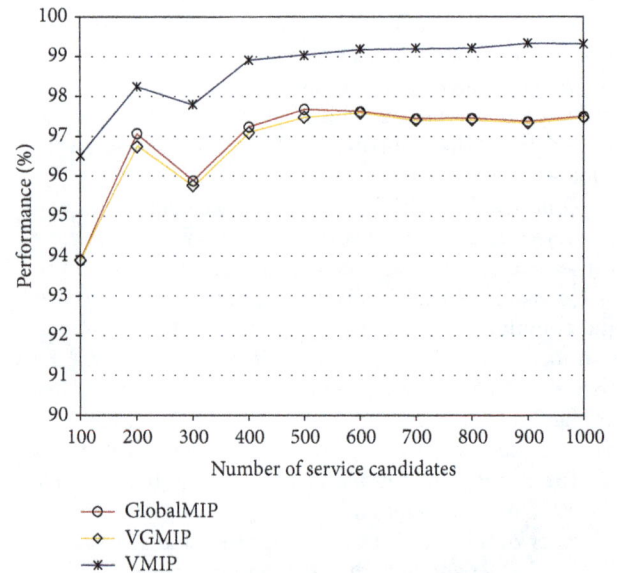

FIGURE 7: Performance in Random dataset.

its reliability is the highest of the three. Hence, for overall comparison, our approach has the highest scores.

4.5. Study of the Parameters

4.5.1. Parameter V. In this experiment we studied how the computation time, reliability, and performance varied with changes in the value of the variance V. When V was set to 30, 30% of services were selected as effective service candidates. In the experiments, we varied the value of V within the range 10 to 90 with a step value of 20.

As shown in Figure 8, the computation time of our approach increases as V is increased from 10 to 90. This demonstrates that the computation time increases with

increasing V because the service selection search space grows rapidly. Figure 9 shows that, for the WSDream dataset, the reliability of our approach also increases with V. As shown in Figure 10, VMP performance always improved with increasing V. Hence, no matter how V is chosen, VMP can always achieve good performance.

4.5.2. Parameters ω. In this experiment, we studied how the user preference weights ω affect the computation time, reliability, and performance of service selection. The weights ω represent the user's requests for each QoS attribute and are important for service selection. We fixed the number of service candidates at 500; the weight of response time was varied from 0.125 to 0.875. Correspondingly, the weight of price was varied from 0.125 to 0.875.

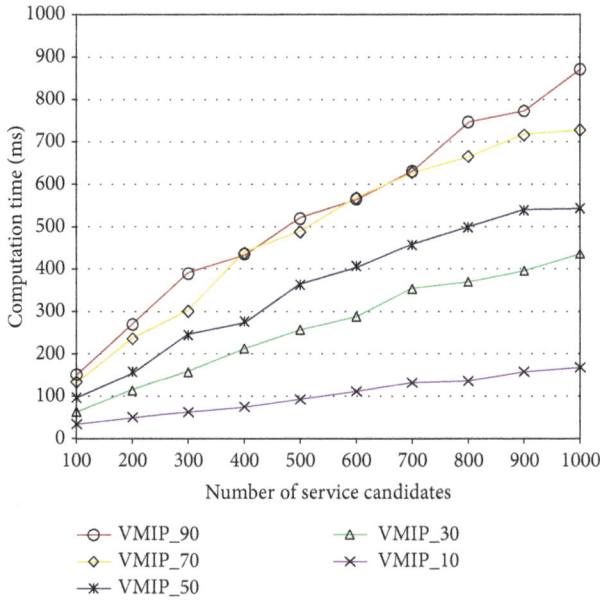

FIGURE 8: Computation time in WSDream dataset with V.

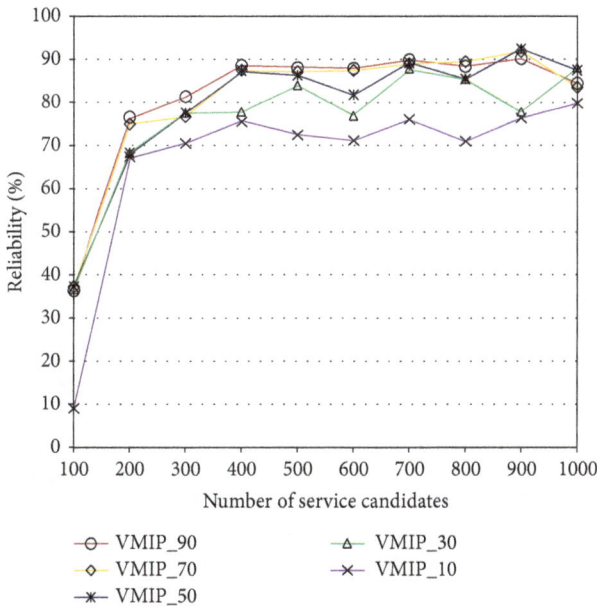

FIGURE 10: Performance in the WSDream dataset with V.

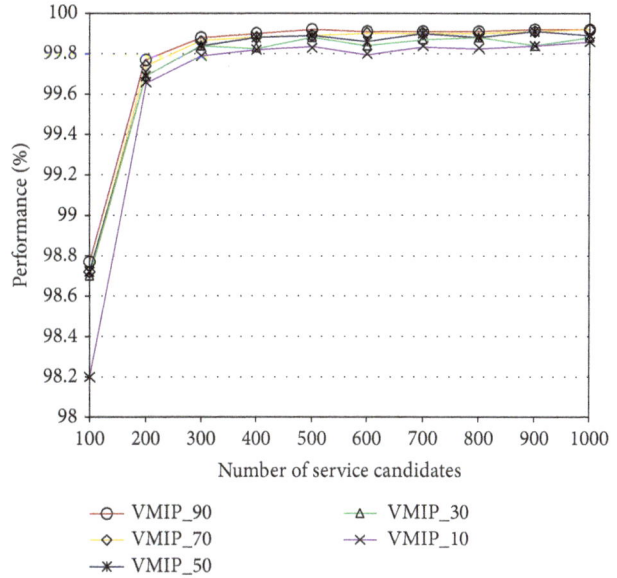

FIGURE 9: Reliability in WSDream dataset with V.

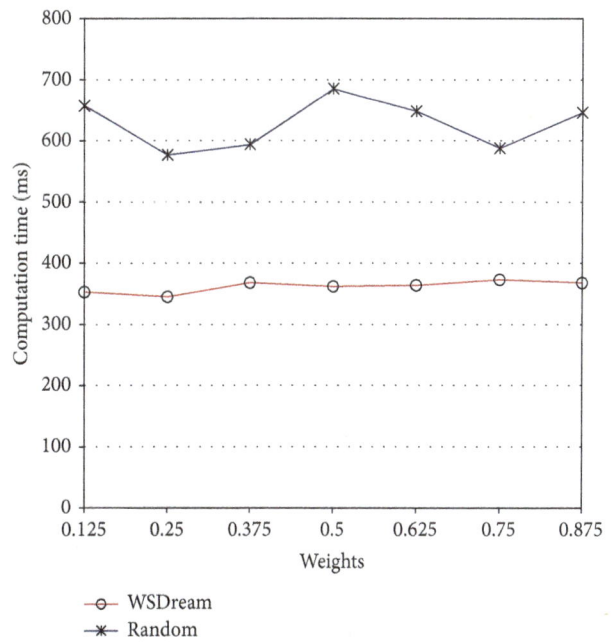

FIGURE 11: Computation time in WSDream dataset with ω. $V = 50$, service candidates = 500.

Figures 11, 12, and 13 show the results with two datasets. We find that no matter how ω were allocated, VMP still obtained good results.

5. Conclusions

We have presented a fast and reliable QoS-aware service selection approach based on the concept of variance in mobile information systems. Our approach first uses variance to compute the QoS uncertainty and to prune services with high uncertainty and reduce the search space of service selection. We then use a Mixed Integer Programming to select a service composition near optimal but more reliable with shorter computation time. We evaluated our approach using both real-world and randomly generated datasets. The experimental results show that our approach needs less computation time to find the best high-reliability, high-performance service composition. This means that our approach can perform service selection on the basis of user preferences more efficiently and effectively for mobile information systems.

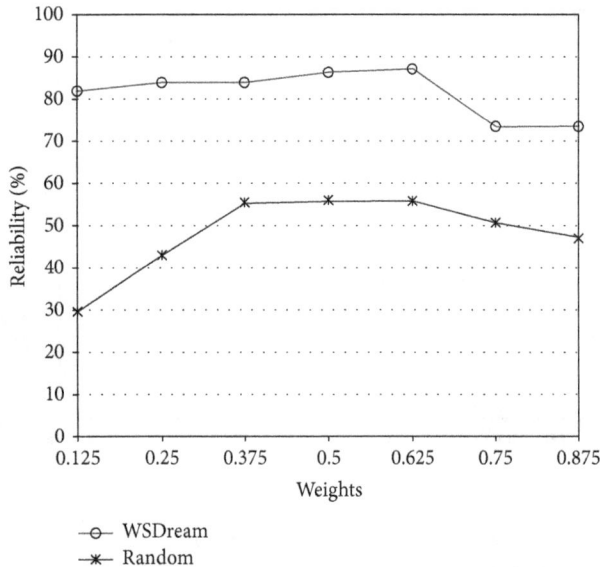

FIGURE 12: Reliability in WSDream dataset with varied ω. $V = 50$, service candidates = 500.

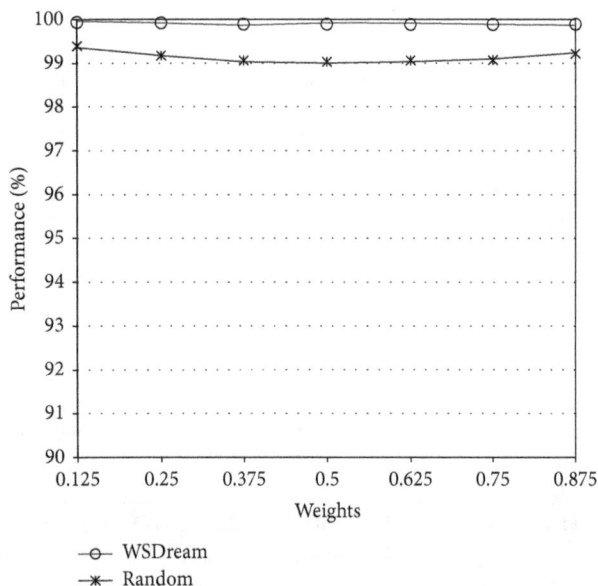

FIGURE 13: Performance in WSDream dataset with varied ω. $V = 50$, service candidates = 500.

Conflict of Interests

The authors declare that there is no conflict of interests regarding the publication of this paper.

Acknowledgments

The work presented in this study is supported by the Natural Science Foundation of Beijing under Grant no. 4132048, NSFC (61472047), and NSFC (61202435).

References

[1] J. Poncela, G. Gómez, A. Hierrezuelo, F. J. Lopez-Martinez, and M. Aamir, "Quality assessment in 3G/4G wireless networks," *Wireless Personal Communications*, vol. 76, no. 3, pp. 363–377, 2014.

[2] D. Amzallag, R. Bar-Yehuda, D. Raz, and G. Scalosub, "Cell selection in 4G cellular networks," *IEEE Transactions on Mobile Computing*, vol. 12, no. 7, pp. 1443–1455, 2013.

[3] H. H. Ju, B. Liang, J. D. Li, and X. N. Yang, "Dynamic joint resource optimization for LTE-advanced relay networks," *IEEE Transactions on Wireless Communications*, vol. 12, no. 11, pp. 5668–5678, 2013.

[4] O. N. C. Yilmaz, J. Hämäläinen, and S. Hämäläinen, "Optimization of adaptive antenna system parameters in self-organizing LTE networks," *Wireless Networks*, vol. 19, no. 6, pp. 1251–1267, 2013.

[5] I. Siomina and D. Yuan, "Analysis of cell load coupling for LTE network planning and optimization," *IEEE Transactions on Wireless Communications*, vol. 11, no. 6, pp. 2287–2297, 2012.

[6] O. Sallent, J. Pérez-Romero, J. Sánchez-González et al., "A roadmap from UMTS optimization to LTE self-optimization," *IEEE Communications Magazine*, vol. 49, no. 6, pp. 172–182, 2011.

[7] F. Curbera, M. Duftler, R. Khalaf, W. Nagy, N. Mukhi, and S. Weerawarana, "Unraveling the Web services Web: an introduction to SOAP, WSDL, and UDDI," *IEEE Internet Computing*, vol. 6, no. 2, pp. 86–93, 2002.

[8] G. Canfora, M. Di Penta, R. Esposito, and M. L. Villani, "A framework for QoS-aware binding and re-binding of composite web services," *Journal of Systems and Software*, vol. 81, no. 10, pp. 1754–1769, 2008.

[9] M. Alrifai, D. Skoutas, and T. Risse, "Selecting skyline services for QoS-based web service composition," in *Proceedings of the 19th International World Wide Web Conference (WWW '10)*, pp. 11–20, Raleigh, NC, USA, April 2010.

[10] D. Ardagna and B. Pernici, "Adaptive service composition in flexible processes," *IEEE Transactions on Software Engineering*, vol. 33, no. 6, pp. 369–384, 2007.

[11] T. Yu, Y. Zhang, and K.-J. Lin, "Efficient algorithms for Web services selection with end-to-end QoS constraints," *ACM Transactions on the Web*, vol. 1, no. 1, article 6, 26 pages, 2007.

[12] M. Alrifai and T. Risse, "Combining global optimization with local selection for efficient QoS-aware service composition," in *Proceedings of the 18th International Conference on World Wide Web (WWW '09)*, pp. 881–890, April 2009.

[13] L. Qi, Y. Tang, W. Dou, and J. Chen, "Combining local optimization and enumeration for QoS-aware web service composition," in *Proceedings of the IEEE 8th International Conference on Web Services (ICWS '10)*, pp. 34–41, Maimi, Fla, USA, July 2010.

[14] S. Wang, Z. Zheng, Q. Sun, H. Zou, and F. Yang, "Cloud model for service selection," in *Proceedings of the IEEE Conference on Computer Communications Workshops (INFOCOM WKSHPS '11)*, pp. 666–671, Shanghai, China, April 2011.

[15] L. Barakat, S. Miles, and M. Luck, "Efficient correlation-aware service selection," in *Proceedings of the IEEE 19th International Conference on Web Services (ICWS '12)*, pp. 1–8, Honolulu, HI, USA, June 2012.

[16] L. Zeng, B. Benatallah, A. H. H. Ngu, M. Dumas, J. Kalagnanam, and H. Chang, "QoS-aware middleware for Web services composition," *IEEE Transactions on Software Engineering*, vol. 30, no. 5, pp. 311–327, 2004.

[17] Z. Zheng, Y. Zhang, and M. R. Lyu, "Distributed QoS evaluation for real-world web services," in *Proceedings of the IEEE 19th International Conference on Web Services (ICWS '12)*, pp. 83–90, Honolulu, Hawaii, USA, 2012.

[18] Z. Zheng and M. R. Lyu, "Collaborative reliability prediction of service-oriented systems," in *Proceedings of the 32nd ACM/IEEE International Conference on Software Engineering (ICSE '10)*, pp. 35–44, Cape Town, South Africa, May 2010.

[19] Z. Zheng, Y. Zhang, and M. R. Lyu, "Investigating QoS of real-world web services," *IEEE Transactions on Services Computing*, vol. 7, no. 1, pp. 32–39, 2014.

[20] S. Wang, Z. Zheng, Z. Wu, Q. Sun, H. Zou, and F. Yang, "Context-aware mobile service adaptation via a Co-evolution eXtended Classifier System in mobile network environments," *Mobile Information Systems*, vol. 10, no. 2, pp. 197–215, 2014.

Permissions

List of Contributors

Yilei Wang
School of Computer Science and Technology, Shandong University, Jinan 250101, China
School of Information and Electrical Engineering, Ludong University, Yantai 264025, China

Chuan Zhao and Qiuliang Xu
School of Computer Science and Technology, Shandong University, Jinan 250101, China

Zhihua Zheng
School of Information Science and Engineering, Shandong Normal University, Jinan 250014, China

Zhenhua Chen
School of Computer Science, Shaanxi Normal University, Xi'an 710062, China

Zhe Liu
Laboratory of Algorithmics, Cryptology and Security (LACS), 1359 Luxembourg, Luxembourg

Shigen Shen
Department of Computer Science and Engineering, Shaoxing University, Shaoxing 312000, China
College of Mathematics, Physics and Information Engineering, Jiaxing University, Jiaxing 314001, China

Keli Hu
Department of Computer Science and Engineering, Shaoxing University, Shaoxing 312000, China

Longjun Huang
Department of Computer Science and Engineering, Shaoxing University, Shaoxing 312000, China
College of Computer Science and Technology, Zhejiang University of Technology, Hangzhou 310014, China

Hongjie Li and Risheng Han
College of Mathematics, Physics and Information Engineering, Jiaxing University, Jiaxing 314001, China

Qiying Cao
College of Computer Science and Technology, Donghua University, Shanghai 201620, China

Hung-Yu Chien
Department of Information Management, National Chi-Nan University, 470 University Road, Puli, Nantou, Taiwan

Giuseppe Vitello and Vincenzo Conti
Faculty of Engineering and Architecture, University of Enna Kore, 94100 Enna, Italy

Salvatore Vitabile
Department of Biopathology and Medical Biotechnologies, University of Palermo, 90127 Palermo, Italy

Filippo Sorbello
Department of Chemical Engineering, Management, Computer Science, and Mechanics, University of Palermo, 90128 Palermo, Italy

Minqing Zhang
School of Computer Science, Northwestern Polytechnical University, Xi'an 710072, China
Key Laboratory of Information and Network Security, Engineering University of Chinese Armed Police Force, Xi'an 710086, China

Xu An Wang and Xiaoyuan Yang
Key Laboratory of Information and Network Security, Engineering University of Chinese Armed Police Force, Xi'an 710086, China

Weihua Li
School of Computer Science, Northwestern Polytechnical University, Xi'an 710072, China

En Zhang
College of Computer and Information Engineering, Henan Normal University, Xinxiang 453007, ChinaState Key Laboratory of Information Security, Institute of Information Engineering, Chinese Academy of Sciences, Beijing 100093, China
Engineering Lab of Intelligence Business & Internet ofThings, Xinxiang, Henan 453007, China

Peiyan Yuan
College of Computer and Information Engineering, Henan Normal University, Xinxiang 453007, China
Engineering Lab of Intelligence Business & Internet ofThings, Xinxiang, Henan 453007, China

Jiao Du
College of Mathematics and Information Science, Henan Normal University, Xinxiang 453007, China

Lingling Xu and Shaohua Tang
School of Computer Science and Engineering, South China University of Technology, Guangzhou 510006, China

Jin Li
School of Computer Science and Educational Software, Guangzhou University, Guangzhou 510006, China

Joonsang Baek
Khalifa University of Science, Technology and Research, P.O. Box 127788, Abu Dhabi, UAE

Baojiang Cui and Ziyue Wang
School of Computer, Beijing University of Posts and Telecommunications, Beijing 100876, China

Bing Zhao and Xiaobing Liang
State Grid Metering Center, Beijing 100192, China

Yuemin Ding
Department of Electronic Systems Engineering, Hanyang University, Ansan 426791, Republic of Korea

Anitha Manikandan and Yogesh Palanichamy
Department of Information Science and Technology, College of Engineering, Anna University, Guindy, Chennai, Tamilnadu 600 025, India

Tarek R. Sheltami
Computer Engineering Department, King Fahd University of Petroleum and Minerals, Dhahran 31216, Saudi Arabia

Won-Suk Kim and Sang-Hwa Chung
Department of Computer Engineering, Pusan National University, Busan 609-735, Republic of Korea

Hui Zhu, Yingfang Xue, Xiaofeng Chen, Qiang Li and Hui Li
State Key Laboratory of Integrated Services Networks, Xidian University, No. 2 South Taibai Road, Yanta District, Xi'an 710071, China

Kam-Yiu Lam and Nelson Wai-Hung Tsang
Department of Computer Science, City University of Hong Kong, Kowloon Tong, Hong Kong

Joseph Kee Yin Ng, Jiantao Wang and Calvin Ho Chuen Kam
Department of Computer Science, Hong Kong Baptist University, Kowloon Tong, Hong Kong

Fei Tang, Hongda Li, Qihua Niu and Bei Liang
The Data Assurance and Communication Security Research Center, Chinese Academy of Sciences, No. 89 Minzhuang Road, Haidian District, Beijing 100093, China State Key Laboratory of Information Security, Institute of Information Engineering, Chinese Academy of Sciences, No. 89 Minzhuang Road, Haidian District, Beijing 100093, China

Luis Gómez-Miralles and Joan Arnedo-Moreno
Internet Interdisciplinary Institute (IN3), Universitat Oberta de Catalunya, Roc Boronat Street 117, 7th Floor, 08018 Barcelona, Spain

Lian-Fen Huang, Sha-Li Zhou and Yi-Feng Zhao
Department of Communication Engineering, Xiamen University, Xiamen, Fujian 361005, China

Han-Chieh Chao
Institute of Computer Science & Information Engineering and Department of Electronic Engineering, National Ilan University, I-Lan, Taiwan
Department of Electrical Engineering, National Dong Hwa University, Hualien, Taiwan

Ashutosh Bhatia and R. C. Hansdah
Department of Computer Science and Automation, Indian Institute of Science, Bangalore 560012, India

Shangguang Wang, Lei Sun, Qibo Sun and Fangchun Yang
State Key Laboratory of Networking and Switching Technology, Beijing University of Posts and Telecommunications, Haidian, Beijing 100876, China

Xuyan Li
Basic Research Service Ministry of Science and Technology the People's Republic of China, Haidian, Beijing 100862, China

www.ingramcontent.com/pod-product-compliance
Lightning Source LLC
Chambersburg PA
CBHW050441200326
41458CB00014B/5030